UCLA Symposia on Molecular and Cellular Biology, New Series

Series Editor, C. Fred Fox

RECENT TITLES

Volume 33
Yeast Cell Biology, James Hicks, *Editor*

Volume 34
Molecular Genetics of Filamentous Fungi, William Timberlake, *Editor*

Volume 35
Plant Genetics, Michael Freeling, *Editor*

Volume 36
Options for the Control of Influenza, Alan P. Kendal and Peter A. Patriarca, *Editors*

Volume 37
Perspectives in Inflammation, Neoplasia, and Vascular Cell Biology, Thomas Edgington, Russell Ross, and Samuel Silverstein, *Editors*

Volume 38
Membrane Skeletons and Cytoskeletal—Membrane Associations, Vann Bennett, Carl M. Cohen, Samuel E. Lux, and Jiri Palek, *Editors*

Volume 39
Protein Structure, Folding, and Design, Dale L. Oxender, *Editor*

Volume 40
Biochemical and Molecular Epidemiology of Cancer, Curtis C. Harris, *Editor*

Volume 41
Immune Regulation by Characterized Polypeptides, Gideon Goldstein, Jean-François Bach, and Hans Wigzell, *Editors*

Volume 42
Molecular Strategies of Parasitic Invasion, Nina Agabian, Howard Goodman, and Nadia Nogueira, *Editors*

Volume 43
Viruses and Human Cancer, Robert C. Gallo, William Haseltine, George Klein, and Harald zur Hausen, *Editors*

Volume 44
Molecular Biology of Plant Growth Control, J. Eugene Fox and Mark Jacobs, *Editors*

Volume 45
Membrane-Mediated Cytotoxicity, Benjamin Bonavida and R. John Collier, *Editors*

Volume 46
Development and Diseases of Cartilage and Bone Matrix, Arup Sen and Thomas Thornhill, *Editors*

Volume 47
DNA Replication and Recombination, Roger McMacken and Thomas J. Kelly, *Editors*

Volume 48
Molecular Strategies for Crop Protection, Charles J. Arntzen and Clarence Ryan, *Editors*

Volume 49
Molecular Entomology, John H. Law, *Editor*

Volume 50
Interferons as Cell Growth Inhibitors and Antitumor Factors, Robert M. Friedman, Thomas Merigan, and T. Sreevalsan, *Editors*

Volume 51
Molecular Approaches to Developmental Biology, Richard A. Firtel and Eric H. Davidson, *Editors*

Volume 52
Transcriptional Control Mechanisms, Daryl Granner, Michael G. Rosenfeld, and Shing Chang, *Editors*

Volume 53
Recent Advances in Bone Marrow Transplantation, Robert Peter Gale and Richard Champlin, *Editors*

Volume 54
Positive Strand RNA Viruses, Margo A. Brinton and Roland R. Rueckert, *Editors*

UCLA Symposia Board

Molecular Entomology

Molecular Entomology

Proceedings of a Monsanto–UCLA Symposium
Held in Steamboat Springs, Colorado
April 6–13, 1986

Editor

John H. Law
Department of Biochemistry
University of Arizona
Tucson, Arizona

Alan R. Liss, Inc. • New York

Address all Inquiries to the Publisher
Alan R. Liss, Inc., 41 East 11th Street, New York, NY 10003

Printed in the United States of America

Library of Congress Cataloging in Publication Data

UCLA Symposium (1986 : Steamboat Springs, Colo.)
Molecular entomology.

(UCLA symposia on molecular and cellular biology ; new ser., v. 49)
Includes bibliographies and index.
1. Insects—Physiology—Congresses. 2. Molecular biology—Congresses. I. Law, John H. II. Title. III. Series [DNLM: 1. Entomology—congresses. 2. Molecular Biology—congresses. W3 U17N new ser. v.49 / QL 461 U17m 1986]
QL495.U64 1986 595.7′088 87-2701
ISBN 0-8451-2648-2

Contents

II. MOLECULAR ENDOCRINOLOGY AND DEVELOPMENT IN INSECTS

III. HEMOLYMPH PROTEINS OF INSECTS

Contributors

Yehia A.I. Abdel-Aal, Departments of Entomology and Environmental Toxicology, University of California, Davis, CA 95616 **[315]**

Ad M.Th. Beenakkers, Department of Experimental Zoology, University of Utrecht, 3508 TB Utrecht, The Netherlands **[247]**

D. Benke, Fachbereich Biologie/ Chemie, Universität Osnabrück, D-4500 Osnabrück, Federal Republic of Germany **[95]**

E.A. Bernays, Division of Biological Control and Department of Entomological Sciences, University of California, Berkeley, CA 94720 **[107]**

Rolf Bodmer, Howard Hughes Medical Institute and Department of Physiology, University of California, San Francisco, CA 94143 **[45]**

Mary Bownes, Department of Molecular Biology, University of Edinburgh, Edinburgh EH9 3JR, Scotland **[391]**

J.Y. Bradfield, Department of Biology, Queen's University, Kingston, Canada K7L 3N6; present address: Department of Entomology, Texas A&M University, College Station, TX 77843 **[275]**

H. Breer, Fachbereich Biologie/ Chemie, Universität Osnabrück, D-4500 Osnabrück, Federal Republic of Germany **[95]**

Victor J. Brookes, Department of Entomology, Oregon State University, Corvallis, OR 97331 **[415]**

R.W. Brueggemeier, College of Pharmacy, Ohio State University, Columbus, OH 43210 **[189]**

Douglas R. Cavener, Department of Molecular Biology, Vanderbilt University, Nashville, TN 37235 **[453]**

R.F. Chapman, Division of Biological Control and Department of Entomological Sciences, University of California, Berkeley, CA 94720 **[107]**

Haruo Chino, Biochemical Laboratory, Institute of Low Temperature Science, Hokkaido University, Sapporo, Japan **[235]**

Kenneth D. Cole, Biochemistry Department, University of Arizona, Tucson, AZ 85721 **[267]**

Wendell L. Combest, Department of Biology, University of North Carolina, Chapel Hill, NC 27514 **[129]**

The numbers in brackets are the opening page numbers of the contributors' articles.

Glenn E. Croston, Departments of Entomology and Environmental Toxicology, University of California, Davis, CA 95616 **[315]**

Christine Dambly-Chaudiere, Laboratoire de Genetique, Université Libre de Bruxelles, 1640 Rhode-St.-Genèse, Belgium **[45]**

D.L. Denlinger, Department of Entomology, Ohio State University, Columbus, OH 43210 **[189]**

Yu-Shin Ding, Department of Chemistry, State University of New York, Stony Brook, NY 11794-3400 **[57]**

D.R. Drake, Department of Entomology, Purdue University, West Lafayette, IN 47907; present address: Department of Microbiology, University of Tennessee, Knoxville, TN 37916 **[381]**

P.E. Dunn, Department of Entomology, Purdue University, West Lafayette, IN 47907 **[381]**

Judy Fabian, Department of Zoology, University of Vermont, Burlington, VT 05405 **[179]**

Dianne K. Fristrom, Department of Genetics, University of California, Berkeley, CA 94720 **[155]**

James W. Fristrom, Department of Genetics, University of California, Berkeley, CA 94720 **[155]**

Masahiko Fujino, Central Research Institute, Takeda Chemical Industry, Osaka, Japan **[119]**

Joel R. Garbow, Physical Sciences Center, Monsanto Company, St. Louis, MO 63167 **[331]**

Alain Ghysen, Laboratoire de Genetique, Université Libre de Bruxelles, 1640 Rhode-St.-Genèse, Belgium **[45]**

Lawrence I. Gilbert, Department of Biology, University of North Carolina, Chapel Hill, NC 27514 **[129]**

Bruce D. Hammock, Departments of Entomology and Environmental Toxicology, University of California, Davis, CA 95616 **[315]**

W. Hanke, Fachbereich Biologie/Chemie, Universität Osnabrück, D-4500 Osnabrück, Federal Republic of Germany **[95]**

Terry N. Hanzlik, Departments of Entomology and Environmental Toxicology, University of California, Davis, CA 95616 **[315]**

George M. Happ, Department of Zoology, University of Vermont, Burlington, VT 05405 **[433]**

Masaaki Hatta, Biological Institute, Faculty of Science, Nagoya University, Nagoya, Japan; present address: Institute for Biophysics, Faculty of Science, Kyoto University, Kyoto, Japan **[119]**

Craig Heim, Department of Entomology, University of Kentucky, Lexington, KY 40546 **[469]**

Gary M. Hellmann, Department of Biochemistry, University of Kentucky, Lexington, KY 40506 **[295]**

Robert Hice, Department of Zoology, University of Washington, Seattle, WA 98195; present address: Department of Entomology, University of California, Riverside, CA 92521 **[201]**

John G. Hildebrand, Arizona Research Laboratories, Division of Neurobiology, University of Arizona, Tucson, AZ 85721 **[21]**

Shivanand T. Hiremath, Department of Biochemistry, University of Kentucky, Lexington, KY 40506 **[295]**

David Hollar, Department of Molecular Biology, Vanderbilt University, Nashville, TN 37235; present address:. Department of Genetics, North Carolina State University, Raleigh, NC 27695 [453]

Theodore L. Hopkins, Department of Entomology, Kansas State University, Manhattan, KS 66506 [331]

Frank Horodyski, Department of Zoology, University of Washington, Seattle, WA 98195 [201]

Joseph Huesing, Department of Entomology, University of Kentucky, Lexington, KY 40546 [469]

Hironori Ishizaki, Biological Institute, Faculty of Science, Nagoya University, Nagoya, Japan [119]

Akira Isogai, Department of Agricultural Chemistry, Faculty of Agriculture, The University of Tokyo, Tokyo, Japan [119]

Gary S. Jacob, Physical Sciences Center, Monsanto Company, St. Louis, MO 63167 [331]

Lily Yeh Jan, Howard Hughes Medical Institute and Department of Physiology, University of California, San Francisco, CA 94143 [45]

Yuh Nung Jan, Howard Hughes Medical Institute and Department of Physiology, University of California, San Francisco, CA 94143 [45]

Davy Jones, Department of Entomology, University of Kentucky, Lexington, KY 40546 [469]

Grace Jones, Department of Entomology, University of Kentucky, Lexington, KY 40546 [295]

Karl-Ernst Kaissling, Max-Planck-Institut für Verhaltensphysiologie, 8131 Seewiesen, Federal Republic of Germany [33]

M.R. Kanost, Department of Biology, Queen's University, Kingston, Canada K7L 3N6 [275,381]

Hiroshi Kataoka, Department of Agricultural Chemistry, Faculty of Agriculture, The University of Tokyo, Tokyo, Japan [119]

N. Katlic, College of Pharmacy, Ohio State University, Columbus, OH 43210 [189]

John K. Kawooya, Department of Biochemistry, University of Arizona, Tucson, AZ 85721 [425]

Joel B. Kirschbaum, Codon Corp., Brisbane, CA 94005 [465]

Chieko Kitada, Central Research Institute, Takeda Chemical Industry, Osaka, Japan [119]

R. Kleene, Fachbereich Biologie/Chemie, Universität Osnabrück, D-4500 Osnabrück, Federal Republic of Germany; present address: Department of Physiological Chemistry, Universität München, 8000 München, Federal Republic of Germany [95]

M. Knipper, Fachbereich Biologie/Chemie, Universität Osnabrück, D-4500 Osnabrück, Federal Republic of Germany [95]

Karl J. Kramer, U.S. Grain Marketing Research Laboratory, Agricultural Research Service, U.S. Department of Agriculture, and Department of Biochemistry, Kansas State University, Manhattan, KS 66502 [331]

John H. Law, Department of Biochemistry, University of Arizona, Tucson, AZ 85721 [257,425]

Arden O. Lea, Department of Entomology, University of Georgia, Athens, GA 30602 [403]

Judith A. Lengyel, Department of Biology and Molecular Biology Institute, UCLA, Los Angeles, CA 90024 **[223]**

Charles E. Linn, Jr., Department of Entomology, Cornell University, Geneva, NY 14456 **[67]**

J. Locke, Department of Biology, Queen's University, Kingston, Canada K7L 3N6; present address: Fox Chase Cancer Center, Philadelphia, PA 19111 **[275]**

Holman C. Massey, Jr., Biology Department, University of Pennsylvania, Philadelphia, PA 19104-6018 **[305]**

H.L. McDonald, Department of Biology, Queen's University, Kingston, Canada K7L 3N6 **[275]**

R. Mechoulam, Faculty of Medicine, Hebrew University, Jerusalem 91 120, Israel **[189]**

Akira Mizoguchi, Biological Institute, Faculty of Science, Nagoya University, Nagoya, Japan **[119]**

John Moore, Department of Genetics, University of California, Berkeley, CA 94720 **[155]**

Thomas D. Morgan, Department of Entomology, Kansas State University, Manhattan, KS 66506 **[331]**

David B. Morton, Department of Zoology, University of Washington, Seattle, WA 98195 **[165]**

Michael Murtha, Department of Molecular Biology, Vanderbilt University, Nashville, TN 37235 **[453]**

Hiromichi Nagasawa, Department of Agricultural Chemistry, Faculty of Agriculture, The University of Tokyo, Tokyo, Japan **[119]**

Shunji Natori, Faculty of Pharmaceutical Sciences, University of Tokyo, Tokyo 113, Japan **[369]**

Jeanette E. Natzle, Department of Genetics, University of California, Berkeley, CA 94720; present address: Department of Zoology, University of California, Davis, CA 95616 **[155]**

Barbara E. Noyes, Department of Psychiatry, University of Chicago, Chicago, IL 60637 **[141]**

Thomas R. Odhiambo, The International Centre of Insect Physiology and Ecology, (ICIPE), Nairobi, Kenya **[1]**

Ellie O. Osir, Department of Biochemistry, University of Arizona, Tucson, AZ 85721; present address: Department of Zoology, University of Washington, Seattle, WA 98195 **[425]**

David Osterbur, Department of Genetics, University of California, Berkeley, CA 94720; present address: Department of Biology, Indiana University, Bloomington, IN 47405 **[155]**

Stephenie Paine-Saunders, Department of Genetics, University of California, Berkeley, CA 94720 **[155]**

Richard N. Pau, Insect Chemistry and Physiology Group, Agricultural and Food Research Council, University of Sussex, Brighton BN1 9RQ, England **[443]**

Sarvamangala V. Prasad, Biochemistry Department, University of Arizona, Tucson, AZ 85721 **[267]**

Glenn D. Prestwich, Department of Chemistry, State University of New York, Stony Brook, NY 11794-3400 **[57]**

Alexander S. Raikhel, Department of Entomology, University of Georgia, Athens, GA 30602 **[403]**

John Rebers, Department of Zoology, University of Washington, Seattle, WA 98195 **[201]**

Robert E. Rhoads, Department of Biochemistry, University of Kentucky, Lexington, KY 40506 **[295]**

Lynn M. Riddiford, Department of Zoology, University of Washington, Seattle, WA 98195; present address: Department of Genetics, University of Cambridge, Cambridge, England **[201, 211]**

David B. Roberts, Genetics Laboratory, Department of Biochemistry, Oxford University, Oxford OX1 3QU, England **[285]**

R. Michael Roe, Departments of Entomology and Environmental Toxicology, University of California, Davis, CA 95616; present address: Department of Entomology, North Carolina State University, Raleigh, NC 27650 **[315]**

Wendell L. Roelofs, Department of Entomology, Cornell University, Geneva, NY 14456 **[67]**

Dorothy B. Rountree, Department of Biology, University of North Carolina, Chapel Hill, NC 27514 **[129]**

Robert O. Ryan, Department of Biochemistry, University of Arizona, Tucson, AZ 85721 **[257]**

Robert D.C. Saunders, Department of Molecular Biology, University of Edinburgh, Edinburgh EH9 3JR, Scotland **[391]**

Jacob Schaefer, Physical Sciences Center, Monsanto Company, St. Louis, MO 63167 **[331]**

Martin H. Schaffer, Department of Psychiatry, University of Chicago, Chicago, IL 60637; present address: Department of Psychiatry, University of Texas Health Science Center at Dallas, Dallas, TX 75235 **[141]**

Christopher Schonbaum, Department of Molecular Biology, Vanderbilt University, Nashville, TN 37235 **[453]**

Thomas K.F. Schulz, Department of Experimental Zoology, University of Utrecht, 3508 TB Utrecht, The Netherlands **[247]**

Paul D. Shirk, Insect Attractants, Behavior, and Basic Biology Research Laboratory, Agricultural Research Service, USDA, Gainesville, FL 32604 **[415]**

Alan D. Shirras, Department of Molecular Biology, University of Edinburgh, Edinburgh EH9 3JR, Scotland **[391]**

Wendy A. Smith, Department of Biology, University of North Carolina, Chapel Hill, NC 27514; present address: Department of Biology, Northeastern University, Boston, MA 02115 **[129]**

Roy D. Speirs, U.S. Grain Marketing Research Laboratory, Agricultural Research Service, U.S. Department of Agriculture, Manhattan, KS 66502 **[331]**

Edward O. Stejskal, Physical Sciences Center, Monsanto Company, St. Louis, MO 63167 **[331]**

M. Sugumaran, Department of Biology, University of Massachusetts at Boston, Boston, MA 02125 **[357]**

Akinori Suzuki, Department of Agricultural Chemistry, Faculty of Agriculture, The University of Tokyo, Tokyo, Japan **[119]**

Saburo Tamura, Department of Agricultural Chemistry, Faculty of Agriculture, The University of Tokyo, Tokyo, Japan **[119]**

P.E.A. Teal, Department of Environmental Biology, University of Guelph, Guelph, Ontario N1G 2W1, Canada **[79]**

William H. Telfer, Biology Department, University of Pennsylvania, Philadelphia, PA 19104-6018 **[305]**

James W. Truman, Department of Zoology, University of Washington, Seattle, WA 98195; present address: Department of Zoology, Cambridge University, Cambridge, England **[165]**

Kozo Tsuchida, Biochemistry Department, University of Arizona, Tucson, AZ 85721 **[267]**

J.H. Tumlinson, Insect Attractants, Behavior, and Basic Biology Research Laboratory, USDA-ARS, Gainesville, FL 32604 **[79]**

Dick J. Van der Horst, Department of Experimental Zoology, University of Utrecht, 3508 TB Utrecht, The Netherlands **[247]**

Miranda C. Van Heusden, Department of Experimental Zoology, University of Utrecht, 3508 TB Utrecht, The Netherlands **[247]**

Richard G. Vogt, Department of Chemistry, State University of New York, Stony Brook, NY 11794-3400 **[57]**

Michael A. Wells, Biochemistry Department, University of Arizona, Tucson, AZ 85721 **[257, 267]**

L. Wieczorek, Fachbereich Biologie/Chemie, Universität Osnabrück, D-4500 Osnabrück, Federal Republic of Germany **[95]**

Thomas G. Wilson, Department of Zoology, University of Vermont, Burlington, VT 05405 **[179]**

Donald Withers, Department of Genetics, University of California, Berkeley, CA 94720 **[155]**

Mietek Wozniak, Department of Entomology, University of Kentucky, Lexington, KY 40546 **[295]**

G.R. Wyatt, Department of Biology, Queen's University, Kingston, Canada K7L 3N6 **[275]**

G.D. Yocum, Department of Entomology, Ohio State University, Columbus, OH 43210 **[189]**

L.B. Yocum, Department of Entomology, Ohio State University, Columbus, OH 43210 **[189]**

Ernö Zador, Department of Entomology, University of Kentucky, Lexington, KY 40546 **[469]**

Preface

This conference, the first of its kind, was predicated on the emergence of a new field—a blend of insect science, molecular biology, and biochemistry. It was particularly significant that it should be held side-by-side, and with considerable cross-over, with a conference on Molecular Strategies for Crop Protection, a title that also covers a new hybrid field of considerable excitement and ferment. While many of the participants in *Molecular Entomology* have emerged from natural products chemistry, insect physiology and endocrinology, genetics, and neurobiology, definite shifts toward the use of molecular techniques were apparent. The natural products chemists now are moving to proteins and tissues, the biochemists to cDNA and genes, and the molecular biologists into more complicated genetic control systems. This engendered an exciting and extremely interactive conference that promises to give the field a new vitality over the next few years.

The conference began with an address by Sidney Brenner, and was shared with the crop protection conference. The first and last plenary sessions also were shared between the conferences, and some of those papers are found in the companion volume, *Molecular Strategies for Crop Protection.* The banquet address by Professor T.R. Odhiambo, Director of the International Centre of Insect Physiology and Ecology, Nairobi, Kenya, is the opening paper in this volume.

Generous financial support to cover expenses for invited speakers was provided by primary sponsorship from Monsanto, with additional support from CIBA-Geigy Corporation, Rohm and Haas Company, Agricultural Biotechnology, Mycogen Corporation, Codon Corporation, FMC Corporation, and Agriculture Chemical Group. Robin Yeaton of the UCLA Symposia staff gave invaluable help and provided a poster presentation on her own research. Hank Harwood was extremely helpful with financial arrangements and Betty Handy expertly managed the editorial arrangements. I am grateful to my colleagues Bill Bowers and John Hildebrand, who provided indispensable help in organizing the program.

Last, but not least, I am indebted to Fred Fox for suggesting the conference and badgering me into serving as its organizer.

John H. Law

Molecular Entomology, pages 1–18

BANQUET ADDRESS

By

THOMAS R. ODHIAMBO
The International Centre of Insect
Physiology and Ecology, (ICIPE),
P.O. Box 30772, Nairobi, Kenya

More than 700 years ago, a dragonfly flitting by pounced upon a gadfly that had stung the arm of the then Emperor of Japan. The Emperor was so overcome by gratitude to the dragonfly that he composed a poem to commemorate this singular event to his person:

> Even a creeping insect
> Waits upon the Great Lord
> Thy form it will bear
> O Yamato, land of the dragonfly.

Other royalty do not necessarily contemplate the dragonfly or any other insect in such a sanguine mood. The Uganda poet, Okot p'Bitek, in Song of Ocol, listens to the sign of a deposed monarch and his encounter with a housefly[1]:

Have you heard
The sigh of a monarch
In Exile?

He squats on a log
In the shadow
Of a disused hut,
It is cold
The keen wind
Knifes through his
Torn trousers
Licking his bruised knee
With rough fenile tongue,

Yesternight!
Yesternight ah!

The smallest toe
On the left foot
Slowly weeps blood,
A fat house-fly
Drones away

But there is more to insects than simply flitting around and droning away. Spare a thought for spring in Italy, for here you will find that at least one insect has aspired to a symbol of national importance. We are told by that intrepid pair, Ronald Taylor and Barbara Carter,[2] of an Italian tradition in May of great character:

> Ascension Day is a day of great festivity in many European countries In Florence, Italy, families celebrate la Festa del Grillo, or Festival of the Cricket, for the cricket is the symbol of spring. A picnic is prepared and the families meet in the Cascino Public Gardens. Here vendors display crickets in brightly colored cages. The children take the crickets home and for those whose crickets are still singing when they arrive home, good luck is promised for the year.

But good luck is not the gift of those who have strayed from the straight-and-narrow path as the Prophet Joel divines:

> What the locust swarm has left
> the great locusts have eaten;
> What the great locusts have left
> the young locusts have eaten;
> What the young locust have left
> other locusts have eaten.

Indeed, it is generally reckoned that about one-third of the world's potential crop production is currently lost to insects, diseases and weeds. The loss is as high as 40% in Africa - where there is the greatest concentration and diversity of insect pests and insect disease vectors than probably anywhere else in the world.

AGRICULTURAL PRODUCTION IN AFRICA

Food production has been declining steadily in Africa during the last two decades or so. Only a few exceptions can be cited in this general, and seemingly inexorable, trend: those of The Ivory Coast, The Cameroon, Malawi, Kenya, and to some extent Zimbabwe. Sets of data compiled by the Food and Agriculture Organization of the United Nations (FAO), the World Food Programme, and several other organizations on the rate of annual growth of the aggregate food production in Africa, when juxtaposed with the rate of annual growth of the human population over the ten-year period 1969-71 to 1977-79, clearly demonstrates the food production is steadily falling behind its base-line level in the 1960s.

What are the causes of this general decline in Africa?

There is no general concensus as yet of the factors behind this general decline, when in other areas of the tropical developing world - Asia and Latin America - agricultural production is very much on the increase, and these areas are in the middle of a "green revolution", or what others prefer to term "yield revolution."[4] One major factor which has been propounded by many authorities, both scientific and geopolitical, is that the recent cycles of drought, and therefore the great variability and unpredictability of rainfall, affecting a large part of Sub-Saharan Africa (the Sahelian zone of West Africa, as well as those areas of eastern, Central and Southern Africa regions showing "Sahelian conditions") is the principal cause of the food production decline. There is a good basis for believing that Sub-Saharan Africa has seen a fluctuating hydrological cycle of drought and wet periods since the last glacial maximum[5]. For instance, there was an increasing aridity in tropical Africa, after the glacial maximum, reaching a peak of aridity approximately 14,000-12,500 B.P. This period was followed by a progressively weter climate; but was then interrupted by two episodes of prolonged drought between 10,200 and 7,400 B.P. The modern pattern and level of lakes in this region became established at around 2,500 B.P.; since when most African lakes have clustered in a band between 2°S and 13°N, and have assumed relatively low water levels or are completely dry. But it is during the recent historical period (since 1500 A.D.) that we have had the most comprehensive studies of the African tropical regimes of climates, most of these data stemming from documentary evidence (e.g. lake levels) and oral evidence (e.g. population migrations). This evidence suggests strongly that there have been major droughts over the last five centuries, and that these drought affected large areas of Sub-Saharan Africa. Africa's dependence of rain-fed agriculture, with its great unreliability in terms of precipitation, has therefore placed great emphasis on _stability_ of agricultural production, rather than on _high yields_.

A second major factor probably responsible for the decline in African agricultural production is the lack of innovative technology - beyond that needed for subsistence food production - to give a push and backstopping for an agricultural production process that will provide a significant surplus even under small-farmer circumstances. For instance, in the African tropics, the pests (insects, ticks, mites, weeds and diseases) of crops and livestock play a very dominant role in the agricultural production system, as well as in the overall socioeconomic development of the region, more than in any other tropical regions of the world[3]; yet, these countries are only now discovering insect science and pest management, and are very far from adopting a strategic programme of research and development (R & D) which would lead them to long-range, sustainable pest and vector management systems. A further example is in the field of intercropping, where African traditional cropping and agroforestry systems have long incorporated several to a large number of plant species planted in the same space and time. Such systems have a structure close to the naturally occurring ecosystem, and has proved itself in many recent studies to be superior to a monocrop farming practice under tropical circumstances in terms of reduced pest pressure, a spreading of risks, higher soil fertility, and more effective exploitation of microclimate and the soil-water profile. A shift from this simulated natural ecosystem to specialised agricultural practices (such as moncrop) disturbs this fragile ecological balance, and creates serious problems in regard to insect pests, plant diseases, and soil conservation. There is no doubt that traditional farmers use resources efficiently within an apparently static technological framework.[6] By the same token, what traditional, resource-poor farmers need - more than the benign neglect they have suffered in Africa this century - is concentrated mission-oriented R & D to produce the kind of technological breakthroughs that would raise farmers' returns from their investment in family labour and other inputs.

A third major factor probably responsible for the food deficit problem in Africa is the almost religious faith in the formula that initiated the green revolution in Asia and Mexico. It is that this sure-fire formula relies on the constellation of four elements that must be provided as a "technological package": the development of high-yielding crop varieties; the provision of irrigation water (in order to assure the farmer of a stable water regime); the provision of adequate fertilizers (for those varieties which usually respond rapidly to such inputs); and the application of pesticides (in order to provide an umbrella against the major pests - insects, weeds, and diseases). Yet, the use of this technological package is not only unlikely to be feasible for the more complex cropping systems practiced in Africa, where the component plant species have diverse and sometimes divergent input requirements, but it is probable that the important yield platform a farmer would wish to reach first is stability (or reliability) of yield. Only then would he experiment with genotypes that give higher yields, but still within his self-perceived higher-level platform of stability of crop yields. It can be demonstrated that it is possible to trigger off a yield revolution in respect of rain-fed sorghum. Indeed, it seems that in India, sorghum is about to join wheat and rice as a green-revolution component crop (Table 1).

TABLE 1. "Green Revolution" Production Increases, India (1967/68 - 1978/79)[7]

CROP	COMPOUND GROWTH RATE		
	TOTAL AREA	TOTAL PRODUCTION	YIELD (KG/HA)
SORGHUM*	1.49	2.07	3.62
WHEAT**	3.16	6.02	2.76
RICE**	0.82	2.64	1.80

*Rain-fed; **Irrigated.

It all began by the introduction, in the 1965/66 season, of an early-maturing (95-100 days) hybrid sorghum CSH-1 in the black soil areas of the Deccan, where otherwise traditional sorghum land races predominated (and took 5-5.5 months to come to maturity). The new hybrid rapidly extended its geographical coverage in India, even though it was susceptible to a number of pests, among them stem-borers, the sorghum shootfly, grain moulds, leaf spot, and _Striga_ weed. Subsequently, hybrids of superior performance (with more resistance to pests, and possessing better grain quality) - such as CSH-5, CSH-6, and CSH-9 - were introduced and became widely grown. All the four early-maturing hybrids provide yield stability of rain-fed sorghum, because their critical growth phases coincide with periods of optimal soil moisture[7]; and breeders can now shoot for a higher level of resistance against insect pests, diseases and weeds, and at the same time test the possibility of harvesting two crops per rainy season in the better rain-fed areas.

The transformation of tropical Africa's traditional agricultural system to a more productive system can only be successful and sustainable if the traditional knowledge base is taken as the initial platform from which to attempt a higher level of agricultural productivity, given the poverty of resources normally available to the majority of the small-scale farming households. In discussing these traditional farming strategies which are the consequence of a long period of trial-and-error and experimentation, Okigbo[8] laments the fact that they are often overlooked or ignored in the rush to make quick agricultural gains by the adoption of exotic farming practices not attuned to the ecological and societal African environment:

> [These traditional farming practices] are often either labelled primitive and inappropriate by those who least understand them or are constantly being assessed when under pressures to achieve objectives for which they were not evolved and in situations to which they are not adapted. On the other hand modern resource management techniques under trial in Africa are either those developed elsewhere to solve different problems under completely different circumstances or they are strategies aimed at achieving maximum short-term or short-sighted returns with serious adverse repercussions on the environment and of decreasing sustainability of benefits to man. Traditional technologies and strategies need to change because the circumstances under which they were evolved are increasingly under pressures of modernization. But as long as the existing ecosystems and societies that live in them cannot be swept overboard..... the only choice we have is to adopt new strategies and develop appropriate technologies that are sensitive to both the old and the new.[8]

This whole-plant, whole-animal approach is only one of many approaches to agricultural production and its attendant problems, including those related to pest management. The great farment in the life sciences - first ushered in some 130 years ago by the publication of On the Origin of Species by Means of Natural Selection by Charles Darwin, then followed closely ten years later by the revelations of genetic inheritance whose foundations were laid down by Gregor Mendel, and lately the revolution of our biological thinking brought about by advances in molecular biology in the last thirty years - has brought these sciences to the leading edge of countemporary science. And it is not surprising then that there are some creative attempts being made to utilize the applied aspects of molecular biology in finding safer, more selective methods of pest management.

THE PROMISE OF BIOTECHNOLOGY

The use of recombinant DNA, which provides the basis for genetic engineering, has opened up two areas of profound potential for biotechnology as applied to the life sciences. First, in the 1970s, methodologies for the rapid analysis of complex biological molecules were developed. This led in 1978 to the compilation of the first computer protein atlas comprising more than 500 proteins. Second, the rapid synthesis of genes was opened up by the discovery of three kinds of enzymes: enzymes capable of slicing the DNA molecule at precise sites; enzymes capable of sealing the loose ends of DNA fragments; and enzymes having the function of synthesizing DNA from messenger RNA. These discoveries led to the rapid synthesis of numerous genes; and by the end of 1980, a computer gene atlas of 350,000 genes had been compiled. Following these two major molecular biological advances, it became possible to synthesize biological molecules to order by implanting specific genes into various kinds of microorganisms.

One of the existing possibilities opened up by biotechnology is the manipulation and amplification of a toxin principle. For instance, Bacillus thuringiensis has been available for commercial use against plant insect pests - as well as certain insect vectors of human tropical diseases, such as mosquitoes and blackflies - since the 1950s. This bacterium accomplishes this result by the production of several toxins, which directly affect specific insect groups. One of these toxins (delta-endotoxin) is deposited as a crystalline product, a protoxin. The protoxin is a single polypeptide chain (of some 80,000-120,000 daltons), and is therefore the product of a single gene. The recent isolation of the gene responsible for protein toxin production, and its cloning in another bacterium (Escherichia coli)[9] has opened up immense possibilities for the manipulation and amplification of this toxin principle for selective pest management, by applying the engineered toxin as a biological pesticides. But the same procedure may make it feasible to transfer the toxin-producing gene into plants, and so enable us to experiment with the strategy of inducing plants to make the toxin in situ for direct insect control: this methodology would be an improvement on the routine use of the toxin as a biological pesticide.

However, we need to approach these new possibilities with the agnostic expectation that the use of microbial pathogens for insect control is not the ultimate weapon, and that the development of resistance to such microbes is a real possibility. Indeed, Boman[10] has drawn up a theoretical scheme of the kinds of mutations needed to give target insects increased resistance to B. thuringiensis.

Ticks, especially Rhipicephalus appendiculatus and its close relatives, are extremely important in much of the eastern and central zone of Africa, as carriers of the protozoan parasite, Theileria parva, the causative agent of East Coast Fever (ECF), a devastating livestock disease, only second in its impact to animal trypanosomiasis, transmitted by tsetse. The conventional method for controlling the tick vectors of ECF and other tick-borne diseases is by close-interval application of acaricides to livestock in dips or sprays. Thus, to control ECF, twice-weekly applications are necessary. This procedure has at least four major disadvantages: the high cost of installing, maintaining and staffing dips and spray races; the high cost of acaricides; the high levels of acaricide residues in beef and dairy products; and the development of resistance to acaricides by the ticks. But perhaps the greatest disadvantage of using acaricides routinely to control ticks is that an inherently unstable situation is created, where regularly treated livestock are completely susceptible to disease and naive to tick infestation. If for any reason acaricide application fails, catastrophies occur. It is at this juncture that the International Centre of Insect Physiology and Ecology (ICIPE), based in Nairobi, Kenya set out to investigate the possibility of an immulogical approach to tick control in the last six years or so.

Two different approaches have been used at the ICIPE to induce resistance in host animals, and are referred to as Type 1 resistance and Type 2 resistance.

Type 1 resistance is stimulated in cattle in response to allergens inoculated with the saliva of the feeding tick. When ticks probe in order to feed on resistant cattle, they excrete allergens in the course of their salivating. These allergens immediately stimulate a hypersensitivity reaction, and marked swelling occurs at the site within 20 min of attachment. This reaction interferes with the ability of the tick to feed properly, and most do not feed to engorgement. The few that do feed, produce small ticks, which have a reduced potential for survival, and the females produce smaller egg batches, mostly sterile. The resistance is produced in the cattle in the first place, by exposing cattle to ticks in paddocks; it is long-lasting, at least for two years. In experiments where cattle made resistant by this procedure have been introduced into a paddock experimentally heavily infested with larval ticks, such paddocks (protected from ingress by other mammals) have been cleared of ticks within a matter of months. This procedure explains the traditional method that the Maasai utilize in clearing their rangelands of ticks, by first keeping goats in a new range (thus greatly reducing the tick burden) before introducing cattle.

Type 2 resistance approach originated at the ICIPE, and depends on the observation that mammalian gamma-globulins ingested by ticks pass unchanged from the blood-meal into the tick's haemolymph. When homogenates of fed female ticks are inoculated into animals, antibodies are produced against the antigens from tick tissues. When later ticks are fed on the immunized host animal, the antibodies ingested with the blood-meal react with the tick target antigens, and thus interfere with their normal function; the result is high tick mortality; those females which survive do not lay eggs, or produce mostly sterile eggs. At least 10 target antigens involved in Type 2 resistance have been identified, one of which has been purified to a high degree. When Type 2 resistance becomes developed into an effective procedure, it will enhance Type 1 resistance for the control of ECF vector ticks.

This novel approach to tick control is paralled by similar approaches at the ICIPE with other target tropical pests. For instance, it has been shown that *Nosema* can be perfected as an instrument for the control of cereal stem-borers otherwise difficult to control by conventional control techniques; further, a newly characterized baculovirus in tsetse, which renders male tsetse sterile, opens up the possibility of bringing a new tool for the control of one of the major scourges of Africa; and, finally, we are beginning to understand part of the story why one of the protozoan causative agents of trypanosomiasis, *Trypanosoma b. brucei,* is usually found in only a very small proportion of tsetse (something like 0.03%), even in circumstances under which trypanosome-infected animals, on which they feed, abound. If *Glossina m. morsitans* is injected with a sub-lethal dose of live *E. coli,* such a procedure confers protection against subsequent lethal doses, and at the same time a dramatic increase of two haemolymph proteins arises. These immune proteins (of 17,000 and 70,000 daltons, respectively) begin to accumulate 18-48 hrs, respectively, after *E. coli* injection. Injection of heat-killed bacteria, on the other hand, had no such effect. If live *T.b. brucei* are injected into the haemocoel of *G.m. morsitans*, they are rapidly eliminated, leaving less than 1% of the trypanosomes 48 hrs after their injection - and these parasites become progressively sluggish. Thus, it is probably that the immune proteins that have been demonstrated soon after the injection episode, have anti-bacterial and anti-trypanosomal properties, as has been demonstrated recently by Boman[10] in similar experiments with the silk moth *Hyalophora cecropia*.

What is being alluded to hear is that in the tropical developing countries, where it has routinely been assumed that there is no choice but to use conventional methods for insect control, mission-oriented research is now being harnessed to tackle these problems in a more rational, systematic way using the most effective basic tools needed for the task. The ICIPE is pioneering this strategic shift.

We know that the simplicity in the use of insecticide has, in the end, proved a mirage; so that, with Picket[11], we are able to say that "...while we can now control almost any specific pest [with insecticides)... nevertheless the problem of controlling pests is more acute than ever." Likewise, plant resistance to pests, although a sound basis for developing an array of other tactics for pest control, can have its own drawbacks. In many cases where these newly engineered crop varieties have been extensively cultivated, the pest condition has eventually developed into epidemic proportions after an initial lull of several seasons or months, particularly where the resistance is conferred by a single gene. For example, previously insignificant pests of rice in South-East Asia have become very important - the whorl maggot, the brown planthopper, the white-backed hopper, and the leaf-folder.[12] Consequently, the one-component approach is no longer effective or sustainable; and the strategy is one of integrated pest management (IPM), by which we take a more holistic approach - through "the selection, integration, and implementation of pest control based on predicted economic, ecological, and sociological consequences".[13] Acceptance of this philosophy implies that the kind of IPM tactics we should experiment with and develop for the management of crop pests could comprise the components summarized in Table 2.

TABLE 2. Major Components of Integrated Pest Management for the Resource-Poor Farmer in the Tropics

1.	VARIETAL RESISTANCE
2.	CHEMICAL CONTROL (selective, strategic)
3.	CULTURAL CONTROL (intercropping phenological synchrony, etc.)
4.	MANIPULATION OF INSECT BEHAVIOUR
5.	BIOLOGICAL CONTROL
6.	GENETIC CONTROL
7.	POPULATION MODELLING
8.	SOCIOECONOMIC CONTEXT

What has been the uniqueness of the ICIPE in its approach to tropical insect science?

THE UNIQUENESS OF THE ICIPE

It is that the founders of the ICIPE realised from its very beginning that the road to the long-range, effective and sustainable pest management in the tropics is likely to be a difficult one, which cannot be approached by the simplistic decision of the sledge-hammer (such as chemical pesticides), nor the militaristic stratagems of some pest eradicators (such as the alteration of the habitat of tsetse, by bush-clearing and game elimination). Our whole experience over the last 50 years, when pest control came into its own as a practitioners calling, is that we cannot agree with Forel[14] when, in commenting on the senses of insects, he concluded that ".... we are asking too much of a poor little insect-brain when we play it such tricks."

We have learnt to know that the only way open to us is to learn all that needs to be known of the basic biology of the selected target pests, including the relevant molecular aspects of their functioning, development, and behaviour, their relations with their hosts and their environment, and then strike at the unstable or fragile episodes and linkages of the life of the target insect. Because of this strategy, the ICIPE scientific community comprises a large number of disciplines, many of them not traditionally associated with insect science - population modelling, social sciences/biosciences interphase, agronomy and veterinary sciences, insect physiology and ecology, natural products chemistry and cell biology, taxonomy and molecular biology, sensory physiology and insect behaviour, and many others - all concentrating on the five or six target arthropods that have been selected for in-depth study because of their African continental or pan-tropical pestiferous importance.

We have tried to understand the insect not making any easy assumptions; and we have come up with priceless information - on the chemical communication world of the grassland mould-building termites, on the sequential diapause of the African armyworm, on the short-range tsetse aphrodisiac, on the rapid evolution of biotypes of the brown planthoper which break down ersturhile resistant rice cultivars, and on the aggregation behaviour and pheromonal regulation of ticks, to mention only a few. We have let our imagination spread its wings, in the belief that we may well understand, and we may then do battle with these tropical insect pests. Alexis Carrel, in his monumental medico-philosophical book, _Man the Unknown_ , sung praises on our sense of curiosity in these words:

> Creative imagination alone is capable of inspiring conjectures and dreams pregnant with the worlds of the future. We must continue asking questions which, from the point of view

> of sound, scientific criticism, are meaningless. And even if we tried to prevent our mind from pursuing the impossible and the unknowable, such an effort would be in vain. Curiosity is a necessity of our nature, a blind impulse that obeys no rule[15]

The scientific community in Africa is small and isolated - more so than in any other developing region of the world. While the primary mission of the ICIPE is to undertake R & D that would lead to the design of novel and effective IPM for the major tropical pests that would be sustainable on a long-term basis, its other major mission is to strengthen the scientific capabilities of the national programmes, especially in Africa. Perhaps, the most innovative means that the ICIPE has pioneered in this respect is the launching of the African Regional Postgraduate Programme in Insect Science (ARPPIS), in which the ICIPE works with a consortium of a dozen African universities to undertake Ph.D. training programmes at the ICIPE, while the universities regularly monitor the academic content of the programme and award the degrees to the successful candidates. At present 32, Ph.D. students frrom 9 African countries are resident at the ICIPE, with 8 of them due to graduate within the next few months. Each year, students are admitted to the course. In time, this cadre of talented young scientists will provide the leadership in insect science in the continent.

Acquiring the capacity for the generation of scientific knowledge and the development of effective technology is a central issue for the sustainable development of Africa, and other tropical developing regions of the world. The ICIPE is intimately engaged in this process in the tropics within the area of insect science and its application to pest management. It has been our privilege to find such dedicated intellectual friends in the major academies of science and advanced research laboratories throughout the world, and donors (including the host country for the ICIPE) to sustain this scientific enterprise, which is so singular in Africa - because it is private.

REFERENCES

1. Okot p'Bitek (1970) Song of Ocol. Nairobi: East Africa Publishing House.
2. Taylor R.L. and Carter B.J. (1976) Entertaining with Insects. Santa Barbara: Woodbridge Press Publishing Company.
3. Odhiambo T.R. (1984) International aspects of crop protection: The needs of tropical developing countries. Insect Sci. Application 5(2), 59-67.
4. Yudelman M. (1985) The World Bank and Agricultural Development - An Insider's View. Washington, D.C.: World Resources Institute.
5. Farmer G. and Wigley T.M.L (1985) Climatic Trends for Tropical Africa. Norwich: Climatic Research Unit, University of East Anglia.
6. Schultz T.W. (1970) Transforming Peasant Agriculture. New Haven: Yale University Press.
7. Rao N.G.P. (1985) Sorghum production in relation to cropping systems, pp. 3-11. In: Proceedings of the International Sorghum Entomology Workshop. Hyderabad: ICRISAT.
8. Okigbo B.N. (1983) Agriculture and agroforestry: Roles in natural resources development in tropical Africa. Paper presented at Consultative Meeting on the United Nations University's Institute for Natural Resources in Africa, Nairobi (Kenya), 26-28 January 1983.
9. Schnepf H.E. and Whiteley H.R. (1981) Cloning and expression of the Bacillus thuringiensis crystal protein gene in Escherichia coli. Proc. Nat. Acad. Sci. 78, 2893-2897.
10. Boman H.G. (1981) Insect responses to microbial infections, pp. 769-783. In: Microbial Control of Pests and Plant Diseases, 1970-1980 (ed. by H.D. Burges). London: Academic Press.

11. Picket A.D. (1949) A critique on insect chemical control methods. Can. Entomol. 81, 67-76.
12. Lim G.B. and Heong K.L. (1984) The role of insecticides in rice integrated pest management, pp. 19.39. In: Proceedings of the FAO/IRRI Workshop on Judicious and Efficient Use of Insecticides on Rice . Los Banos: International Rice Research Institute.
13. Bottrell D.G. (1979) Integrated Pest Management. Washington, D.C.: Council on Environmental Quality.
14. Forel A. (1908) The Senses of Insects. London: Methuen.
15. Carrel A. (1935) Man the Unknown. London: Harper and Brothers, Publishers.

I. MOLECULAR ASPECTS OF THE INSECT NERVOUS SYSTEM

Molecular Entomology, pages 21–31

FROM SEMIOCHEMICAL TO BEHAVIOR: OLFACTION IN THE SPHINX MOTH *MANDUCA SEXTA*[1]

John G. Hildebrand

Arizona Research Laboratories
Division of Neurobiology
University of Arizona, Tucson, AZ 85721

ABSTRACT The olfactory system of the sphinx moth *Manduca sexta* has been studied extensively by the methods of neuroanatomy, neurophysiology, and neurochemistry. These studies, together with behavioral studies of oviposition, mating, and feeding, have revealed much about the organization and functions of chemical senses in controlling important behaviors of insects.

INTRODUCTION

Olfaction plays a major role in the regulation of insect behavior. Orientation and movement toward, and interaction with, sources of food, receptive mating partners, appropriate sites for oviposition, and hosts for parasitism usually involve olfactory signals that initiate, sustain, and guide the behaviors. We study insect olfaction both to contribute to understanding of the structure, function, and ontogeny of chemical-sensory systems in general and to advance understanding of the earth's most numerous and biologically successful fauna. Our research on the experimentally favorable insect, the sphingiid moth *Manduca sexta*, probes the functional organization, cellular neurophysiology, neurochemistry, postembryonic development, and behavioral roles of the antennal olfactory system and other related

[1]This work was supported by NIH grants AI-16150, AI-17711, and AI-23253, and by NSF grants BNS 77-13281, BNS 80-13511, and BNS 83-12769.

sensory pathways. These efforts focus on the male-specific olfactory subsystem responsible for detection of, and integration of information about, the female's sex pheromones, as well as on elements involved in sensing host plants. We are especially interested in cell-by-cell processing of olfactory signals in the central nervous system (CNS). This brief review of our studies emphasizes the functional organization and cellular neurophysiology of the antennal olfactory pathway in the CNS of *Manduca*. Other recent reviews have treated certain aspects of this research in more detail than is possible here (1-4).

FUNCTIONAL ORGANIZATION OF THE ANTENNAL OLFACTORY PATHWAY

Each antenna of *Manduca* consists of 2 basal segments and a long third segment, the flagellum, which is divided into about 80 annuli (5). The basal segments, the scape and pedicel, possess mechanosensory receptor cells associated with known mechanosensory organs, as well as the insertions of the muscles that move the antenna. The male flagellum has an orderly array of about 10^5 sensory organules or sensilla. A few appear to be mechanosensory (*sensilla chaetica)*, but the vast majority are chemosensory (*sensilla basiconica, coeloconica*, and *styloconica*, as well as male-specific *sensilla trichodea*). To date there is no evidence that any of these is gustatory; all appear to be olfactory and may include receptors for water vapor, CO_2, and heat.

The antennal nerve (AN), which runs from the lumen of the antenna to the brain, carries sensory axons (about 2.6×10^5 from the flagellum, in a male, plus hundreds more from the basal segments) into the brain and motor axons (probably no more than 20) out to the antennal muscles. The AN enters the brain at the level of the antennal lobe (AL), which is the most prominent and anteriormost structure in the ipsilateral deutocerebrum (6).

The AL consists of a central region of coarse neuropil (largely principal neurites of AL neurons) surrounded by an orderly array of spheroidal glomeruli (50-100 μm in diameter and made up of terminals or primary olfactory axons, arborizations of CNS neurons, all of the recognized synapses in the AL, and a glial investment) and 3 groups of cell bodies of AL neurons (a large lateral group, a smaller medial group, and a still smaller anterior group) (7-9). In the male, but not in the female, there is a large "macroglomerular complex"

(MGC) located near the entrance of the AN into the AL but outside the array of "ordinary" glomeruli (9-11).

Intracellular staining with cobalt sulfide and Lucifer Yellow has revealed a number of features of the antennal pathway, including the following. [A] Nearly all of the axons from the flagellum -- the olfactory fibers -- project to the glomeruli in the ipsilateral AL. A few, probably mechanosensory axons project to the adjacent mechanosensory and motor center of the deutocerebrum (9,10). [B] Mechanosensory axons from the basal segments of the antenna project to the mechanosensory and motor center of the deutocerebrum and not to the AL (9,10). [C] No antennal sensory fibers project to the contralateral AL (10). [D] Many, and perhaps a majority, of the *ca.* 1000 AL neurons are local, amacrine interneurons, which participate in extensive dendro-dendritic synaptic interactions (in the glomeruli) with other AL neurons (11). [E] The rest of the AL neurons are projection neurons that send their axons out of the AL to other parts of the CNS -- usually to the calyces of the mushroom bodies and/or the lateral protocerebrum, but sometimes to the contralateral AL or other targets (3,12). [F] All local interneurons (LNs) have multiglomerular dendritic arborizations in the AL neuropil, and all projection neurons (PNs) have either uniglomerular dendritic arborizations or ramifications in a small number of glomeruli (3,11, and Montague & Hildebrand, in preparation). [G] Male ALs have male-specific LNs and PNs with dendritic arborizations in the sexually dimorphic MGC (9,11). [H] Axons of primary sensory cells project to single glomeruli without branching before entering their target glomerulus (unpublished observations of Harrow, Kent, Schneiderman & Hildebrand).

Ultrastructural studies (13) have shown that the chemical synapses in the AL, which are confined to the glomeruli, fall into at least 3 morphological classes characterized by different types of synaptic vesicles in the presynaptic elements. The most prominent type, with round, electron-lucent vesicles, represents the primary-afferent terminals and possibly some AL-neuron presynaptic elements as well. All of the synapses reconstructed to date have been "multiplads" in which 2-7 presynaptic elements synapse upon a single postsynaptic element. No gap junctions have been observed between neurites in AL glomeruli.

PHEROMONE-PROCESSING SUBSYSTEM

As has been found in a number of other insect species

(e.g. cockroaches and silkmoths, 14-17), the MGC in the AL of male *Manduca* is responsible for processing afferent information about the female's sex pheromones (11). The afferent input to the MGC is relayed by axons of the receptor cells that innervate the male-specific trichoid sensilla of the male's antennal flagellum. Each of these sensilla houses the receptor dendrites of a pair of olfactory receptor cells (6, Hildebrand & Kaissling, unpublished observations, and Keil, personal communication). One of these receptors responds sensitively and specifically to *E*,*Z*-10,12-hexadecadienal (bombykal), which is the principal pheromone in the female's lure-gland pheromone blend (18 and Hildebrand & Kaissling, unpublished results). The second receptor cell detects a second pheromone in the female's blend, the identity of which is as yet unclear. The axons of these pheromone-detecting, male-specific receptor cells project exclusively to the MGC in the AL (9, 10, and Harrow, Schneiderman & Hildebrand, unpublished observations). All of the AL neurons the receive primary-synaptic input from pheromone-detecting olfactory receptors have dendritic arborizations in the MGC, and all AL neurons with ramifications in the MGC respond postsynaptically when the ipsilateral antenna is stimulated with female pheromones (11).

Gynandromorphs.

Transplantation of antennal imaginal disks between male and female 5th-instar larvae leads to the metamorphic development of gynandromorphic adult moths in which the transplanted disk has given rise to an antenna of the donor's sex and the sensory axons from that antenna project into an AL that belongs to, and thus has the "sex" of, the recipient animal. The AL tissue is thus gynandromorphic (19). When we transplant a male antennal disk into the head of a female larva in place of one of her own disks prior to pupation (several days before the mitotic birth of the antennal sensory neurons), the graft develops into an apparently normal male antenna and its developing sensory neurons send axons into the host female's AL, where they contact and form synapses with processes of AL neurons. The development of at least some of the neurons in the gynandromorphic AL is strongly influenced by the gender of the antennal-sensory axons innervating them (19). A "MGC" develops only in an AL, genetically either male or female, that is innervated by sensory axons from a male antenna. In "female gynandromorphs" -- female moths with ALs innervated by fibers from grafted male antennae -- the "MGC" that arises as a consequence of male-afferent ingrowth is also innervated by dendritic processes of AL neurons that resemble typical male-specific, sexually dimorphic AL neurons (19). Such neurons never are found in normal female ALs.

PHYSIOLOGY OF AL NEURONS

Coordinated physiological and morphological characterization of AL neurons in our studies involves intracellular recording and injection of either cobaltous ions (for subsequent staining by precipitation of cobaltous sulfide and silver intensification) or Lucifer Yellow (11,12,21). Among our physiological findings are the following. [A] In all AL neurons tested to date, sensory and/or electrical stimulation of the ipsilateral antennal afferent input elicited both postsynaptic potentials and action potentials (11,20,21). [B] Exclusively nonspiking responses, often observed in other insect CNS neurons, have never been detected in AL neurons of *Manduca*. [C] Both hyper- and depolarizing synaptic potentials -- inhibitory postsynaptic potentials (IPSPs) and excitatory postsynaptic potentials (EPSPs), respectively -- have been recorded, often in the same AL neuron (11). [D] Primary-afferent synapses are excitatory; IPSPs always have longer latencies and cannot follow at frequencies of primary-afferent stimulation greater than 5-10 hz. [E] LNs usually (or perhaps always) respond to sensory inputs with EPSPs, which often excite the LNs, which typically have very low rates of spontaneous firing, to fire action potentials (11). [F] PNs typically have higher rates of spontaneous, or "background", firing and respond to sensory inputs with a mixture of inhibitory and excitatory postsynaptic potentials that tend to "modulate" the background activity (11).

Male-Specific PNs

The male-specific PNs, or "output neurons", have dendritic arborizations confined to the MGC and can be referred to as "MGC projection neurons" (MPNs). MPNs can be categorized on the basis of their physiological responsiveness to stimulation of the ipsilateral antennal-sensory inputs either with female pheromones or with electric shocks. When odors are delivered to the antennal flagellum, the only effective stimulants eliciting postsynaptic responses in MPNs are female sex pheromones (11,21). Three types of MPNs have been recognized so far: [i] MPNs exhibiting pure excitation in response to female pheromone blend or to electrical stimulation of the ipsilateral AN. About half of the examples to date have responded to only one or the other of the two female pheromones; the other half of our sample have responded equally to both female pheromones. The axons of these cells generally project via the "inner antenno-cerebral tract" to the calyces of the mushroom bodies and the lateral protocerebrum (Christensen & Hildebrand, in preparation). [ii] MPNs tonically inhibited in response to electrical stimulation of the ipsilateral AN. These cells have not yet been studied with olfactory stimuli. [iii] MPNs showing excitation by one female sex pheromone, inhibition by the other, and a mixed response to the natural blend of the two pheromones. Electrical stimulation of the ipsilateral AN elicits similar on-off or off-on responses in these MPNs. These cells apparently integrate pheromonal information in a manner more complicated than that for the other types of MPNs, and their mixed excitatory/inhibitory responses imply more complicated polysynaptic inputs than those driving cells of the type described above under [i]. Indeed, this third type of MPNs seem to be "feature detectors" whose full, normal response is elicited only by the female's natural blend of pheromones (21).

Synaptic Transmission.

To understand how olfactory information is processed and integrated in the ALs, we have begun a comprehensive study of synaptic transmission between neurons in the AL.

Neurotransmitter candidates in the AL. Neurotransmitter "screening" experiments have revealed that the principal transmitter candidates in the AL are acetylcholine (ACh), gamma-aminobutyric acid (GABA), serotonin (5-hydroxytryptamine, 5HT), histamine (HA), tyramine (TA), and a number of apparent neuropeptides including substances immunochemically similar to substance P (SP), locust adipokinetic hormone (AKH), FMRFamide, and corticotrophin-releasing factor (CRF) (12, 22-26). Earlier studies indicated that the primary sensory cells of the antenna are cholinergic and make nicotinic synapses upon their CNS target neurons in the AL (7,8,27). A few AL neurons may also be cholinergic (7,28). About 230 neurons in each AL exhibit GABA-like immunoreactivity; most of these cells are LNs, while a few are PNs (12). 5HT-like immunoreactivity is confined to 1 neuron in each AL (29). Putative neuropeptides, revealed by immunocytochemistry, are distributed as follows: SP (>20 cells); FMRFamide (≥1 cell, possibly a LN but not one containing GABA); CRF (>80 cells, probably LNs that do not contain GABA); and AKH (>10 LNs) (25,26, and Homberg & Hildebrand, in preparation). From these findings it is clear that the LNs, although mophologically largely similar to each other, are of several different types neurochemically and probably therefore exert different physiological effects upon other cells in the AL.

Inhibitory synaptic transmission. Inhibitory synaptic activity plays a prominent role in information processing in the AL (11,30,31). For example, in response to electrical stimulation of the ipsilateral AN, many PNs exhibit complex postsynaptic responses typically comprising a relatively early, compound IPSP (graded with different stimulus intensities) followed by a compound EPSP and burst of action potentials. The compound IPSP (up to 10 mV) can follow afferent stimuli only when they are delivered at relatively low frequency (<10 hz), and the latency of the IPSP, measured from the stimulus artifact, is relatively long (ca. 10 msec or more). These observations suggest that the inhibitory input to the PN is indirect, via at least one inhibitory interneuron (probably in the AL). Injection of hyperpolarizing current simultaneously with afferent stimulation reverses the sign of the PSP, indicating that the IPSP is due to chemical synaptic transmission (31). Moreover, the reversal potential is negative with respect to the cell's resting potential, suggesting that an increase in Cl^- or K^+ conductance underlies the IPSP (31). Preliminary results of Cl^--substitution experiments (involving KCl-filled microelectrodes) suggest that the IPSPs are Cl^--mediated. Treatment of the desheathed preparation with picrotoxin or bicuculline reversibly blocks the IPSP, suggesting that GABA (acting on a receptor that gates a Cl^- channel) mediates the inhibitory effects (31). Injection of GABA into the AL neuropil can hyperpolarize PNs and the response reverses near the apparent cellular resting potential; thus GABA mimics the neurally-evoked inhibitory influences.

ACCESSORY OLFACTORY PATHWAY

In addition to the principal, antennal olfactory pathway, many insects possess an "accessory" olfactory pathway originating at a receptor organ recessed in a depression or pit on the distalmost segment of the labial palp (32,33). The sensilla in this organ are clearly chemosensilla, and probably olfactory receptors, based upon their appearance in the scanning and transmission electron microscopes. Preliminary physiological recording experiments confirm that the receptors of this labial-pit organ (LPO) respond to volatile materials (odors, CO_2, water vapor, or the like) (32). Anterograde staining of the LPO axons with cobaltous sulfide reveals that the sensory neurons of the LPO sensilla send their axons via one of the labial nerves into the subesophageal ganglion. These fibers then continue beyond the subesophageal ganglion to project bilaterally to the two ALs, where the LPO axons terminate in one particular and identifiable glomerulus (the LPOG) in each AL. Double-staining experiments in which the LPO fibers were stained with Lucifer Yellow and the antennal sensory fibers, with cobalt, revealed that the LPOG apparently receives sensory input *only* from the LPO (29). It appears that olfactory inputs from the antennal sensilla and the LPO sensilla are integrated in the AL, and it is likely that there are LNs in the AL that receive synaptic inputs from both types of olfactory afferents.

CONCLUSIONS

Progress toward understanding the "wiring" of the primary olfactory center in the insect brain -- the AL -- has been considerable. It should soon be possible to construct a reasonable model for olfactory information processing in that glomerular center in the brain. It has also begun to become clearer how the output elements of the AL -- the PNs -- encode information about biologically significant odors (at least in the case of the sex pheromones). The next challenge is to understand what happens to that processed olfactory information in the higher centers of the brain, the mushroom bodies and the lateral protocerebrum; where premotor, descending interneurons that are influenced by olfactory information are situated in the brain and how these improtant cells are affected by olfactory inputs; and finally how such descending, olfactorily modulated neural activity is translated into behavioral changes in the free-living insect. The methods are at hand to meet this challenge, and intriguing beginnings have been recorded (34,35). The "black box" of the CNS may soon yield its secrets to persistent investigators determined to understand how semiochemicals control insect behavior.

REFERENCES

1. Hildebrand JG (1985). Metamorphosis of the insect nervous system: Influences of the periphery on the postembryonic development of the antennal sensory pathway in the brain of *Manduca sexta*. In Selverston A (ed): "Model Neural Networks and Behavior," New York: Plenum, p. 129.
2. Schneiderman AM, Hildebrand JG (1985). Sexually dimorphic development of the insect olfactory pathway. Trends in Neurosci 8:494.
3. Hildebrand JG, Montague RA (1986) Functional organization of olfactory pathways in the central nervous system of *Manduca sexta*. In Payne TL, Birch MC, Kennedy JS (eds): "Mechanisms in Insect Olfaction," Oxford: Oxford Univ. Press, in press.
4. Christensen TA, Hildebrand JG (1986) Olfactory information processing in insects -- Functions, organization, and physiology of the olfactory pathways in the lepidopteran brain. In Gupta AP (ed): "Arthropod Brain: Its Evolution, Development, Structure and Functions," New York: John Wiley, in press.
5. Sanes JR, Hildebrand JG (1976) Structure and development of antennae in a moth, *Manduca sexta*. Devel Biol 51:282.
6. Sanes JR, Hildebrand JG (1976) Origin and morphogenesis of sensory neurons in an insect antenna. Devel Biol 51:300.
7. Sanes JR, Prescott DJ, Hildebrand JG (1977) Cholinergic neurochemical development of normal and deafferented antennal lobes in the brain of the moth, *Manduca sexta*. Brain Research 119:389.
8. Hildebrand JG, Hall LM, Osmond BC (1979) Distribution of binding sites for ^{125}I-labeled α-bungarotoxin in normal and deafferented antennal lobes of *Manduca sexta*. Proc Nat Acad Sci USA 76:499.
9. Hildebrand JG, Matsumoto SG, Camazine SM, Tolbert LP, Blank S, Ferguson H, Ecker V (1980) Organisation and physiology of antennal centres in the brain of the moth *Manduca sexta*. In "Insect Neurobiology and Pesticide Action (Neurotox 79)," London: Soc Chem Ind, p 375.
10. Camazine SM, Hildebrand JG (1979) Central projections of antennal sensory neurons in mature and developing *Manduca sexta*. Soc Neurosci Abstr 5:155.
11. Matsumoto SG, Hildebrand JG (1981) Olfactory mechanisms in the moth *Manduca sexta*: Response characteristics and morphology of central neurons in the antennal lobes. Proc Roy Soc (Lond) B213:249.

12. Hoskins SG, Homberg U, Kingan TG, Christensen TA, Hildebrand JG (1986) Immunocytochemistry of GABA in the antennal lobes of the sphinx moth *Manduca sexta*. Cell Tiss Res, in press.
13. Tolbert LP. Hildebrand JG (1981) Organization and synaptic ultrastructure of glomeruli in the antennal lobes of the moth *Manduca sexta*: A study using thin sections and freeze-fracture. Proc Roy Soc (Lond) B213:279.
14. Bretschneider F (1924) Über die Gehirne des Eichenspinners und des Seidenspinners (*Lasiocampa quercus* und *Bombyx mori* L.). Jena Z Naturforsch 60:563.
15. Jawlowski H (1954) Über die Struktur des Gehirnes bei *Saltatoria*. Ann Univ M Curie (Sklowdowska) 8:403.
16. Boeckh J, Boeckh V (1979) Threshold and odor specificity of pheromone-sensitive neurons in the deutocerebrum of *Antheraea pernyi* and *A. polyphemus* (Saturniidae). J Comp Physiol 132:235.
17. Prillinger L (1981) Postembryonic development of the antennal lobes in *Periplaneta americana* L. Cell Tiss Res 215:563.
18. Starratt AN, Dahm KH, Allen N, Hildebrand JG, Payne TL, Roller H (1979) Bombykal, a sex pheromone of the sphinx moth *Manduca sexta*. Z Naturforsch 34c:9.
19. Schneiderman AM, Matsumoto SG, Hildebrand JG (1982) Transsexually grafted antennae influence development of sexually dimorphic neurones in moth brain. Nature 298:844.
20. Tolbert LP, Matsumoto SG, Hildebrand JG (1983) Development of synapses in the antennal lobes of the moth *Manduca sexta* during metamorphosis. J Neurosci 3:1158.
21. Christensen TA, Hildebrand JG (1984) Functional anatomy and physiology of male-specific pheromone-processing interneurons in the brain of *Manduca sexta*. Soc Neurosci Abstr 10:862.
22. Prescott DJ, Hildebrand JG, Sanes JR, Jewett S (1977) Biochemical and developmental studies of acetylcholine metabolism in the central nervous system of the moth, *Manduca sexta*. Comp Biochem Physiol 56c:77.
23. Maxwell GD, Tait JF, Hildebrand JG (1978) Regional synthesis of neurotransmitter candidates in the CNS of the moth *Manduca sexta*. Comp Biochem Physiol 61c:109.
24. Kingan TG, Hildebrand JG (1985) γ-Aminobutyric acid in the central nervous system of metamorphosing and mature *Manduca sexta*. Insect Biochem 15:667.
25. Hoskins SG, Homberg U, Kingan TG, Hildebrand JG (1985) Neurochemical anatomy of the brain of the sphinx moth *Manduca sexta*. In "Neuropharmacology and Pesticide Action

(Neurotox 85)," London: Soc Chem Ind, p 84.
26. Homberg U, Hoskins SG, Hildebrand JG (1985) Immunocytochemical mapping of peptides in the brain and subesophageal ganglion of *Manduca sexta*. Soc Neurosci Abstr 11:942.
27. Sanes JR, Hildebrand JG (1976) Acetylcholine and its metabolic enzymes in developing antennae of the moth, *Manduca sexta*. Devel Biol 52:105.
28. Hoskins SG, Hildebrand JG (1983) Neurotransmitter histochemistry of neurons in the antennal lobes of *Manduca sexta*. Soc Neurosci Abstr 9:216.
29. Kent KS (1985) "Metamorphosis of the antennal center and the influence of sensory innervation on the formation of glomeruli in the hawk moth *Manduca sexta*." PhD Dissertation, Harvard University.
30. Harrow ID, Hildebrand JG (1982) Synaptic interactions in the olfactory lobe of the moth *Manduca sexta*. Soc Neurosci Abstr 8:528.
31. Christensen TA, Waldrop BR, Hildebrand JG (1985) GABA-mediated inhibition in the antennal lobes of the moth *Manduca sexta*. Soc Neurosci Abstr 11:163.
32. Harrow ID, Quartararo P, Kent KS, Hildebrand JG (1983) Central projections and possible chemosensory function of neurons in a sensory organ on the labial palp of *Manduca sexta*. Soc Neurosci Abstr 9:216.
33. Kent KS, Harrow ID, Quartararo P, Hildebrand JG (1986) An accessory olfactory pathway in Lepidoptera: The labial pit organ and its central projections in *Manduca sexta* and certain other sphinx moths and silk moths. Cell Tiss Res, in press.
34. Olberg RM (1983) Pheromone-triggered flip-flopping interneurons in the ventral nerve cord of the silkworm moth, *Bombyx mori*. J Comp Physiol 152:297.
35. Light DM (1986) Central integration of sensory signals, as exploration of processing of pheromonal and multimodal information in lepidopteran brains. In Payne TL, Birch MC, Kennedy JS (eds): "Mechanisms in Insect Olfaction," Oxford: Oxford Univ Press, in press.

Molecular Entomology, pages 33–43

TRANSDUCTION PROCESSES IN OLFACTORY RECEPTORS OF MOTHS

Karl-Ernst Kaissling

Max-Planck-Institut für Verhaltensphysiologie
8131 Seewiesen, FRG

ABSTRACT This paper summarizes present knowledge of olfactory transduction in insect pheromone receptors as obtained from morphological, electrophysiological and biochemical studies including measurements of radiolabelled pheromone compounds. Main topics are the adsorption of pheromone molecules on the antenna, the stimulus transport on and within the olfactory hairs, the generation of receptor potentials and nerve impulses and, eventually, the rapid inactivation and metabolism of the stimulus molecules. The possible role of the pheromone binding protein and of the sensillar esterase is discussed.

INTRODUCTION

Transduction of an olfactory stimulus into nervous excitation involves a complex of several molecular processes, discussed here for the antennae of male silkworm moths of *Bombyx mori* and *Antheraea polyphemus*. First the odor (pheromone) molecules are adsorbed by the olfactory hairs, extended cuticular tubes which contain the sensillum lymph and one or more receptor cell dendrites (fig. 1). The adsorbed molecules penetrate the hair wall and most likely interact with specific molecular receptors at the dendritic cell membrane. This interaction leads to alterations of electrical properties of the receptor cell which generate the extracellularly recorded responses, such as receptor potentials and nerve impulses. Finally, the stimulus molecules have to be inactivated in order to maintain the sensitivity of the receptor cells. Each of the subprocesses of transduction may influence the output signal of the receptor cells, the number of nerve impulses and their distribution over time, both of which carry information about the intensity and time characteristics of the odor stimulus. This paper summarizes the present state of knowledge about these processes, which is based on morphological, electrophysiological and biochemical data and on measurements of radiolabelled pheromone compounds. For a detailed review and more complete references see (1).

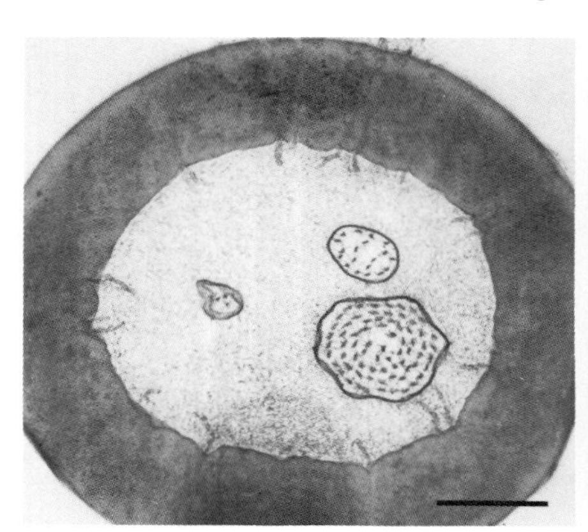

a

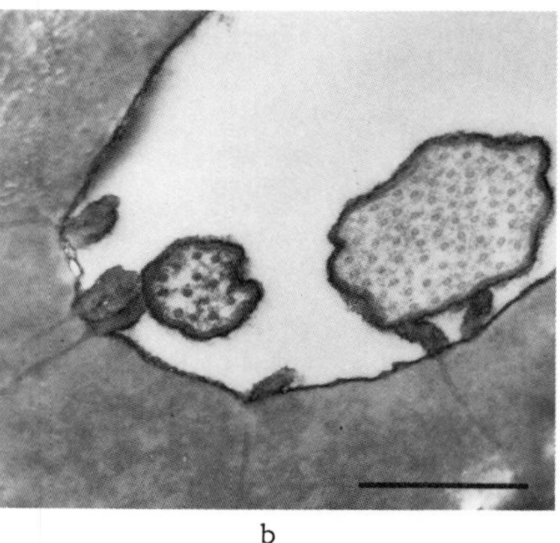

b

FIGURE 1. Cross sections of olfactory hairs (Sensilla trichodea) of Antheraea polyphemus with three (a) and two (b) receptor cell dendrites. a) Freeze substituted preparation. Three of the pore tubules seem to contact the thickest of the three dendrites (by courtesy of T.A. Keil), b) preparation treated with ruthenium red (13). Length of horizontal bars 0.5 μm.

STIMULUS ADSORPTION AND TRANSPORT

The antennae of male saturniid moths have a comb-like structure with an "outline area" of up to 1 or 2 cm^2 (2). About 30% of odour molecules in a free airstream passing a cross sectional area equal to the antennal outline area were adsorbed on the antennae of male moths of Antheraea polyphemus (3). This fraction or "adsorption quotient" was determined using ^{3}H-labelled E-6,Z-11 hexadecadienyl acetate (HDA), a pheromone component of the female moth. A similar value (27%) was measured with ^{3}H-labelled bombykol (E-10,Z-12 hexadecadienol) in male moths of the silkworm

Bombyx mori (4,5).

Only 20% of the adsorbed molecules desorb within 30 min from an antenna in still air, whereas 40% desorb if fresh air is blown over the antenna (3). Therefore, desorption cannot explain the fact that the receptor potential and the nerve impulse response start diminishing immediately after the end of an external stimulus and terminate after a few seconds at physiological stimulus intensities.

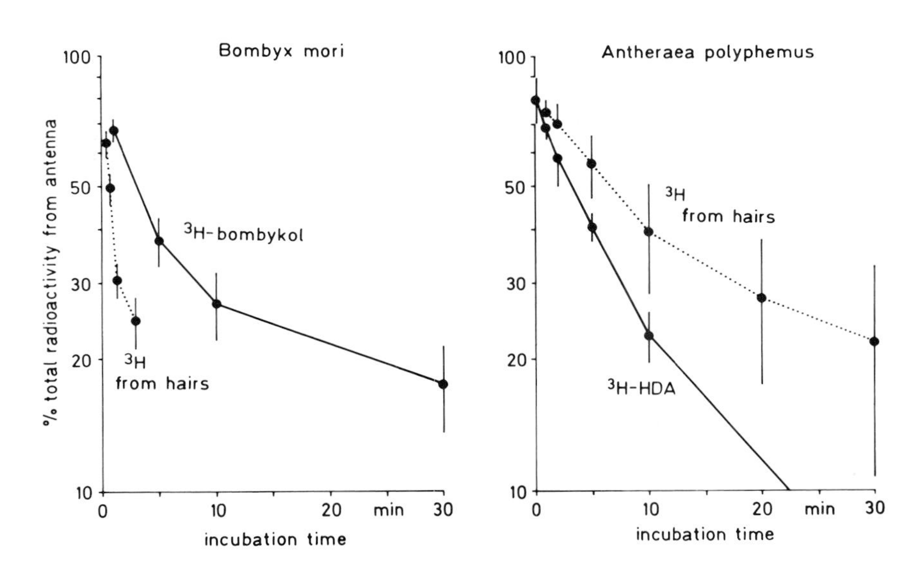

FIGURE 2. Decrease of ^{3}H-activity associated with olfactory hairs (Sensilla trichodea) due to transport from the hairs to the antennal body (dotted lines) (3,6) and decrease of pheromone associated with the antenna, incl. hairs (continous lines) due to enzymatical degradation of bombykol (n=6 to 8, 5 * 10^{10} molecules per antenna, see (27)) and HDA (n=4, 9 * 10^{12} molecules per antenna, by courtesy of G. Kasang). In Bombyx mori considerable degradation must take place after transport from the hairs. In Antheraea polyphemus the transport is delayed due to longer hairs and substantial degradation occurs in the hairs.

Of the adsorbed molecules 80% were found on the long hairs belonging to the sensilla trichodea, which have a length of 100 µm in Bombyx mori and of 300 µm in Antheraea *polyphemus* (fig. 2) (3,6). For the latter species the odour concentration in the hairs increased by a factor of 10^5 over the concentration in air within one second of stimulation at an airstream velocity of 2.5 m/s. The remarkable accumulation of odour molecules by the antenna can be understood on the basis of the laws of convective diffusion and was, in fact, predicted from these laws and from the geometry and arrangement of the sensilla on the antennae of moths (7).

The adsorbed stimulus molecules migrate from the hairs towards the antennal branches, obviously following a concentration gradient (fig. 2). The velocity of this transport corresponds to diffusion coefficients of $5 * 10^{-7}$ cm^2/s as estimated for bombykol in *Bombyx mori* (6) and $3 * 10^{-7}$ cm^2/s for HDA in *Antheraea polyphemus* (3). These values were determined after extremely strong stimuli of about 10^8 molecules loaded on each hair within 10 s. Such high densities of stimulus molecules -- maximally about 10 molecules per nm^2 of hair surface -- may alter the diffusion velocity.

One hundred times lower diffusion coefficients would be sufficient according to the average latencies of the receptor cell responses observed at low stimulus intensities (several hundred ms) (1,4,8). Such values are in the range of diffusion coefficients of molecules in lipid membranes (9) and may not be implausible for diffusional transport on the hairs -- under certainly non-ideal conditions. It can, however, not be excluded that the molecules arrive at the dendritic membrane after only a few ms, as expected from the above diffusion coefficients, and that other processes are time limiting -- e.g., the recognition of the stimulus molecule at the cell membrane.

As for desorption, the migration of the stimulus molecules from the hairs is also much too slow to explain the decrease of the cell responses after the end of exposure to an odor stimulus. Therefore, an inactivation of the stimulus molecules must occur on or within the olfactory hairs.

The stimulus molecules are thought to interact with highly specific molecular receptors in the membrane of the receptor cell (10). It is tempting to speculate that the lipophilic odor molecules do not enter the extracellular sensillum fluid on their way towards the receptor cell, but reach the dendritic membrane directly via the inner extensions of the pore tubules (11). However, in many electron-microscopic preparations few or no contacts between pore tubules and receptor cell membrane are visible (fig. 1a) (12,13).

It is clear that the pore tubules are not extensions

of the receptor cell dendrite. The hair wall including the pore tubules is formed by the trichogen cell, one of the three accessory cells. Only after this cell withdraws from the hair lumen do the receptor cell dendrites invade the hair, where they may form contacts with the pore tubules (14,15). Cationic markers (cationized ferritin, ruthenium red) of the cell coat on the dendritic membrane also accumulate at the inner extensions of the pore tubules, indicating the presence of negatively charged material which may be involved in the formation of contacts (13).

Actually, the cationic agents seem to increase the number of contacts in the preparation (fig. 1b). Recently, it has been demonstrated that certain lectins bind specifically to the dendritic membrane and others to the pore tubules (16).

As the alternative to the contact model, the stimulus molecules would enter the sensillum lymph before reaching the sites of interaction at the cell membrane. In this case they might immediately bind to the pheromone-binding protein present in the sensillum lymph in an extraordinary high concentration (~10mM (17)). This protein may then carry the pheromone molecule towards the cell membrane (18). In the contact model, the pheromone-binding protein could bind and thereby inactivate the odour molecule after the stimulatory interaction (see below).

Both the contact model and the carrier model have to cope with the fact that, eventually, a considerable amount of pheromone (or its metabolites) moves into the hair lumen. Sensillum lymph collected 1-2 min after stimulation of male Antheraea polyphemus with ^{3}H-labelled HDA contained 40% of the radioactivity associated with the hairs (3)

GENERATION OF ELECTRICAL SIGNALS

A combined electrophysiological, behavioral and radiometric study of Bombyx mori led to the conclusion that one molecule of bombykol is sufficient to trigger a nerve impulse (4,5). The first electrical response to a single pheromone molecule is the so-called elementary receptor potential preceding the nerve impulse (1,8,19). These extracellularly recorded events are negatively directed potential fluctuations of an amplitude up to 0.5 mV and a duration of 10-50 ms or more. Sometimes these potentials are rectangular in shape, but ordinarily they are rounded, possibly due to capacitive coupling in the sensillar circuit. According to an analysis of static electrical properties of the sensillum, the elementary potentials might reflect a change of conductance in the dendritic membrane in the range of 30 pS. Therefore, these events may indicate opening of single ionic channels in the cell membrane (8).

This model of chemo-electrical transduction does not require amplification mechanisms other than ion gating in a charged cell membrane such as has been found, for instance, in muscle cells sensitive to acetylcholine (20).

The static model of the sensillum includes the assumption of a relatively high specific resistance of the dendritic membrane (above $3 * 10^3\ \Omega cm^2$). An analysis of dynamic electrical properties of the sensillum (21) puts the resistance of the dendritic membrane above $2000\ \Omega cm^2$. This author, however, questions the current idea of nerve impulse initiation in the soma region of the receptor cell (22). Instead he proposes that the impulses are initiated in the receptor cell dendrite as has been suggested for certain mechanoreceptive sensilla (23). There is no model available which simulates the static and dynamic electrical behavior of the sensillum in a satisfactory way. Therefore, other mechanisms of generation of electrical responses cannot be excluded. Any model of transduction has to consider the unusual ionic conditions in the sensillum lymph: 200 mM of K^+ and 25 mM of Na^+ (8).

At weak stimulus intensities the elementary receptor potentials are irregularly distributed, as expected for random molecular stimuli (1). The reaction time of the responses is several hundred ms on average and varies between several tens of ms and several seconds. This reaction time may reflect the transport of the molecule from the adsorption site at the hair surface through the pore tubule towards the cell membrane as discussed above. Alternatively, the transport could be quicker and other stochastic processes, such as the assumed interaction between stimulus and receptor molecule may be responsible for the delay of the responses. With increased stimulus intensities, the elementary responses superimpose and form the typical fluctuating receptor potential.

SPECIFICITY OF THE CELL RESPONSE

Minimal alterations of pheromone molecules -- for example, removal, shifting or change of configuration of a double bond -- reduce the stimulus effectiveness 10 to 1000 fold if determined for a given mean amplitude of the receptor potential. In addition, certain pheromone derivatives produce altered fluctuations of the receptor potential (8,19). Shortening of the pheromone molecule by one or two CH_2-groups or removal of a double bond can produce much smoother receptor potentials. This response type may result from superimposition of large number of comparatively small elementary responses. The single events may have a much smaller duration, beyond the limits of resolution of the recording method. Ligand-dependent life times of the open state of ion channels are known from the acetylcholine receptor (20).

These observations show that the structure of the stimulus molecule does not only influence the probability of an excitatory effect but also induces different types of effects -- e.g., different types of conformational changes of a receptor molecule. Therefore, the effectiveness of a stimulus compound cannot adequately be described by one single number as usually done in studies on structure-activity relationships.

ODOR INACTIVATION

Two possible mechanisms of inactivation have been discussed for pheromone molecules adsorbed on moth antennae: a) enzymatic degradation of the pheromone molecules associated with the hairs (18,24,25), and b) binding of the stimulus molecules to the abundant pheromone binding protein found in the sensillum lymph (17,26).

Enzymatic degradation of the pheromone component bombykol (E-10,Z-12 hexadecadienol) has been shown to occur on antennae of male and female *Bombyx mori* and also in other parts of the body. Two types of metabolic products, acids and esters, were eluted from intact antennae previously exposed to airborne pheromone stimuli. Therefore, at least two pheromone-degrading enzymes are associated with the antennae. Marked conversion of pheromone into acids but not into esters was found on the scales, and on the legs tightly covered with scales. One function of the pheromone degradation is, obviously, to avoid secondary stimulation by odour adsorbed on the insect body (27). Pheromone metabolism was also observed in other moths including *Antheraea polyphemus* (18,25). However, the half life of pheromone associated with intact antennae was about 3 min for bombykol over more than 100-fold range of stimulus loadings on *Bombyx* antennae (27) and in a similar range for HDA loaded on the antennae of Antheraea polyphemus (fig. 2). The pheromone concentrations used in these experiments elicited electrical responses ceasing within seconds after termination of the stimulus (3). Therefore, a rapid nonenzymatic inactivation mechanism has been postulated in addition to the slow enzymatic degradation.

The possible reduction of free pheromone concentration (S_{free}) in the sensillum lymph due to pheromone binding was calculated for Antheraea polyphemus (1). In all experiments the total initial concentration (S_{tot}) of pheromone in the hair was far below the total concentration of binding protein ($B_{tot} = 10^{-2}$ M). Furthermore, the dissociation constant ($K_D = 6 * 10^{-8}$ M) (26) of the pheromone-binding protein and HDA was far below B_{tot}. Therefore, the mass action equation

$S_{free} / S_{bound} = K_D / B_{free}$ can be converted into
$S_{free} / S_{bound} = K_D / B_{tot} = 6 * 10^{-6}$. The formula shows a strong reduction of the free pheromone concentration under equilibrium conditions. This inactivation mechanism would function in the contact model (see above) if the stimulus molecules reach the assumed receptors in the dendritic membrane before binding to the protein occurred. The pore tubules may have a protecting function besides serving as lipophilic pathways for the odour molecules.

In contrast, the carrier model requires that the odour molecules bind to the protein before arriving at the cell membrane. This means that binding of the odour molecule to the protein should not interfere with its excitatory effect at the cell membrane. Therefore, another rapid inactivation mechanism has to be postulated for the carrier model.

The astonishingly high rate of pheromone degradation found for the sensillar esterase in the sensillum lymph of male Antheraea polyphemus moths was the reason to reconsider the idea that the rapid inactivation may be enzymatic. The electrophoretically isolated enzyme degraded the female pheromone component HDA with a maximal velocity v_{max} of $5 * 10^{-9}$ M/s, for one antennal unit of enzyme in one ml (18). In the sensillum lymph of 55,000 hairs per antenna, v_{max} would be F = ml/55,000 pl = $1.82 * 10^{4}$ times higher if the enzyme is located exclusively in the 1pl volume of the hair lumen. With $v_{max\ adj} = F * v_{max}$ one calculates a half life (τ) of the pheromone compound of 15 ms under first order conditions, i.e. at free pheromone concentrations below K_M (= $2 * 10^{-6}$ M), from the equation

$\tau = \ln 2 * K_M / v_{max\ adj}$ (18). Therefore, the enzyme alone would be sufficiently fast for the rapid inactivation in the carrier model. However, the enormous discrepancy between the calculated value and the pheromone half life of 3 min found in intact antennae (see above) needs explanation before a model can be accepted.

The discrepancy may be solved if the pheromone concentration (S_{free}) available for the enzyme is reduced due to the presence of the pheromone-binding protein. Indeed addition of binding protein considerably slowed down the enzymatic reaction in vitro (28). In vivo, the half life of HDA within the hair could maximally increase by the factor $S_{tot} / S_{free} = 1/6 * 10^6$, to as much as 45 min, if the binding to the protein equilibrates faster than the enzymatic reaction (1). The half life of the pheromone reduces to 4.5 min if one considers a 10% yield of the enzyme isolation (estimated from (18) and (29)). This value agrees with the half life of pheromone on intact antennae (fig. 2). The conclusion seems clear, that the enzymatic breakdown of the pheromone cannot account for the postulated rapid inactivation in either model. Since for logical reasons binding of the stimulus to the protein cannot serve for inactivation in the carrier model, this model so far lacks a mechanism for rapid inactivation. Therefore, the contact model with the binding protein as inactivator is more complete.

The small number of contacts observed in electron-microscopic preparations is puzzling. However, these contacts are very sensitive to preparation artifacts. They may be of temporary nature. Considering the average latencies of the elementary receptor potentials (see above), there is plenty of time for the molecule eventually to "find" a tubule attached to the cell membrane.

REFERENCES

1. Kaissling K-E (1986). Chemo-electrical transduction in insect olfactory receptors. Ann Rev Neurosci 9:121-45.
2. Boeckh J, Kaissling KE, Schneider D (1960). Sensillen und Bau der Antennengeißel von Telea polyphemus. Zool Jb Anat 78:559-84.
3. Kanaujia S, Kaissling K-E (1985). Interactions of pheromone with moth antennae: adsorption, desorption and transport. J Insect Physiol 31:71-81.
4. Kaissling K-E, Priesner E (1970). Die Riechschwelle des Seidenspinners. Naturwiss 57:23-28.
5. Kaissling K-E (1971). Insect olfaction. In Beidler LM

(ed): "Handbook of Sensory Physiology" Berlin: Springer Verlag, p 351-431.
6. Steinbrecht RA, Kasang G (1972). Capture and conveyance of odour molecules in an insect olfactory receptor. In Schneider D (ed): "Olfaction and Taste IV" Stuttgart: Wiss Verlagsges, p 193-199.
7. Adam G, Delbrück M (1968). Reduction of dimensionality in biological diffusion processes. In Rich A, Davidson N (eds): "Structural Chemistry and Molecular Biology," San Francisco: Freeman, p 198-215.
8. Kaissling K-E, Thorson J (1980). Insect olfactory sensilla: structural, chemical and electrical aspects of the functional organization. In Sattelle DB, Hall LM, Hildebrand JG (eds): "Receptors for Neurotransmitters, Hormones and Pheromones in Insects," Amsterdam: Elsevier/North-Holland, p 261-282.
9. Träuble H, Sackmann E (1972). Studies of the crystalline-liquid crystalline phase transition of lipid model membranes. III. Amer Chem Soc 94:4499-4510.
10. Kafka WA (1970). Molekulare Wechselwirkungen bei der Erregung einzelner Riechzellen. Z vergl Physiol 70:105-43.
11. Steinbrecht RA (1973). Der Feinbau olfaktorischer Sensillen des Seidenspinners (Insecta, Lepidoptera): Rezeptorfortsätze und reizleitender Apparat. Z Zellforsch Mikrosk Anat 139: 533-565.
11. Keil TA (1982). Contacts of pore tubules and sensory dendrites in antennal chemosensilla of a silkmoth: demonstration of a possible pathway for olfactory molecules. Tissue & Cell 14:451-462.
12. Keil TA (1984). Reconstruction and morphometry of silkmoth olfactory hairs: a comparative study of sensilla trichodea on the antennae of male Antheraea polyphemus and Antheraea pernyi (Insecta, Lepidoptera). Zoomorphol 104:147-56.
13. Keil TA (1984). Surface coats of pore tubules and olfactory sensory dendrites of a silkmoth revealed by cationic markers. Tissue & Cell 16:705-17.
14. Ernst KD (1972). Die Ontogenie der basiconischen Riechsensillen auf der Antenne von Necrophorus (Coleoptera). Z Zellforsch mikrosk Anat 129:217-36.
15. Sanes JR, Hildebrand JG (1976). Origin and morphogenesis of sensory neurons in an insect antenna. Dev Biol 51:300-19.
16. Keil TA (1986). Lectin binding studies on olfactory sensilla of the silkmoth Antheraea polyphemus. Verh

Dtsch Zool Ges 79, in press.
17. Vogt RG, Riddiford LM (1981). Pheromone binding and inactivation by moth antennae. Nature (Lond) 293:161-63.
18. Vogt RG, Riddiford LM, Prestwich GD (1986). Kinetic properties of a pheromone degrading enzyme: the sensillar esterase of Antheraea polyphemus. Proc Nat Acad Sci 82:8827-31.
19. Kaissling K-E (1974). Sensory transduction in insect olfactory receptors. In Jaenicke L (ed): "Biochemistry of Sensory Functions," Berlin: Springer, p 243-73.
20. Neher E, Sakmann B (1976). Single-channel currents recorded from membrane of denervated frog muscle fibers. Nature 260:779-802.
21. De Kramer JJ (1985). The electrical circuitry of an olfactory sensillum in Antheraea p. J Neurosci 5:2484-93
22. Morita H, Yamashita S (1959). The back-firing of impulses in a labellar chemosensory hair of the fly. Mem Fac Sci Kyushu Univ, Ser E (Biol) 3:81-87.
23. Erler G, Thurm U (1981). Dendritic impulse initiation in an epithelial sensory neuron. J Comp Physiol 142:237-49.
24. Kasang G (1971). Bombykol reception and metabolism on the antennae of the silkmoth Bombyx mori. In Ohloff G, Thomas AF (eds): "Gustation and Olfaction," London/New York: Acad Press, p 245-50.
25. Ferkovich SM, Mayer MS, Rutter RR (1973). Sex pheromone of the cabbage looper: reactions with antennal proteins in vitro. J Insect Physiol 19:2231-43.
26. Kaissling KE, Klein U, de Kramer JJ, Keil TA, Kanaujia S, Hemberger J (1985). Insect olfactory cells: electrophysiological and biochemical studies. In Changeux JP, Hucho A, Maelicke A, Neumann E (eds): "Molecular Basis of Nerve Activity," Berlin/New York: W. de Gruyter & Co, p 173-83.
27. Kasang G, Kaissling K-E (1972). Specificity of primary and secondary olfactory processes in Bombyx antennae. In Schneider D (ed): "Int Symp Olfaction and Taste IV," Stuttgart: Wiss Verlagsges, p 200-6.
28. Vogt RG, Riddiford LM (1986). Pheromone reception: a kinetic equilibrium. In Payne T, Birch M, Kennedy S (eds): "Mechanisms in Insect Olfaction," Oxford University Press, in press.
29. Klein U, Keil TA (1984). Dendritic membrane from insect olfactory hairs: Isolation method and electron microscopical observations. Cell Molec Neurobiol 4:385-96.

Molecular Entomology, pages 45–56

MUTATIONS AFFECTING THE EMBRYONIC DEVELOPMENT OF THE PERIPHERAL NERVOUS SYSTEM IN DROSOPHILA[1]

Yuh Nung Jan*, Rolf Bodmer*, Lily Yeh Jan*, Alain Ghysen+ and Christine Dambly-Chaudiere+

*Howard Hughes Medical Institute and
*Department of Physiology, University of California
San Francisco, California 94143
+Laboratoire de Genetique, Universite Libre de Bruxelles, 67, rue des Chevaux, 1640,Rhode-St.-Genese, Belgium

ABSTRACT In a genetic approach to problems concerning the formation of the nervous system, we began by systematically screening chromosomal deletions and embryonic lethal mutations for alterations of embryonic neural development. About one third of the genome has been scanned this way. With the help of several neuronal specific antibodies, we examined the anatomy of the nervous system of these mutant embryos and focused our attention primarily on the peripheral nervous system. More than 20 regions or loci appear to be primarily involved in the early development of the embryonic peripheral nervous system, while other mutations with other alterations also cause neuronal defects. We discuss the question as to what a specific neuronal defect could encompass.

INTRODUCTION

How a nervous system is built during development remains a major unsolved problem in biology. It would be particularly intriguing to learn how intrinsic factors and external cues influence the determination and differentiation of individual neurons. If we consider the early development of the peripheral nervous system, the problem can be divided into three basic questions:

[1] This study was supported by NIH grant no. NS19191.

1) What causes an ectodermal cells to become a neuron at a particular location?
2) How does a neuron acquire its identity?
3) What factors determine the guidance of the axonal (and dendritic) pathway of a neuron?

An attractive approach in Drosophila to address these questions at the molecular level is to make use of its well established genetics. This approach was very successful in studying problems of developmental biology. For example, in recent years, Nusslein-Volhard, Wieschaus and Jurgens (1,2,3) systematically and exhaustively searched for genes involved in cuticular patterning in Drosophila and found a limited number of genes which were responsible for the basic segmental body plan. Analysis of many of those genes have now progressed to the molecular level. Combination of molecular and genetic studies of these genes suggest that they interact with one another in regulating their regional expression and that they may act in a combinatorial way to produce positional information.

A systematic but less extensive search for genes involved in early neural development of the central nervous system (CNS) was attempted by Campos-Ortega and his coworkers (4,5). They described 7 genes that give a neuralized CNS phenotype when mutated (i.e. cells of the neurogenic region that normally become epidermal cells turn into neurons). One such gene, Notch, has been cloned. Interestingly, the molecular analysis revealed that the Notch gene product has striking homology to mammalian epidermal growth factor as well as the gene product of lin-12, a nematode gene controlling cell lineage (6,7). At the time when Campos-Ortega and his colleagues did the search, methods for visualizing the entire nervous system with single cell resolution were not yet available and they described mutations causing fairly general abnormalities in the central nervous system. More recently, antibodies specific for neuronal structures became available (8,9). These tools allowed the resolution of the highly stereotyped pattern of the peripheral nervous system (PNS) in the Drosophila embryo at identified cell level. The number, position and identity of those neurons and their dendritic and axonal processes have been fully characterized in the periphery (10,11, Bodmer & Jan in preparation). Thus, the anatomy of the embryonic nervous system provides a very sensitive assay system for identifying mutations affecting early neural development. This assay enabled us to attempt a search for genes that may be specifically involved in the early neuronal development in Drosophila. Having screened through one-third of the genome, we found 10-20 regions or loci which probably are fairly directly involved in neural development. Besides these, a fair number of regions or loci (maybe less than 30) clearly have effects on the ectodermal differentiation in addition to the underlying neuronal structures. A detailed description of this work will be published elsewhere. In the following we give a brief account of the range of phenotypes observed, and a rough estimate of the number of regions (loci) involved for the different processes.

METHODS

Staining of Embryonic Nervous System.

The embryonic nervous system was visualized by staining the embryos in whole mount with either a neuronal-specific antibody originally raised against horseradish peroxidase (8) or a monoclonal antibody (21A4) which stains all peripheral and a small subset of central neurons in Drosophila (9,10), and second antibodies labeled with horseradish peroxidase. Normarski optics was used in visualizing the stained embryos.

Stocks.

Stocks used in this study include most of the deficiencies and some of the point mutations previously described by Nusslein-Volhard, Wieschaus and Jurgens (1,2,3).

RESULTS

The Embryonic Nervous System in Wild-Type Drosophila.

The entire embryonic nervous system (both PNS and CNS) of Drosophila can be visualized by using an antibody specific for the nervous system (Fig. 1). The embryonic PNS consists of sensory and motor components in a stereotyped arrangement that repeats itself from segment to segment. The sensory neurons have their cell bodies located in the periphery and their axons projecting to the CNS. The cell bodies of motor neurons are located within the CNS and their axons extend peripherally to innervate the muscles. The entire sensory nervous system has been characterized in detail (10,11, Bodmer & Jan in preparation). A brief summary is provided in the following.

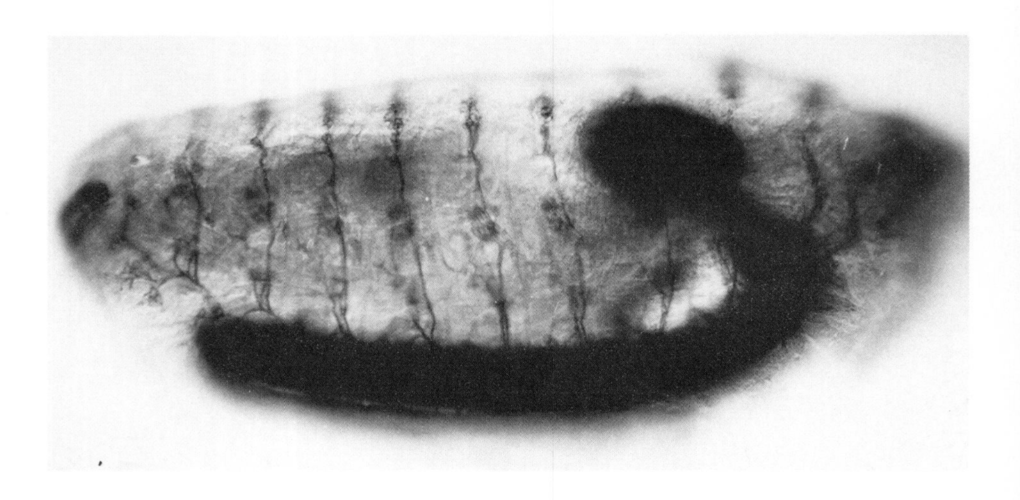

FIGURE 1. Central and peripheral nervous system in a 14 h wild-type Drosophila embryo stained in whole mount with antibodies against horseradish peroxidase (HRP), which binds to the cell surface of all neurons in Drosophila. The same antibody was used in the subsequent figures.

The peripheral sensory neurons can be divided into five classes according to their morphology (and putative function):

(1) ES neurons which innervate External Sensory structures (mostly trichoid and campaniform sensilla)
(2) CO neurons which innervate internal presumptive proprioreceptors (Chorodotonal Organs)
(3) DA neurons which have elaborate Dendritic Arborization and probably serve as proprio- or touch receptors
(4) BD neurons which have Bipolar Dendrites. Their function is unknown
(5) TD neurons with Dendrites that wrap around a Tracheal branch

The peripheral nervous system is bilaterally symmetric. The most anterior segments (head and 1st thoracic)and the most posterior segments (abdominal 8th and 9th segments) have unique patterns. The second and third thoracic segments exhibit the same pattern, while the first through seventh abdominal segments share another common pattern. For example, in each half of the abdominal segment there are 44 neurons organized into four clusters, one dorsal, one lateral and two ventral. The dorsal cluster contains only ES, BD and DA neurons. The lateral and ventral clusters contain all five types of neurons. The thoracic cell type distribution is similarly mixed.

Deficiencies and Mutations Affecting Embryonic Neural Development.

So far, we screened about 70 deficiencies covering approximately 30% of the genome and a number of embryonic lethal point mutations. We limit our attempt to classify the wide variety of different phenotypes essentially to those affecting the PNS. Many mutant PNS defects were, however, accompanied by various degrees of CNS defects. Head defects were very common and have been essentially disregarded since we concentrated on the PNS in the body segments. The problem of extrinsic, non-neuronal structures affected in the deficiency or point mutants had to be dealt with individually, since the occurrence of other alterations during gastrulation, segmentation, or later stages does not preclude a specific neuronal defect: 1) the different phenotypes may be attributed to different genes present in the deficiency; 2) a particular gene may have more than one tissue specific action; 3) the "non-neuronal" mutations might affect neuronal development by altering positional cues for neuronal differentiation or pathfinding cues for axons or dendrites. In these cases we compared the type and severity of the gastrulation or segmentation defect with the specificity and severity of the neuronal defects to decide tentatively if the neuronal defect was likely to be the simple consequence of a more general defect.

The PNS abnormalities are classified as follows:

Class 1) Gross derangement of the entire nervous system. CNS and PNS are generally equally affected.

We found more than 10 deficiencies and point mutations within this category. An example is Df(1)RA2 uncovering stardust (std) (2) shown in Fig. 2.

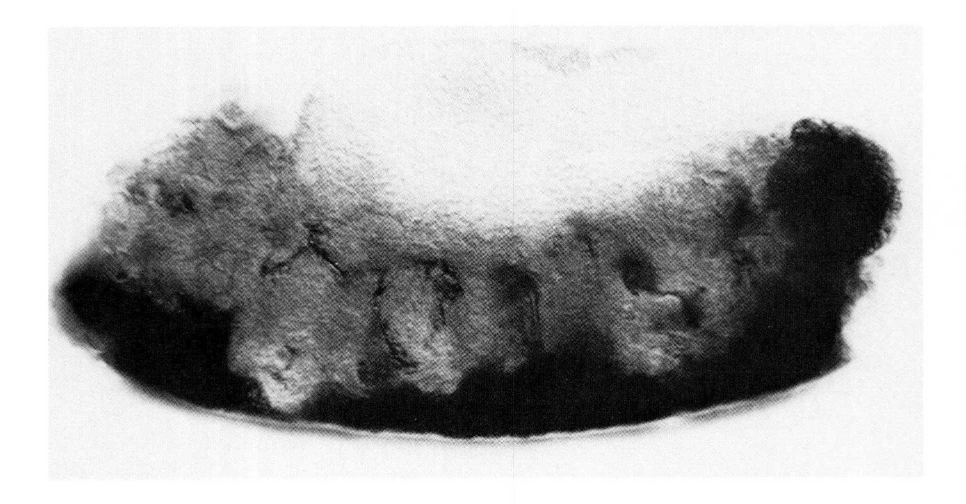

FIGURE 2. Effects of the deficiency Df(1)RA2 on the central and peripheral nervous system. This embryo is homozygous for the deficiency.

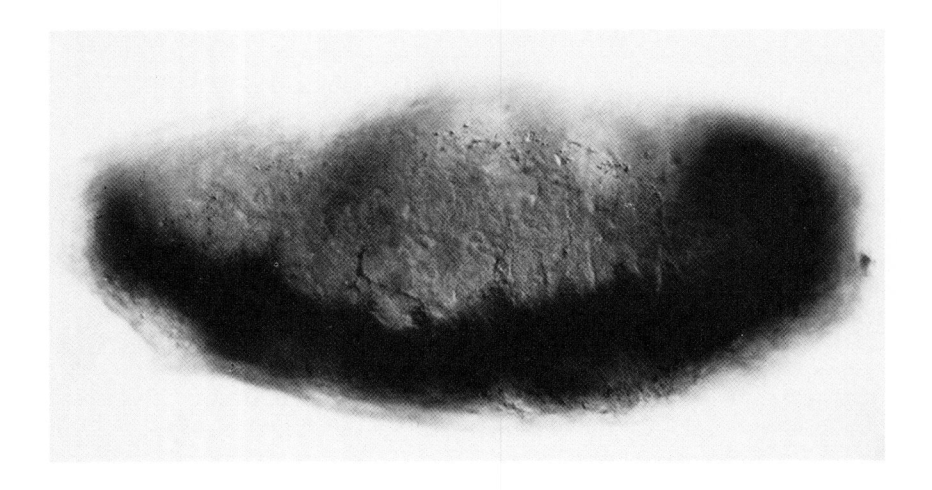

FIGURE 3. Effects of the deficiency Df(1)NI9 on the embryonic nervous system. Note that most peripheral neurons are missing while the central nervous system is only moderately deranged.

Class 2) General reduction of the PNS. Phenotypes range from a moderate reduction of neurons to the complete absence of the PNS. Often we saw misrouting (class 6 defect, see below) associated with the reduction of neurons.

There were a number of deficiencies and point mutations found in this category. Df(1)NI9 is shown in Fig. 3 as an example. The CNS is only moderately deranged while the PNS is almost entirely absent with only occasional fibers or cell bodies. Even though segmentation seems somewhat deranged it cannot be the cause of the absence of the entire PNS.

Class 3) Specific alterations in the number of cells of a particular type. These together with class 4 are probably the most specific mutations affecting neuronal determination. Several mutations would specifically delete, duplicate or reduce cells of a subgroup or a cell type.

More than 10 deficiencies and point mutations were found. For example Df(1)svr deletes all the ES neurons (12), while Df(3L)vi^3 deletes the CO neurons but probably none of the other types of neurons (Fig. 4). Another example is Df(2L)E55 which reduces specifically the lateral abdominal five CO neurons to three (Fig. 5).

Class 4) Specific transformation of a set of neurons into another. A number of mutations of an identified locus cause transformation of ES neurons into CO neurons without affecting any other sensory neurons. The existence of this type of transformation suggests that ES and CO neurons have a common pathway of determination and that this gene may serve as a molecular switch for an ES neuron to acquire its identity (manuscript in preparation).

Class 5) Strongly misrouted or reduced pathways. The phenotypes ranged from moderately to completely erratic misrouting. Often fibers did not respect segmental boundaries and join neighboring nerve trajectories. Other phenotypes show aborted pathways and profuse branching. Such mistakes are rare in the wild-type. This category may include mutations affecting extrinsic guidance cues as well as mutations causing failure of axons in reading these cues.

More than 10 regions or loci were found in this category. Five cases are particularly severe. Many of these mutants also cause some disorganization of the peripheral neuronal arrangement. Df(1)HA32 provides an example of considerable misroutings (Fig. 6)

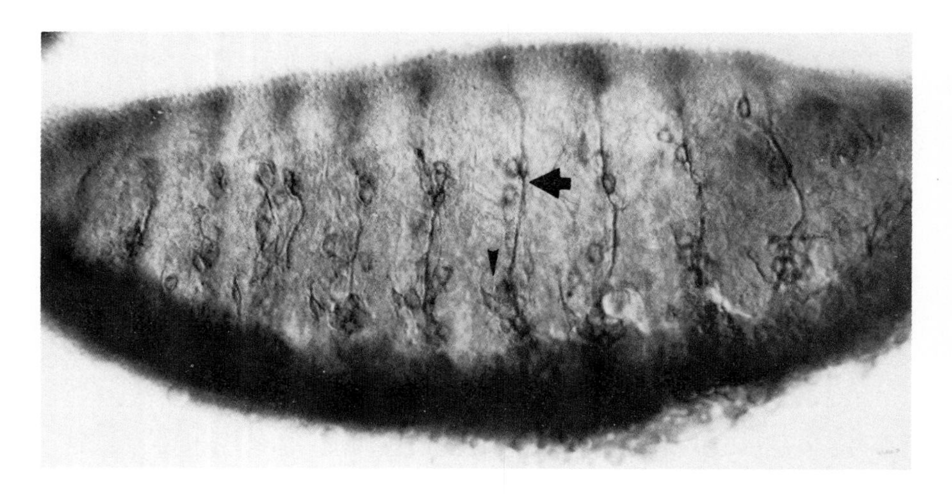

FIGURE 4. Effects of the deficiency Df(3L)vi^3 on the embryonic nervous system. The CO neurons are deleted (arrows).

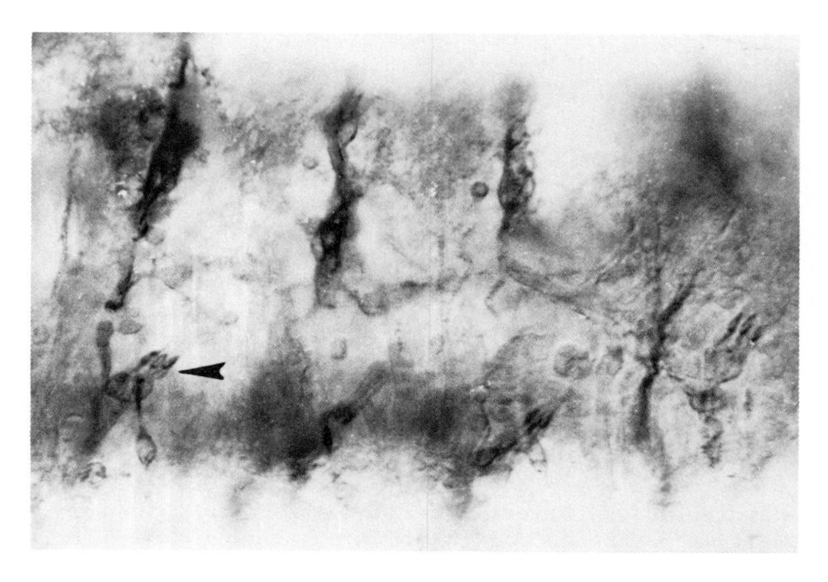

FIGURE 5. Effects of the deficiency Df(2L)E55 on the peripheral nervous system. Note that in these embryos there are three, instead of five, CO neurons in each lateral cluster of the abdominal segment.

FIGURE 6. Effects of the deficiency DF(1)HA32 on the embryonic nervous system. Note the massive misrouting of the axonal pathways in the periphery. There is also a CNS associated with this deficiency.

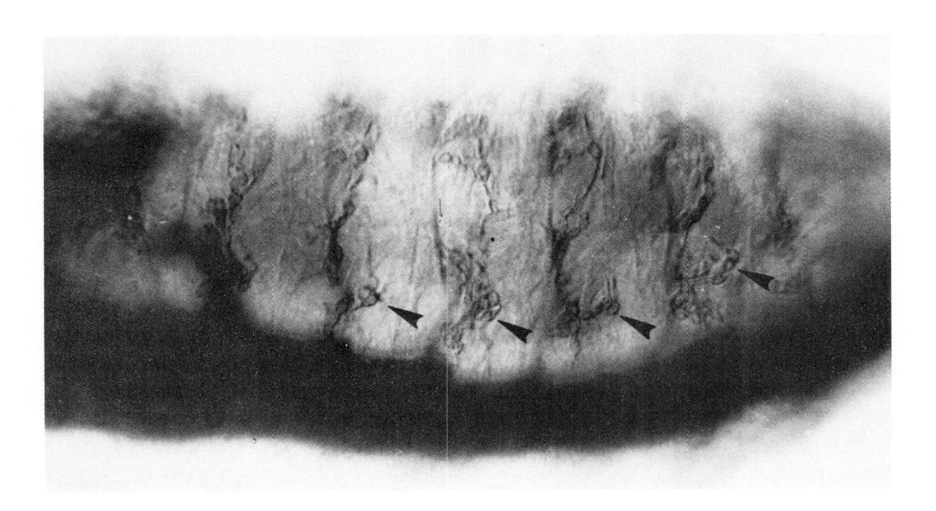

FIGURE 7. Effects of the deficiency DF(1)N12 on the embryonic nervous system. A number of deficiencies gave similar phenotype: a mild general derangement of the nervous tissue. Note the variable effects on neurons from one segment to the other (arrows).

Class 6) Weakly misrouted fibers and/or disorganized clusters. Quite a few loci or regions of genome (more than 15) seem to cause only weakly deranged nervous system. The Df(1)N12 shown in Fig. 7 exhibits that kind of defect. In the wild type embryo, such mistakes are rare. It is less clear whether such defects in the mutant embryos are direct consequences of the mutation.

DISCUSSION

In this paper, we briefly outlined our attempt in systematically identifying genes that are involved in early neural development in the Drosophila embryo. Our search has thus far covered approximately 30% of the genome or about 1500 genes. So far, we have found a limited number (less than 40) of deficiencies or genes which are involved in peripheral neural development. In several instances the specific neural defects initially detected in a deficiency were later found to be due to a single gene within the deficiency. This is consistent with the notion that there are only a small number of genes involved in neural development so that most deficiencies contain zero or one gene important for neural development. However, our estimate of the number of genes involved in neural development is very rough because deletion of a number of regions cause very severe disruption in overall development of the embryo. If within these regions are genes important for the development of the nervous system, they would have escaped detection.

Deletion of a number of regions cause obvious defects in other tissues in addition to defects in the nervous system. In these cases, it is not easy to decide whether the genes are involved directly both in neural and non-neural development or whether the effect on neural development is secondary. In the nematode, studies of a number of cell lineage mutations reveal that both neurons and non-neurons may be specifically affected if they are derived from similar lineage pedigrees (13). Thus, the exclusive action of a gene in neurons may not be a reasonable criterion for genes important for neural development. Further experiments on these identified regions or loci are necessary to assess their functional roles. For the regions of particular interest, mutations of the relevant genes may be obtained and localized more precisely on salivary gland chromosomes, so that both genetic and molecular analyses of these genes may be undertaken for better understanding of the genetic control of neural development.

ACKNOWLEDGEMENTS

We would like to acknowledge Mr. Larry Ackerman for his superb technical assistance in immunocytochemistry and Ms. Phyllis Cameron for preparation of this manuscript. We wish to thank the CalTech stock center, Bowling Green stock center, and Dr. Eric Wieschaus for providing the deficiency and lethal mutation stocks. This study was supported in part by the Howard Hughes Medical Institute. R. Bodmer has a Swiss National Foundation Fellowship.

REFERENCES

1. Nusslein-Volhard C, Wieschaus E, Kluding H (1984) Mutations affecting the pattern of the larval cuticle in Drosophila melanogaster. I. Zygotic loci on the second chromosome. Roux Arch Dev Biol 193: 267.
2. Wieschaus E, Nusslein-Volhard C, Jurgens G (1984). Mutations affecting the pattern of the larval cuticle in Drosophila melanogaster. III. Zygotic loci on the X-chromosome and fourth chromosome. Roux Arch Dev Biol 193: 296.
3. Jurgens G, Wieschaus E, Nusslein-Volhard C, Kluding H (1984). Mutations affecting the pattern of the larval cuticle in Drosophila melanogaster. II. Zygotic loci on the third chromosome. Roux Arch Dev Biol 193: 283.
4. Campos-Ortega JA, Jimenez F (1980). The effect of X-chromosome deficiencies on neurogenesis in Drosophila. In Siddigi O, Babu P, Hall LM, Hall JC (eds): "Development and Neurobiology of Drosophila," Plenum.
5. Lehmann R, Jimenez F, Dietrich U, Campos-Ortega JA (1983). On the phenotype and development of mutants of early neurogenesis in Drosophila melangaster. Roux Arch Dev Biol 192: 62.
6. Wharton K, Johansen K, Xu T, Artavanis-Tsakonas S (1985). Nucleotide sequence from the neurogenic locus notch implies a gene product that shares homology with proteins containing EGF-like repeats. Cell 43: 567.

7. Greenwald I (1985). lin-12, a nematoid homeotic gene, is homologous to a set of mammalian proteins that includes epidermal growth factor. Cell 43: 583.
8. Jan LY, Jan YN (1982). Antibodies to horseradish peroxidase as specific neuronal markers in Drosophila and in grasshopper embryos. Proc Nat Acad Sci USA 72: 2700.
9. Jan YN, Ghysen A, Christoph I, Barbel S, Jan LY (1985). Formation of neuronal pathways in the imaginal discs of Drosophila melanogaster. J Neurosci 5: 2453.
10. Ghysen A, Dambly-Chaudiere C, Aceves E, Jan LY, Jan YN (1986). Sensory neurons and peripheral pathways in Drosophila embryos. Roux Arch, in press.
11. Campos-Ortega JA, Hartenstein V (1985). "The embryonic development of Drosophila melanogaster." Springer Verlag, in press.
12. Dambly-Chaudiere C, Ghysen A (1986). The pattern of sense organs in Drosophila embryos and larvae. Roux Arch, in press.
13. Sternberg PW, Horvitz HR (1984). The genetic control of cell lineage during nematode development. Ann Rev Genet 18: 489.

Molecular Entomology, pages 57–66

CHEMICAL STUDIES OF PHEROMONE CATABOLISM AND RECEPTION[1]

Glenn D. Prestwich,* Richard G. Vogt, and Yu-Shin Ding

Department of Chemistry
State University of New York
Stony Brook, New York 11794-3400

ABSTRACT Biochemical studies of the binding and catabolism of pheromone molecules by proteins in antennal and other tissues of insects can be best studied with high specific activity ^{3}H-labeled pheromones and pheromone analogs. Inhibitors of catabolic enzymes and affinity labels for putative receptor proteins provide new data on pheromone processing. Results are presented for the processing of E6,Z11-16:Ac by *Antheraea polyphemus*, Z11-16:Al and Z9-14:Al by *Heliothis virescens*, and *cis*-7,8-epoxy-2-methyloctadecane by *Lymantria dispar*.

INTRODUCTION

In contrast to the wealth of information on the chemical structures and the behavioral effects of insect pheromones, much less is known of their biogenesis, molecular action, neurobiology, receptor binding, or catabolism *in vivo*. The basic biochemistry of pheromone metabolism must be more fully comprehended if effective strategies for mating disruption are to be implemented.

[1]Supported by NSF grant DMB-8316931 and USDA Competitive Research Grant 85-CRCR-11736. G.D.P. is a Fellow of Alfred P. Sloan Foundation and a Camille and Henry Dreyfus Teacher-Scholar (1981-86).

Our approach integrates the synthesis of chemically reactive pheromone analogs as enzyme inhibitors and as receptor affinity labels with the preparation of tritiated isotopomers of the natural insect pheromones. With these chemical tools, we examine the catabolic and binding proteins of insect sensory hairs and other tissues *in vitro*; we also include studies of the effects of our reactive analogs *in vivo* (antennal recordings; behavior).

We currently employ three insects to cover the major chemical functionalities in lepidopteran sex pheromones. The wild silkmoth *Antheraea polyphemus* uses a mixture of E6,Z11-16:Ac and the corresponding aldehyde. The moths have large feathery antennae with easily isolated sensory hairs. We describe properties of a sensillar esterase and of sensillar binding proteins. Second, the economically important pest *Heliothis virescens* employs a six-component aldehyde blend of which Z11-16:Al and Z9-14:Al are the major components. We present new data on aldehyde dehydrogenase and two other enzyme activities in these moths. New fluorinated analogs which mimic pheromones have been prepared to examine binding to pheromone-specific proteins. Finally, the catabolism of the epoxide disparlure by tissues of the gypsy moth *Lymantria dispar* is being reinvestigated using chiral, ^{3}H-labeled pheromone and several novel epoxide analogs. A comprehensive review of chemical studies of pheromone biochemistry will be published separately (1).

ANTHERAEA POLYPHEMUS

Sensory hair lymph from antennae of males of the wild silkmoth *Antheraea polyphemus* contains several soluble proteins involved in pheromone processing, including an aggressive esterase (MW 55,000) and an abundant binding protein (MW 15,000) (2,3). The interaction of these two soluble proteins in the deactivation of E6,Z11-16:Ac, and the interaction between the degradatory system and the receptor system has been extensively discussed (4,5). In the presence of the esterase ($K_M = 2 \times 10^{-6}$) at physiological concentration, the pheromone has a maximum estimated half-life of 15 msec. This suggests a molecular model for reception in which the binding protein acts as a carrier and the enzyme acts as a rapid inactivator to maintain a low stimulus noise level in the sensillum (2).

The study of the binding protein and esterase at physiologically relevant pheromone concentrations, i.e., 10^{-6} to 10^{-10} M, was only possible using high specific activity tritium-labeled pheromone. Earlier studies of esterases of *Trichoplusia ni* used low specific activity Z7-12:Ac labeled in the acetate with tritium at 0.4 - 0.8 Ci/mmol (6,7). In contrast, the synthesis of [11,12-$^3H_2$]-6E, 11Z-hexadecadienyl acetate by partial reduction of the 11-alkyne with carrier-free tritium gas in the presence of a poisoned catalyst provided labeled pheromone of specific activity >40 Ci/mmol (8) (Figure 1).

FIGURE 1. Synthesis of labeled *A. polyphemus* pheromone

Degradation of tritiated E6,11Z-16:Ac (≈100 nM) by the purified sensillar esterase was subject to competitive inhibition by the corresponding diazoacetate, E6,11Z-16:Dza (I_{50} ≈10 μM), but not by the 6Z stereoisomer or the 11-alkynyl analog (2) (Figure 2). The serine esterase transition-state inhibitor 1,1,1-trifluorotetradecan-2-one (9) showed an I_{50} of 5 nM; most serine esterase inhibitors were ineffective even at 100 μM. Clearly the enzyme preferentially binds fatty alcohol esters, particularly those possessing the correct side chain geometry.

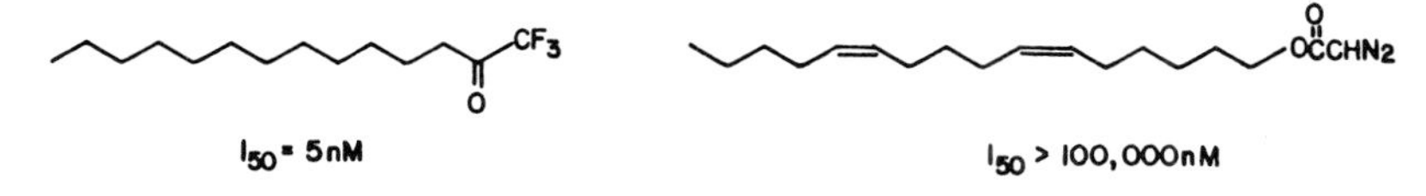

FIGURE 2. Two sensillar esterase inhibitors.

Enzymatic hydrolysis of the acetate or diazoacetate was also examined for three pheromone analogs each labeled with two tritium atoms: (a) E6,Z11-16:Dza, the diazoacetate analog of the acetate pheromone, (b) 16:Ac, the saturated analog of the pheromone, and (c) Z9-14:Ac, a shorter kinked-chain acetate (10). The first two are poor substrates over four decades of concentration. The

Z9-14:Ac however was the best alternative substrate tested for this *in vitro* pheromone metabolism system, which included the buffering binding protein. Unlabeled Z9-14:Ac was also the best competitive inhibitor of the hydrolysis of labeled pheromone.

The diazoacetate analog of the *A. polyphemus* pheromone was first prepated in 1978 in unlabeled form and was shown to possess approximately 10% of the electrophysiological activity of the parent acetate (11). We prepared this photoactivatable analog of the pheromone labeled at >40 Ci/mmol (8) to modify covalently the binding protein located in the receptor lymph and the putative receptor protein located in the dendritic membrane (Figure 3).

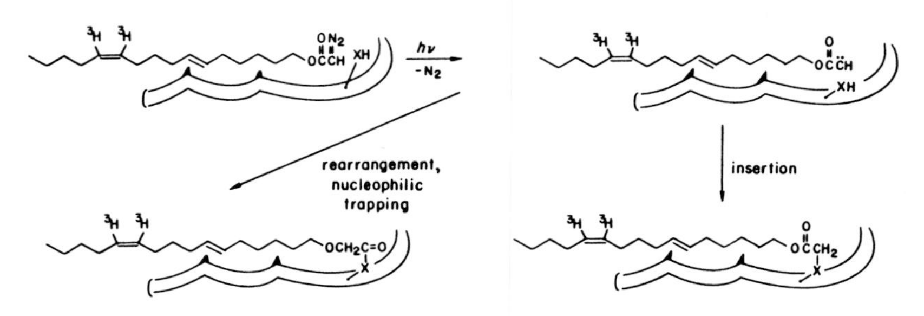

FIGURE 3. Photoaffinity labeling of a pheromone receptor

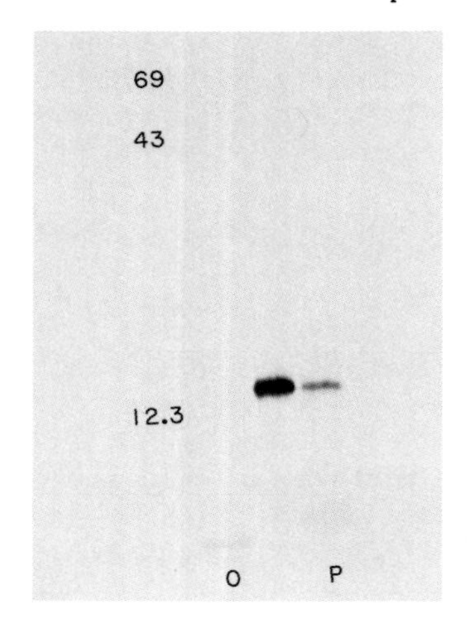

FIGURE 4. Autoradiogram of photoaffinity labeled soluble proteins from male *A. polyphemus* sensory hairs: O = no light; P = protected with 10 μM 6E,11Z-16:Ac; [Protein] = 30 μM.

Indeed, irradiation of sensory hair homogenates containing 100 nM of [^{3}H]-E6,Z11-16:Dza resulted in selective photolabeling of male antennal proteins in both soluble and membrane-associated fractions. Photoaffinity labeling of these proteins is competable with unlabeled acetate pheromone, and does not occur in the absence of irradiation. Autoradiograms of the electrophoretically separated soluble proteins are shown above (Figure 4).

HELIOTHIS VIRESCENS

We synthesized the two unsaturated aldehydes, Z9-14:Al and Z11-16:Al, labeled with two tritiums per molecule in order to investigate the pheromone processing in the economically important noctuid *Heliothis virescens* (12). The corresponding alkynyl acetates (0.1 mmol) were semihydrogenated at 720 torr with carrier-free tritium gas using a poisoned catalyst; this was performed at the National Tritium Labeling Facility of the Lawrence Berkeley Laboratory. Use of THF as solvent gave specific activity >52 Ci/mmol, while CH_3OH gave 20-35 Ci/mmol. High specific activity pheromones and analogs were stored below -20°C in hexane-toluene mixtures to retard radiolytic decomposition. The acetates were the best storage form with <10% decomposition per year. Of the techniques known for partial oxidation of a primary alcohol, only pyridinium dichromate gave the desired aldehyde when working with 0.5 ml solutions in dichloromethane containing 10 mCi or less of labeled alcohol (200 nmole).

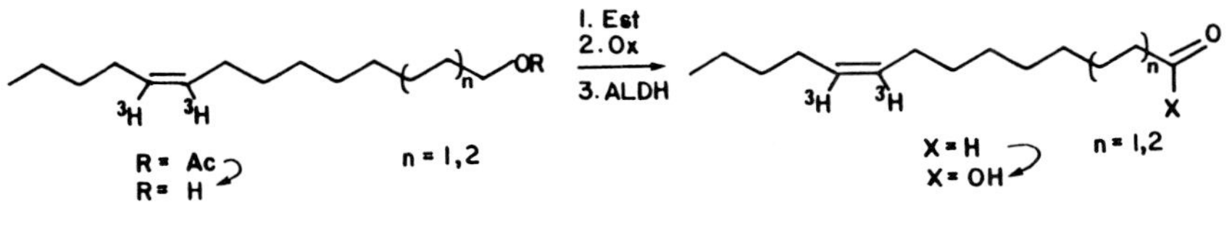

^{3}H ^{3}H OR
R = Ac
R = H
n = 1,2
1. Est
2. Ox
3. ALDH
^{3}H ^{3}H O X
X = H
X = OH
n = 1,2

FIGURE 5. Interconversions of the *H. virescens* pheromone components.

We investigated the tissue specificity of acetate esterase, alcohol oxidase and aldehyde dehydrogenase in both male and female moths (Figure 5). These three activities have also been reported in the spruce budworm *Choristoneura fumiferana* using radiochemical and luminescence assays (13). Soluble and membrane-associated enzyme activities in *Heliothis* were determined by radio-TLC assays. Key observations include:

1. Aldehyde dehydrogenase is the primary enzyme activity found in both leg and antennal tissues of males and females. The male antennal enzyme is probably important in pheromone catabolism in this moth.
2. Oxidase activity is low in antennae, but higher in male legs and glandular tissues of both sexes.
3. Esterase activity is low in antennae, higher in legs, and highest in glandular tissues of both sexes.
4. Little difference is observed between enzymic conversions of the tetradecenal and the hexadecenal.

In addition, two potential mechanism-based inhibitors of the pheromone-specific aldehyde dehydrogenase are shown in Figure 6. The α,α-difluoroaldehyde is completely hydrated in water, and could act as a transition-state analog inhibitor. The substituted cyclopropanone itself cannot be readily prepared (half-life < 30 min), but the cyclopropanol is a stable *in vivo* precursor, if the oxidase is capable of oxidative activation.

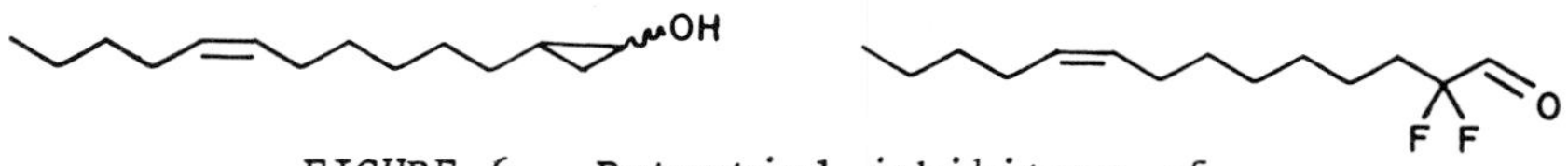

OH

O

F F

FIGURE 6. Potential inhibitors of *H. virescens* ALDH

We postulated that the acyl fluorides would act as reactive mimics of the corresponding aldehydes. By analogy with the visual transduction mechanism involving a protonated Schiff base of retinal and opsin, we proposed that aldehyde perception might involve a transient imino intermediate important in ion channel opening. An acyl fluoride could irreversibly lock the protein in an open or closed state, leading to hyperactivation or anosmia, respectively. Indeed, the Z9-14:Acf and Z11-16:Acf are potent hyperagonists, causing aphrodisia and disorientation in male moths (14).

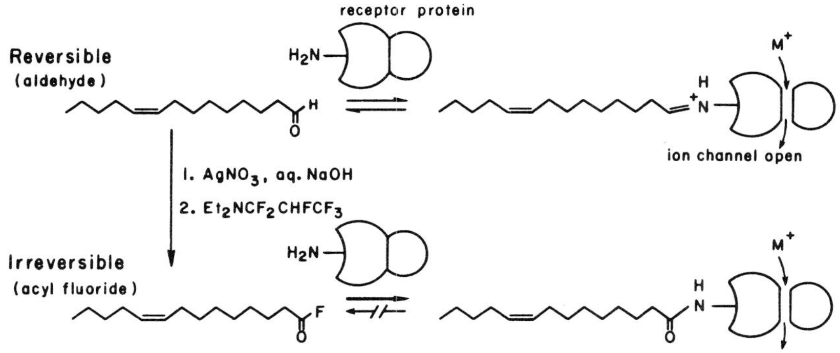

FIGURE 7. Model of acyl fluoride mimicry of aldehydes

In order to determine if a pheromone-specific Schiff base was being formed with soluble or membrane-associated proteins, we used a two-step covalent modification technique (15). Male antennae were sonicated and the eluted proteins were centrifuged; the 100,000 x g supernatant and pellet fractions were incubated with labeled Z9-14:Al in three separate treatments; (a) no fixer, (b) $NaBH_4$ as fixer, (c) $NaBH_3CN$ as fixer. Only the last treatment selectively reduces protonated Schiff bases in the presence of aldehydes. Electrophoretic separations of the soluble and membrane proteins both showed a single major protein band which could be a specific binding protein for pheromone aldehydes.

LYMANTRIA DISPAR

Radiolabeled racemic disparlure was first prepared over ten years ago by epoxidation of the [7,8-$^3H_2$]-Z7-2-methyloctadecene (16). The alkene was used in very preliminary studies of disparlure biosynthesis in pre-eclosion female *L. dispar* (17), while the labeled pheromone was used to study uptake into the male antenna and conversion to metabolites. At an exposure of about 5×10^{11} molecules/antenna of racemic 40 Ci/mmol disparlure, about 2.5 min was required for disappearance of half of the epoxide and the appearance of two uncharacterized polar metabolites (18). We were unsatisfied with these data and initiated a reinvestigation of disparlure binding and catabolism. We felt that it would be important to distinguish binding and catabolism of the two enantiomeric forms, since the proteins involved reside in separate sensory cells (19).

Racemic disparlure (>40 Ci/mmol) was prepared by us by semihydrogenation of 2-methyloctadec-7-yne followed by epoxidation of the Z-alkene. Recently, we completed a synthesis of both enantiomers of disparlure (>95% enantiomerically pure) using an asymmetric epoxidation of an allylic alcohol (20), chain extession, and homogeneous tritiation (21) of an alkenyl oxirane to give the [5,6-$^3H_2$] compounds as shown in Figure 8.

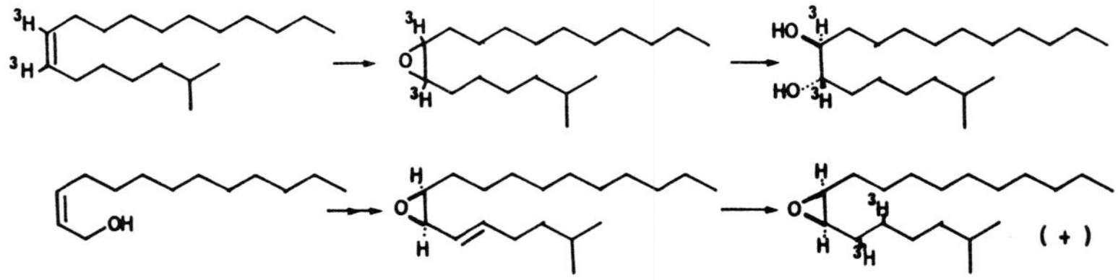

FIGURE 8. Synthesis and hydrolysis of labeled racemic disparlure (top) and (+)-disparlure (bottom).

Initial pheromone degradation studies have deomonstrated antennal-specific epoxide hydrolase activity in both male and female moths. Ultracentrifugation indicates that this activity is membrane associated. The pheromone product co-migrates with the *threo*-alcohol produced by alkaline hydrolysis of the *cis*-disparlure epoxide. The enantioselectivity and enzyme kinetics of the EH are currently under study, and several specific epoxide hydrolase inhibitors have been synthesized.

The sensory hairs of the male gypsy moth have proven relatively easy to isolate in high yield, free of hemolymph and epidermal contamination. Initial electrophoretic studies of isolated sensory hairs have shown two predominant proteins of high mobility on non-SDS polyacrylamide gels. Both proteins were observed in electrophoresed homogenates of male as well as female antennae. Our current studies are focusing on pheromone/analog interactions with the sensory hair proteins.

ACKNOWLEDGMENTS

Antheraea: We thank Prof. L.M. Riddiford for laboratory facilities and Prof. N. Andersen (both Univ. of Washington) for supplying synthetic intermediates. Lymantria: Mr. J.-W.Kuo and Mr. S. Graham (Stony Brook) helped prepare labeled pheromone and unlabeled inhibitors; pupae were provided by C. Schwalbe (USDA). Heliothis: Ms. M. Tasayco, Dr. J. Carvalho, and Dr. D. Hendricks provided data; pupae were supplied by the USDA and the Cotton Foundation.

REFERENCES

1. Prestwich GD (1986). Chemical studies of pheromone reception and catabolism. In Prestwich GD, Blomquist GJ (eds): "Pheromone Biochemistry," New York: Academic Press, submitted.
2. Vogt RG, Riddiford LM, Prestwich GD (1985). Kinetic properties of a pheromone degrading enzyme: the sensillar esterase of Antheraea polyphemus. Proc Natl Acad Sci USA 82:8827.
3. Vogt RG, Riddiford LM (1981). Pheromone binding and inactivation by moth antennae. Nature 293:161.
4. Vogt RG, Riddiford LM (1986). Pheromone reception: a kinetic equilibrium. In Payne T, Birch M, Kennedy J (eds): "Mechanisms of Perception and Orientation to Insect Olfactory Signals," Oxford: Pergamon, p 201.
5. Kaissling K-E, Thorson J (1980). Insect olfactory sensilla: structural, chemical and electrical aspects of the functional organizations. In Satelle DB, Hall LM, Hildebrand JG (eds): "Receptors for Neurotransmitters, Hormones and Pheromones in Insects," Amsterdam: Elsevier/North Holland Biomedical, p 261.
6. Ferkovich SM (1982). Enzymatic alteration of insect pheromones. In Morris DM (ed): "Perception of Behavioral Chemicals," Amsterdam: Elsevier/North Holland, p 165.
7. Ferkovich SM, Oliver JE, Dillard C (1982). Pheromone hydrolysis by cuticular and interior esterases of the antennae, legs, and wings of the cabbage looper moth, Trichoplusia ni (Hubner). J Chem Ecol 8:859.
8. Prestwich GD, Golec FA, Andersen NH (1984). Synthesis of a highly tritiated photoaffinity labelled pheromone analog for the moth Antheraea polyphemus. J Labelled Compd Radiopharmacol 21:593.
9. Hammock BD, Wing KD, McLaughlin J, Lovell V, Sparks TC (1982). Trifluoromethyl ketones as possible transition state analog inhibitors of juvenile hormone esterase. Pest Biochem Physiol 17:76.

10. Prestwich GD, Vogt RG, Riddiford LM (1986). Binding and hydrolysis of radiolabeled pheromone and several analogs by male-specific antennal proteins of the moth Antheraea polyphemus. J Chem Ecol 12:323.
11. Ganjian I, Pettei MJ, Nakanishi K, Kaissling K-E (1978). A photoaffinity-labeled insect sex pheromone for the moth Antheraea polyphemus. Nature 271:157.
12. Ding Y-S and Prestwich GD (1986). Metabolic transformation of tritium-labeled pheromone by tissues of Heliothis virescens moths. J Chem Ecol 12:411.
13. Morse D, Meighen E (1984). Detection of pheromone biosynthetic and degradative enzymes in vitro. J Biol Chem 259:475.
14. Prestwich GD, Carvalho JF, Ding, Y-S, Hendricks DE (1986). Acyl fluorides as reactive mimics of aldehyde pheromones: hyperactivation and aphrodisia in Heliothis virescens: submitted.
15. Mason JR, Leong FC, Plaxco KW, Morton TH (1985). Two-step covalent modification of proteins. Selective labeling of Schiff base-forming sites and selective blockade of the sense of smell in vivo. J Am Chem Soc 107:6075.
16. Sheads RE, Beroza M (1973). Preparation of tritium-labeled disparlure, the sex attractant of the gypsy moth. J Agric Food Chem 21:751.
17. Kasang G, Schneider D (1974). Biosynthesis of the sex pheromone disparlure of olefin-epoxide conversion. Naturwiss 61:130.
18. Kasang G, Knauer B, Beroza M (1974). Uptake of the sex attractant ^{3}H-disparlure by gypsy male moth antennae (Lymantria dispar) [Porthetria dispar]. Experientia 30:147.
19. Hansen K (1984). Discrimination and production of disparlure enantiomers by the gypsy moth and the nun moth. Physiol Entomol 9:9.
20. Rossiter BE, Katsuki T, Sharpless KB (1981). Asymmetric epoxidation provides shortest routes to four chiral epoxy alcohols which are key intermediates in syntheses of methymycin, erythromycin, leukotriene C-1, and disparlure. J Am Chem Soc 103:464.
21. Prestwich GD, Wawrzenczyk C (1985). High specific activity enantiomerically enriched juvenile hormones: synthesis and binding assay. Proc Natl Acad Sci USA 82:1663.

Molecular Entomology, pages 67–77

NEUROACTIVE SUBSTANCES AS PROBES OF CENTRAL THRESHOLDS IN SEX PHEROMONE PERCEPTION[1]

Wendell L. Roelofs and Charles E. Linn, Jr.

Department of Entomology, Cornell University
Geneva, New York 14456

ABSTRACT The sex pheromone-mediated behavior of male moths is a complex response that involves integration of visual and chemical inputs with a circadian rhythm that also regulates normal flight activity. Research in laboratory flight tunnels involving the full behavioral response sequence of oriented upwind flight and landing at the chemical source has shown that males are very sensitive to slight changes in the quantity and quality of the chemical signal. The normal profile of responses to a series of pheromone component blends and release rates is affected by pre-exposure to individual components, by varying the temperature or the assay time in the photoperiod, and by injection of neuroactive substances. Our initial studies have shown that octopamine, as well as its agonists and antagonists, greatly affects the male moth's pheromone response behavioral thresholds.

INTRODUCTION

Male moths demonstrate incredible odor perception and discrimination when flying long distances to "calling" female moths through turbulent air laden with nature's smells. In many cases, this communication system is composed of a specific blend of compounds involving geometrical or positional isomers. For example, with the oak leafroller moth, Archips semiferanus, field studies (1) showed that traps baited with the natural blend of 67:33 E/Z-11-

[1]This research was supported by NSF Grant BNS 82-16752.

tetradecenyl acetates captured significantly more male moths than the 60:40 and the 70:30 mixtures. The specific blend also is tightly controlled in pheromone production in female moths. With redbanded leafroller moths, Argyrotaenia velutinana, strong artificial selection pressure in the laboratory for three years (2,3) was not successful in shifting the 92:8 Z/E blend beyond the natural narrow range of 4-12%E.

Threshold Hypothesis

In addition to specificity to a blend ratio, male moths also were found to be sensitive to release rate. In a large field trapping experiment involving a range of ratios and release rates (Novak and Roelofs, unpublished), redbanded leafroller males were captured best to the natural ratio of 8% E with load rates of 0.5 - 30 mg (polyethylene cap). Male captures with load rates above this range were very poor. Treatments containing 2% and 15% E also were good in the range of 1-10 mg, but the 20% E treatment did not capture males at any dosage. This type of trapping data was used to develop a threshold hypothesis (4) that postulates a lower behavioral threshold for activation of upwind flight and an upper threshold for in-flight arrestment. These thresholds define an area of peak attraction, with males most sensitive to the natural female-produced blend and release rate, but able to respond to some slightly altered blends at increased release rates.

Further support for the existence of the behavioral thresholds was obtained in a flight tunnel study (5) with the Oriental fruit moth (OFM), Grapholita molesta. The sex pheromone-mediated responses of these males to an array of blends and dosages of the Z- and E8-12:0Ac's were analyzed. The range of treatments for which males completed flights to the source appeared to be bounded by dosages too low or too high to result in significant attraction, with the higher dosages causing arrestment of flight at some distance from the source.

Shifting the Behavioral Thresholds

Pre-exposure. We viewed the thresholds as a CNS phenomenon that could be affected by a variety of factors. We first studied (6) the effect of pre-exposure of male moths to the E8-12:0Ac component, which was present in the pheromone at 6% of the Z isomer. The previous studies had shown that the ability of males to orient to the odorous plume and exhibit continuous flight to the source was particularly affected at high dosages of all blends and at lower dosages containing high %E blends. The common element in all of these blends producing arrested flight was a high relative amount of E isomer. Males were pre-exposed to the E isomer and then after 1 min of clean air were assayed to the array of ratios and dosages used in previous studies. Surprisingly, the pre-exposure caused enhanced flight response to treatments above the arrestment threshold, thus expanding the range of blend-dosage combinations that elicited high numbers of completed flights to the source. Only the arrestment threshold appeared to be pushed back in these tests. The selective nature of the pre-exposure effect implied alteration of central processes in discrimination of blend rather than disturbance of olfactory receptor function.

Blend Composition. Further studies (7) with the OFM males showed that complete flights to the source were observed only to blend combinations containing all three pheromone components: Z8-12:0Ac, 6% E8-12:0Ac, 10% Z8-12:OH Although responses to the natural 6%E blend at the two best dosages were relatively unaffected by changes in the proportion of the OH component, responses to treatments surrounding the peak area of attraction were strongly influenced by the proportion of OH in the blend. Another study (8) showed that the 3-, 6-, and 7-component pheromone blends of OFM, cabbage looper (CL), and redbanded leafroller moths, respectively, elicited male upwind flights at concentration levels much below that of single components or partial blends. These data support the concept that the minor pheromone components enhance male sensitivity and specificity to the pheromone signal, and thus, are important for initiating upwind flight of a distant male to a female moth.

Temperature. In testing (9) OFM males to the array of blends and dosages at a higher temperature (26°C) relative to the normal 21°C, it was discovered that the peak area of attraction was greatly expanded at the higher temperature. The data indicated large changes in the behavioral thresholds at the higher temperature. This phenomenon was observed previously in trap studies (10) in California with the omnivorous leafroller, but catches to an expanded range of component ratios in the summer, relative to the spring and fall, had been assumed to be due to higher release rates, rather than changes in the male moth's perception. The new flight tunnel data indicate that the warm summer temperatures actually changed the male's responses to the array of blends at all release rates.

Sublethal effects of neuroactive compounds. Our next study with OFM (11) showed that the complex precopulatory sequence of behaviors exhibited by males is very sensitive to sublethal doses of insecticides and other neuroactive compounds. Permethrin, carbaryl and chlordimeform induced their own unique effects at specific phases of the sequence. Octopamine induced a hypersensitivity to the olfactory signal, whereas yohimbine and cyproheptadine significantly decreased moth activation to the chemical signal and reversed the effects induced by octopamine. These data were obtained by topically treating OFM males on the ventral thoracic region 5 hrs prior to the test period in the flight tunnel. The interesting results with the biogenic amine, octopamine, provided the stimulus to conduct an in-depth study on the effects of biogenic amines, as well as their agonists and antagonists, on the pheromone response thresholds. We switched to the larger CL moth, and injected the test materials instead of using topical applications. A brief summary of this study will be presented in this chapter.

METHODS

The test procedures were similar to those reported previously (11). Generally, 4-day-old CL males were injected into the lateral thoracic region near the neck with 1 μl saline solution of the test material. Males were treated 3 hrs prior to scotophase. Effects on pheromone response were observed with individual males in the flight tunnel during the 5th and 6th hr of the 8 hr scotophase. Males were allowed 30 sec to respond and were scored for 3 key behaviors typically exhibited in the flight tunnel: taking flight, initiation of upwind oriented flight, and source of contact after flying 1.5 m to the source. Tests for effects on the optimum time of pheromone response were conducted by testing the male's response in the flight tunnel every 2 hrs through the scotophase to a dosage series of pheromone.

The pheromone used in this study was a 6-component blend (13) in the following proportions: 12:OAc (6.8), Z5-12:OAc (7.6), Z7-12:OAc (100), 11-12:OAc (2.3), Z7-14:OAc (0.9), and Z9-14:OAc (0.6).

RESULTS AND DISCUSSION

Octopamine (OA) and Serotonin

Male cabbage looper moths treated with increasing amounts of the biogenic amines were tested in the flight

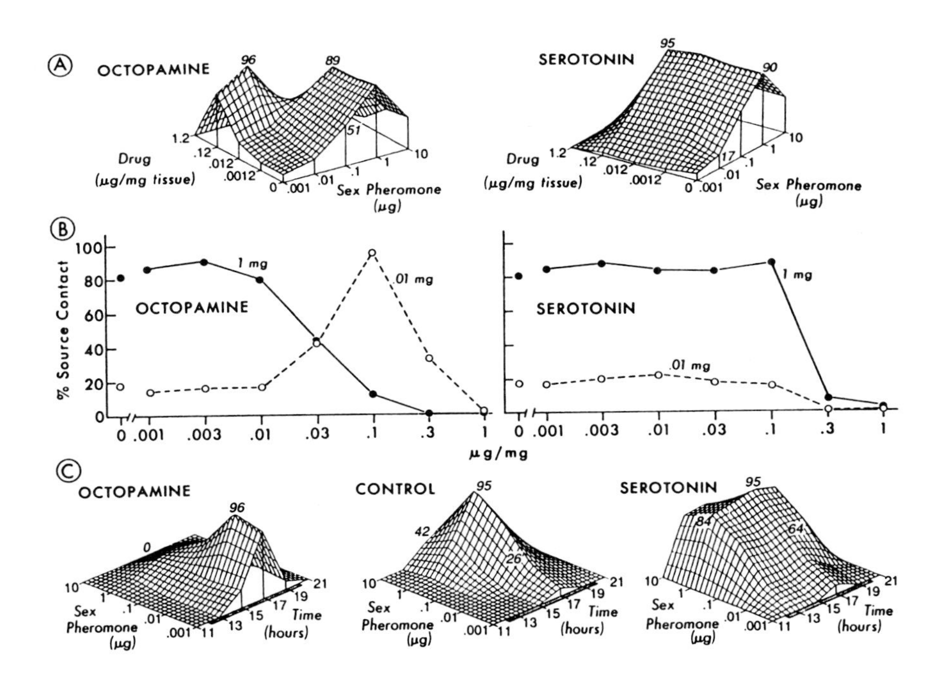

FIGURE 1. A. Response profile of CL males for source contact to a dosage series of pheromone after treatment with octopamine or serotonin. B. Same as A, but showing responses to only 2 dosages of pheromone. C. Response profiles over scotophase period after injection of 0.12 μg/mg octopamine or serotonin.

tunnel at the mid-point of the scotophase to a series of pheromone dosages (Fig. 1A & B). At the 0.12 μg/mg dose of OA, the response thresholds for completed flights were significantly lowered, such that maximal response was achieved with only 0.01 mg of pheromone instead of the 1 mg dosage found to be most attractive to untreated males. At the higher 1.2 μg/mg dose of OA, male response was lowered to all pheromone treatments, the result of significant effects on flight motor behaviors rather than on thresholds for perception of the signal.

Serotonin, on the other hand, had no effect on pheromone sensitivity (Fig. 1A). In this case, when males were treated with 0.12 μg/mg, there was a significant effect on the time of pheromone response activity (Fig. 1C) and random activity in a cage. Males were maximally responsive to the normally optimal 1 mg dosage of pheromone throughout the scotophase when treated with serotonin. Octopamine did not have this effect on their time of maximum response activity (Fig. 1C), but again the effect on pheromone sensitivity is seen as the treated males responded to extremely low levels of pheromone.

Agonists

A series of potential agonist compounds for OA was tested by flying drug-treated males to the low dosage pheromone sources. For example, 10^{-1} μg/mg of OA increased male response to the 10^{-2} mg pheromone dosage from 20% in untreated males to over 80% in treated males. Agonists were found that were much more potent than OA in eliciting over 90% male response to lower amounts of pheromone and with much lower doses of compound. Figure 2 shows the dose of compound eliciting the highest male response, after which toxic effects on flight coordination and motor function sharply lowered completed flights to the source. With the phenylethylamines, dopamine is inactive and synephrine is slightly better than octopamine, presumably because the methylated amine can pass more easily through the brain barrier. Chlordimeform and XAMI are the most potent from the standpoint of eliciting peak male responses to 10^{-3} mg pheromone with only a 10^{-4} μg/mg dose. However, tolazoline and naphazoline produced over 90% male response to only 10^{-4} mg pheromone with slightly higher doses.

Although some of the differences in activity could be due to the ability of the compound to reach the CNS, there is convincing evidence that octopamine and some of its known

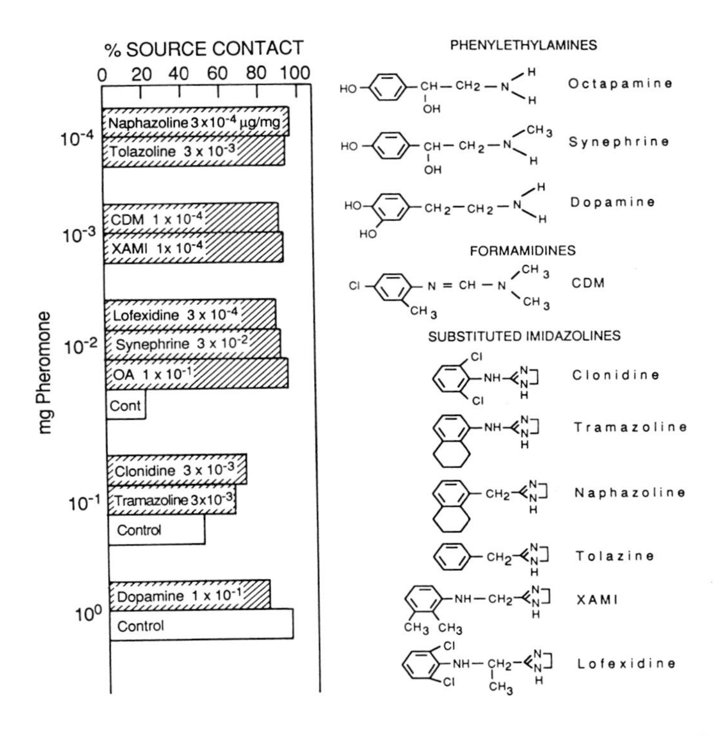

FIGURE 2. Percentage CL males reaching source (mg pheromone) 4-5 hrs after injection of indicated dose of OA or agonist compound (µg/mg) (N=50). The percentages indicate the dosage of pheromone eliciting peak levels of response over the dosage series tested.

agonists greatly lowered the male's pheromone response threshold.

Antagonists

Tests with octopamine antagonists were conducted by observing the male moth's responses to the normally optimal 1 mg pheromone treatment in the flight tunnel after treatment with a dose series of the compounds. Under control conditions over 80% of the untreated males arrive at the 1 mg source (Fig. 1B), but the antagonists decreased these numbers by affecting the whole response sequence, beginning with activation. The treated males did not exhibit the arrested flight of males responding to too much pheromone, but rather appeared to be in a lower motivational state, or below the activation response threshold.

The antagonists used were selected from those tested on other OA systems (14-16). These included the alpha-adrenergic blockers phentolamine, yohimbine, and metochlopromide; the (generally considered) serotonin antagonists cyproheptadine, mianserin, and gramine; the phenothiazines chlorpromazine and promethazine, which generally are classed as dopamine antagonists; and the beta-adrenergic blocker propranolol. The data (Fig. 3) show that the beta blocker and the specific serotonin antagonist, gramine, are inactive in our OA assay.

The high activity of mianserin, metochlopromide and phentolamine as OA antagonists is similar to the data on OA receptors characterized by Evans (14) as type 2 receptors, affecting twitch tension and relaxation in the locust extensor tibia muscle. Our rankings differ from others for

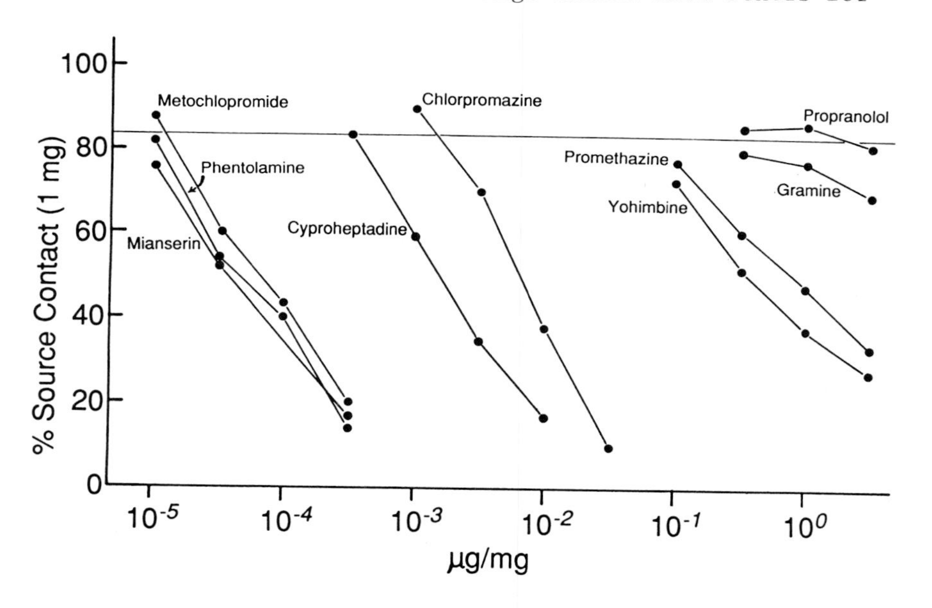

FIGURE 3. Percentage male CL reaching source (1 mg pheromone) 4-5 hrs after injection of antagonists. N=80 for each dose of antagonist. Solid line represents control response.

antagonists of OA-activated adenylate cyclase in the firefly lantern (15) and tobacco hornworm CNS (16) in that they found chlorpromazine to be a much better blocker than phentolamine and metochlopromide.

In order to show that the above antagonists were associated with OA effects, a test was conducted in which the antagonists were added along with OA. The treated males all had an injection of 0.1 μg/mg of OA and were flown to the high 1 mg pheromone lure. Males thus treated respond below 20% (Fig. 1B) to this pheromone source because the OA treatment has lowered the arrestment threshold and the males exhibit arrested flight. When these males were co-injected with various quantities of antagonists, the data showed that the most effective dose or lower of antagonist from Fig. 3 would completely reverse the OA effect and the number of males reaching the source was again over 80%. The order of activity for this reversal effect was the same as seen with the compounds alone in Fig. 3.

Role of cAMP in Pheromone Response Sensitivity

The above studies support the hypothesis that OA is an endogenous modulator of a male moth's response thresholds to sex pheromone. The data show that the effects are specific to OA, and not produced by serotonin, dopamine, and histamine (unpublished). Also, the activity of several agonists and antagonists is similar to that found in previous studies with receptor sites associated with OA-induced actions. One general scheme by which OA exerts its regulatory actions is outlined in (16,17). Stimulation of OA receptors yields an increase in adenylate cyclase activity, which raises the level of cAMP. This, in turn, affects the activity of protein kinases involved with the phosphorylation of proteins, which can alter the physiological status of the receptive tissue. Physiological response is produced in this scheme by elevated levels of cAMP, which can be affected in several ways, such as with chemicals that interact with adenylate cyclase, OA receptors, enzymes involved in OA biosynthesis or metabolism (such as N-acetyl transferase), or by inhibiting phosphodiesterase, which converts cAMP to AMP.

In addition to studying effects of OA agonists and antagonists on this system, we have initiated a study with the phosphodiesterase inhibitor IBMX. The inhibitor alone at 1 μg/mg did effect an increase to 30% male response to a low dosage of pheromone (0.01 mg) relative to 15% with untreated males or males injected a low amount of OA (0.01 µg/mg). However, treating males with both IBMX and OA at those same rates resulted in a synergistic effect of 78% males successfully reaching the source. IBMX clearly potentiated the effect of a low dose of OA on male sensitivity to pheromone. This appears to be similar to the synergistic effects of IBMX on octopamine agonists in *Manduca* anti-feeding studies (15).

Further studies will include a screening of other types of compounds. Since the shift in male moth response thresh-lds appears to be selectively and sensitively affected by changes in the cAMP system, this assay should provide an excellent research tool for investigating the involved physiological processes and for screening for new neuro-active compounds that can be used as a behaviorally active insect control agent.

ACKNOWLEDGMENTS

We gratefully acknowledge Kathy Poole and Marlene Campbell for rearing the moths, and Dr. Robert Hollingworth for the generous gift of agonist compounds.

REFERENCES

1. Miller JR, Baker TC, Carde' RT, Roelofs WL (1976). Re-investigation of oak leafroller sex pheromone components and the hypothesis that they vary with diet. Science 192: 140.
2. Du JW, Linn CE, Roelofs WL (1984). Artificial selection for new pheromone strain of redbanded leafroller moths *Argyrotaenia velutinana*. Contr Shanghai Inst Entomol 4: 21.
3. Roelofs WL, Du JW, Linn CE, Glover TJ, Bjostad LB (1986). The potential for genetic manipulation of the redbanded leafroller moth sex pheromone gland. In Huettel, MD (ed): "Evolutionary Genetics of Invertebrate Behavior," New York: Plenum Pub Corp.
4. Roelofs WL (1978). Threshold hypothesis for pheromone perception. J Chem Ecol 4: 685.
5. Baker TC, Meyer W, Roelofs WL (1981). Sex pheromone dosage and blend specificity of response by Oriental fruit moth males. Ent Exp & Appl 30: 269.
6. Linn CE, Roelofs WL (1981). Modification of sex pheromone

blend discrimination in male Oriental fruit moths by pre-exposure to (E)-8-dodecenyl acetate. Physiol Entomol 6: 421.

7. Linn CE, Roelofs WL (1983). Effect of varying proportions of the alcohol component on sex pheromone blend discrimination in male Oriental fruit moths. Physiol Entomol 8: 291.
8. Linn CE, Campbell MG, Roelofs WL (1986). Male moth sensitivity to multicomponent pheromones: Critical role of female-released blend in determining the functional role of components and active space of the pheromone. J Chem Ecol 12: 659.
9. Linn CE, Campbell MG, Roelofs WL (1987). Temperature modulation of thresholds controlling moth sex pheromone response specificity. Physiol Entomol 12: submitted.
10. Baker JL, Hill AS, Roelofs WL (1978). Seasonal variations in male omnivorous leafroller moth trap catches with pheromone component mixtures. Environ Entomol 7: 399.
11. Linn CE, Roelofs WL (1984). Sublethal effects of neuroactive compounds on pheromone response thresholds in male Oriental fruit moths. Arch Insect Biochem Physiol 1: 331.
12. Linn CE, Roelofs WL (1986). Modulatory effects of octopamine and serotonin on male sensitivity and periodicity of response to sex pheromone in the cabbage looper moth, Trichoplusia ni. Arch Insect Biochem Physiol 3: 161.
13. Linn CE, Bjostad L, Du JW, Roelofs WL (1984). Redundancy in a chemical signal: behavioral responses of male Trichoplusia ni to a 6-component sex pheromone blend. J Chem Ecol 10: 1635.
14. Evans PD (1980). Multiple receptor types for octopamine in the locust. J Physiol 318: 99.
15. Nathanson JA (1985). Phenyliminoimidazolidines: characterization of a class of potent agonists of octopamine-sensitive adenylate cyclase and their use in understanding the pharmacology of octopamine receptors. Mol Pharmacol 28: 254.
16. Hollingworth R, Johnstone E, Wright N (1984). Aspects of the biochemistry and toxicology of octopamine in arthropods. ACS Symp Ser 255: 103.
17. Nathanson JA (1984). Cyclic AMP and protein phosporylation as a transducing mechanism for certain neurohormones and neurotransmitters. In Borkovec AB, Kelly TJ (eds): "Insect Neurochemistry and Neurophysiology," New York: Plenum Press, p 135.

Molecular Entomology, pages 79–94

INDUCED CHANGES IN SEX PHEROMONE BIOSYNTHESIS OF HELIOTHIS MOTHS (LEPIDOPTERA: NOCTUIDAE)

P.E.A. Teal, and J.H. Tumlinson[1]

Department of Environmental Biology, University of Guelph, Guelph, Ontario N1G 2W1

ABSTRACT Capillary gas chromatography (GC) and GC-mass spectroscopy of extracts of hairpencil glands or scales and of volatiles released from these glands by males of *Heliothis virescens*, revealed the presence of acetates, alcohols and acids which correspond to the aldehyde pheromone components released as the sex pheromone by females of this species. None of the aldehydes were found in samples obtained from males. *In vivo* application of 0.5 ug amounts of various primary alcohols or acetates, which were not present in the glands to the surface of the de-scaled hairpencil gland, in a drop of dimethyl sulfoxide showed that biosynthesis proceeds to the alcohol via the acetate. Similar application of primary alcohols and the corresponding acetates to the pheromone gland of females of *H. virescens* and a close relative *H. subflexa* indicated that the acetates were converted to the alcohols which are subsequently converted to the aldehydes via an alcohol oxidase.

INTRODUCTION

The sex pheromone communication systems of Lepidoptera have been studied for the past 25 years. While both behavioral and chemical analyses have led to the identification of pheromones for several hundred species, (1) studies on the biosynthesis of pheromones are relatively new. Investigations on pheromone biosynthesis

[1]Insect Attractants, Behavior and Basic Biology Research Laboratory, USDA-ARS, Gainesville, Fl. 32604

are of considerable importance, not only for developing an understanding of the whole semiochemical communication system, but also they will provide insights into the regulation of pheromone production that will allow us to exploit weak points for developing more effective control strategies.

A number of studies have demonstrated the de novo synthesis of the sex pheromones of moths from acetate (2,3,4,5,6). In most cases, biosynthesis involves the production of fatty acid analogs with the double bond position resulting from a unique Δ-11-desaturase and subsequent chain shortening via beta-oxidation (3,4,5,7). However, in one case some evidence suggests the biosynthesis of the long chain ester precursors directly from acetate (6). Fewer studies have been conducted on the terminal step in biosynthesis which results in production of the actual pheromone components. This step is of considerable importance because the potential is produced by the use of molecular engineering to alter the actual pheromone blend (8).

Recently, we identified a number of C_{14} and C_{16} saturated and mono-unsaturated primary alcohols in the female sex pheromone gland of Heliothis virescens (F.) and a mono-unsaturated C_{16} primary alcohol in H. zea (Boddie) (9,10). These alcohols correspond to the major aldehyde pheromone components used by these species but are either without function or are inhibitors to the capture of males (10,11). Studies on the role of these alcohols in pheromone biosynthesis by these species has revealed that they are the immediate precursors of the aldehyde pheromone components (12). A third species of this genus, H. subflexa (Gn.), uses a blend of aldehydes and acetates as a pheromone but maintains the corresponding alcohols within the pheromone gland (13,14). As occurs with H. zea incorporation of the alcohols into pheromone lures resulted in a reduction of the number of males captured in pheromone traps (13). This strongly suggests a role in pheromone biosynthesis for these alcohols in H. subflexa.

Recently, Jacobson et al. (15) reported the identification of very large amounts (1 mg/insect) of (Z)-9-tetradecenal (Z9-14:AL) from both male hairpencil extracts and homogenates of whole abdomens of male H. virescens. Based on electroantennogram and field trapping studies using the synthetic compound, these authors suggested that Z-9-14:AL was the male hairpencil pheromone and that it functioned to inhibit the response of other males. The production and use of Z-9-14:AL by male *H. virescens* was intriguing to us because this compound is of critical importance to female to male sexual communication (9,11) albeit at a much reduced concentration. Further, the production of mg amounts of this compound was of considerable interest because of the tremendous biosynthetic capability implied. (*Z*)-9-tetradecenol was not identified even though this alcohol has been identified from female sex pheromone gland extracts (9) and has been shown to be the immediate biosynthetic precursor of the aldehyde (12,16). Therefore, a different biosynthetic pathway from that present in the female sex pheromone gland was suggested.

For these reasons, we investigated the terminal steps in pheromone biosynthesis by *H. subflexa* using *in vivo* techniques and re-investigated the pheromone produced by the hairpencil glands of male *H. virescens*, including *in vivo* studies on biosynthesis of the compounds produced by males. The following reports the conversion of topically applied acetates to the corresponding alcohols by females of both *H. subflexa* and *H. virecens* and that the alcohols serve as the immediate precursors of the aldehyde pheromone components. We also discuss the terminal steps in biosynthesis of the pheromone blend produced by male *H. virescens*.

METHODS

Insects.

Heliothis virescens used in this study were obtained as pupae from the Bioenvironmental Insect Control Laboratory, USDA, Stoneville, M.S. *H. subflexa* were obtained from colonies maintained at the University of Guelph and Insect Attractants Laboratory. Insects were sexed as pupae and were maintained under a reversed 16:8 light/dark cycle at 25°C and 55% relative humidity. Newly emerged insects were transferred to 30 x 30 x 30 cm clear plastic cages daily and were provided with a 10% sucrose solution soaked onto cotton. In order to ensure that insects were capable of pheromone biosynthesis, all insects were aged 2-4 days prior to use.

Isolation and Identification.

In initial studies, rinses of the whole male hairpencil gland complex were obtained by removing the abdominal segments of males during the 4th-6th h of the dark period, everting the gland by applying pressure to the anterior abdominal segments using forceps, and immersed the exposed gland into 200 ul of iso-octane (Fisher 99 mole %) for ca 30 sec. Up to 25 rinses were collected in each sample and the extracts were analyzed without concentrating the iso-octane. Extracts of the elongate scales, which serve to disseminate the pheromone, were prepared by carefully cutting the scales from the surface of the gland leaving a short portion attached, and placing the scales in a 0.5 ml conical microvial. The scales were then rinsed with 10 ul of iso-octane containing the appropriate internal standards, and the total volume of the extract was analyzed.

Two methods were used to collect the actual volatile compounds released from the hairpencil glands. In the first method, individual males were placed in an enlarged holding apparatus similar to that described by Teal *et al.* (12) which was linked to another smaller chamber containing a rubber septum impregnated with 1 mg of the 6 component pheromone blend (12). Purified air was passed over the lure and into the chamber holding the males. Volatiles released during hairpencil displays were collected on charcoal microentrainment filters and were recovered in a small amount of dichloromethane and iso-octane as described elsewhere (12). In the second method, the hairpencils were everted using forceps and the volatiles were collected under a stream of dried N_2 (250 mls/min) above a conical microvial containing 250 ul of iso-octane. The volatiles from up to 50 insects were collected in the same sample and the iso-octane was never concentrated to less than 25 ul under a N_2 stream.

Methyl esters of acids suspected to be present in the hairpencil pheromone were formed both by acid methanolysis and methanolysis using boron trifluoride in methanol (17,18). The iso-octane was completely evaporated under N_2 from extracts of the hairpencil scales of groups of 5 males or volatile collections obtained from groups of 10 males using the second method described above. In acid methanolysis studies, the procedure outlined by Bjostad and Roelofs (19) was followed while that of Morrison and Smith (18) was used with boron trifluoride in methanol.

Biosynthesis.

In vivo studies on the terminal step in pheromone biosynthesis by male H. virescens were conducted using glands that were clamped in a fully exposed position using a smooth jawed alligator clamp. The hairpencil scales of some preparations were then carefully removed with forceps which exposed the actual gland. The removed scales were placed in a connical microvial. Other preparations were left intact. Test alcohols or acetates were then applied at a concentration of 500 ng in 1 ul of dimethyl sulfoxide (DMSO) to the denuded gland surfaces, the removed scales or the intact scales using a 1 ul syringe. Treated and DMSO control preparations were then incubated for 1 h at 25° prior to extraction and GC analysis. Experiments were conducted during the reproductive period (ca. 4 h after dark) and 12 h later during the light period.

Experiments on the biosynthesis of the female sex pheromone of H. subflexa and H. virescens were conducted 1-2 h prior to the reproductive period. The terminal abdominal segments were extended and clamped as described in other studies (4,12) and test alcohols and acetates were applied in a 1.0 ul drop of DMSO. Preparations were allowed to incubate for 1 h at 25° and were stopped by excising the glands and soaking the tissue for 2 min in 5 ul of iso-octane. Experimental controls for studies on both species included preparations treated with DMSO alone.

Chemical Analysis

Gas chromatographic (GC) analyses were conducted using Hewlett Packard 5792(R) and 5890(R) GC's equipped with splitless and cool on column capillary injectors and flame ionization detectors. The detectors of the 5792 GC were interfaced to a Hewlett Packard 3390(R) reporting integrator. Data from the 5890 GC were aquired at a rate of 20 points/sec through a Chromadapt(R) interface and Adalab(R) data aquisition system (Interactive Microware Inc.) and processed using an Apple IIe(R) Computer with Chromatochart(R) Software (Interactive Microware Inc.). Fused silica capillary columns used routinely for GC analysis included a 15 m x 0.25 mm (id) DB 225(R) (J&W), a 30 m x 0.25 mm (id) SPB1(R) (Supelco), and a 30 m x 0.25 mm (id) SPB10(R) (Supelco). Only the DB 225 column was used in conjunction with the cool on column injector. In this case the initial temperature of 80° was maintained for 0.5

min and then increased at 15°/min to a final temperature of 135°. Conditions of chromatography when using the splitless injectors were as follows: initial temperature = 80° for 1 min, splitless purge at 0.5 min, temperature increase at 25°/min to 165° (SPB1) or 180° (SPB10). Hydrogen was used as the carrier gas at a linear flow velocity of 38 cm/sec. The primary saturated acetates of tridecanol and pentadecanol (10 ng each) were used as internal standards for both synthetic and natural product samples and were used to calculate relative retention indices and to quantitate the amounts of compounds present in natural product samples. Further GC confirmation was obtained by co-chromatography of the natural product samples with the individual synthetic compounds on all 3 columns.

Both electron impact and chemical ionization mass spectra (MS) were obtained using a VG1212F$^{(R)}$ MS interfaced to a Hewlett Packard 5792 equipped with a cool on column capillary injector. Helium was used as the carrier gas at a linear flow velocity of 18 cm/sec. Both iso-butane and methane were used as ionization gases. Samples were chromatographed on both the SPB1 column used in GC studies and a 50 m x 0.25 mm (id) DB5 column (J&W) under the following conditions: initial temperature = 80°, temperature program = 20°/min after 1 min, final temperature = 225. The retention times and fragmentation patterns of the both synthetic compounds and those present in natural products were compared.

Synthetic Chemicals.

Synthetic chemicals used were obtained from the Sigma Chemical Company. All compounds were purified by high performance liquid chromatography (HPLC) using a 25 cm x 0.46 cm (id) column packed with 5 um silica (Altech) eluted with 20% ether in hexane (Fisher, HPLC grade). Unsaturated compounds were further purified by HPLC on a 25 cm x 0.46 cm (id) _in situ_ A_gNO_3 impregnated silica (5 um) column eluted with toluene (Fisher, HPLC grade) (20). GC analysis using the SPB1 and SPB10 capillary columns indicated that all compounds were at least 99% pure and were free of the corresponding alcohols, acetates and aldehydes and of the corresponding geometrical isomers.

RESULTS

Male Pheromone Components.

Initial GC analyses using all 3 capillary columns of 1 male equivalent (ME) of the whole gland extracts indicated the presence of 3 peaks which were present in a consistent ratio. These peaks had retention indices corresponding to hexadecanol (16:OH) (28.8 $\pm$ 6.4 ng) (n = 10), hexadecanol acetate (16:AC) (313.0 $\pm$ 56.2 ng) and (*Z*)-11-hexadecenol acetate (Z11-16:AC) (8.2 $\pm$ 2.3 ng). Two other peaks were also present in chromatograms obtained using the SPB1 column. This column was capable of chromatographing free fatty acids, although peak shape was poor. The 2 novel peaks had both relative retention indices and peak shapes which suggested that they were tetradecanoic (14:Acid) and hexadecanoic (16:Acid). Methanolysis of the extracts and subsequent chromatography on the 3 capillary columns indicated that the acid assignments were correct because the methyl esters of both acids were detected at concentrations of 2.7 ($\pm$ 0.9) ng (14:Acid) and 22.6 ($\pm$ 4.3) ng (16:Acid).

GC analysis, of extracts of the scales which disseminate the pheromone, indicated that all of the above components were present in the same ratio found in whole gland extracts. Extracts of groups of 5 ME also contained compounds which had relative retention indices corresponding to tetradecanol (14:OH), tetradecanol acetate (14:AC), (Z)-11-hexadecenol (Z11-16:OH), (Z)-7-hexadecenol acetate (Z7-16:AC) and (Z)-9-hexadecenol acetate (Z9-16:AC) (Figure 1). The ratio of these compounds was 0.88 ng 14:AC: 0.14 ng 14:OH: 0.49 ng Z7-16:AC: 0.24 ng Z9-16:AC: 2.80 ng Z11-16:AC: 190.70 ng 16:AC :0.30 ng Z11-16:OH: 23.73 ng 16:OH per male (n= 10) (Figure 1). In addition, each ME of these samples contained 2.1 ng of 14:Acid and 23.3 ng of 16:Acid as indicated by analysis of the methyl esters of the extracts formed using both techniques. Full mass spectra (60-400 AMU) and relative retention indices of the compounds present in the scale extract using both capillary columns coupled with EI, CI (methane) and CI (isobutane) ms confirmed the presence of 16:OH, Z11-16:AC, 16:AC, 14:Acid and 16:Acid. Adequate spectra of the other compounds were not obtained and their identities cannot be confirmed at present by ms.

Volatiles collected on the charcoal entrainment filter of the pheromone released by males responding to the female sex pheromone contained an 11:1 ratio of the 16 AC : 16:OH. The greatest amount of 16:AC recovered from a male was 12.2 ng after 6 hairpencil exposures. Minor components were not detected. The 11:1 ratio present in these samples is comparable to the 8:1 ratio found in scale extracts when considering the recovery efficiencies of the alcohols and acetates from the charcoal filter (21). In order to obtain greater quantities of the volatile components we developed the second technique listed earlier. GC and GC-mass spectral analysis of volatiles confirmed the presence of 6:1 ratio of 16:AC and 16:OH. GC analyses also indicated the presence of 14:AC and Z11-16:AC however backround was too high to obtain useful mass spectra of these compounds. GC and GC-mass spectral analyses of the esterified volatile samples confirmed the presence of 14:Acid and 16:Acid. The ratio of all of the components was consistent with that found in scale extracts.

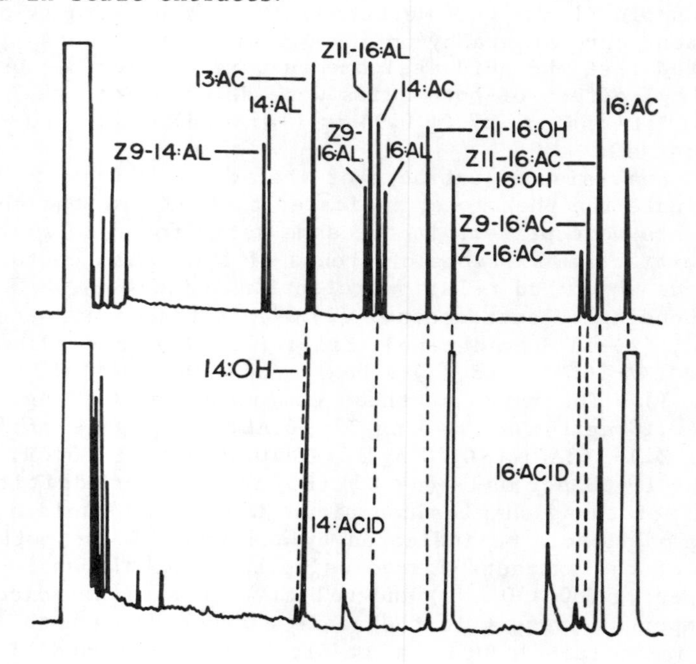

FIGURE 1. Comparison of synthetic standards (upper) and 2 ME of the hairpencil scale extract of male *H. virescens* on the SPB1 capillary column. Captions are as in the text.

Male Pheromone Biosynthesis.

Application of 500 ng of 14:OH or Z11-16:OH to intact hairpencil glands, glands denuded of scales, and to the scales, did not result in changes in the ratios of the corresponding acetates present in gland extracts. Similarly, other alcohols applied to the gland surface including (E) and (Z)-11-tetradecenol (E11-14:OH, Z11-14:OH) or (Z)-9-tetradecenol (Z9-14:OH) were not converted to the corresponding acetates. Scales incubated with 14:AC, E11-14:AC or Z9-14:AC did not preferentially convert the acetates to the alcohol analogs. However, substantial amounts of the alcohols were produced when the acetates were applied to the surface of the denuded gland (Figure 2). The enzyme responsible for the conversion showed no specificity for either presence or geometry of double bonds as is indicated by the production of 4.0 ($\pm$ 1.2) ng more of 14:OH than was present in glands of untreated preparations (n = 10) and 3.7 ($\pm$ 0.7) ng (n = 10) of E11-14:AC after a 1 h incubation. Studies conducted 12 h after the peak of the reproductive period indicated that the temporal periodicity governing male response to the female pheromone had no effect on the biosynthetic capability of the gland because 3.9 ($\pm$ 0.6) ng of E11-14:OH was produced by preparations treated with the corresponding acetate during this period. Similar amounts of Z9-14:OH and Z11-14:OH were produced when glands were treated with the corresponding acetates.

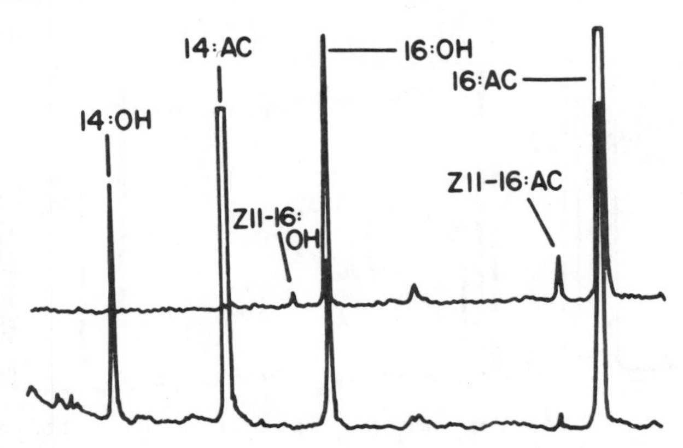

FIGURE 2. Chromatograms on the SPB1 column of a 1 ME extract of the male H. virescens gland (upper) compared with a gland incubated with 14:AC for 1 h (lower). Captions as in text.

Female Pheromone Biosynthesis.

In initial experiments 1 ug of either Z11-16:OH (n = 5) or Z9-16:OH (n = 5) was applied to the surface of the pheromone gland of H. subflexa females 2 h prior to the onset of pheromone release. Extraction and analysis of the tissue 1 h later revealed the presence of 8 times the amount of Z11-16:AL and 9 times the amount of Z9-16:AL present in untreated preparations. This dramatically altered the blend of aldehydes present. No change in the ratio of the corresponding acetates was found. Application of 500 ng of 14:OH (Figure 3), Z9-14:OH, or E11-14:OH resulted in the production of 21.9 (± 6.3) ng 14:AL, 23.6 (± 2.2) ng Z9-14:AL and 22.1 (± 2.4) ng of E11-14:AL respectively (n = 5, each treatment). No detectable amounts of either 14:AL or Z9-14:AL were detected in control preparations. The acetate analogs of these alcohols were not detected in any of the samples.

Application of 1 ug of 14:AC to the surface of the pheromone gland of females of H. subflexa (n = 10) resulted in the production of 4.5 (± 2.3) ng of 14:AL and 3.2 ng (± 2.1) ng of 14:OH after a 1 h incubation period (Figure 4). Similar amounts of both E and Z11-14:AL and E and Z11-14:OH were produced by glands incubated for 1 h after treatment with the corresponding acetates. The amounts of the alcohols were quite variable in all cases but were always less than the corresponding aldehydes.

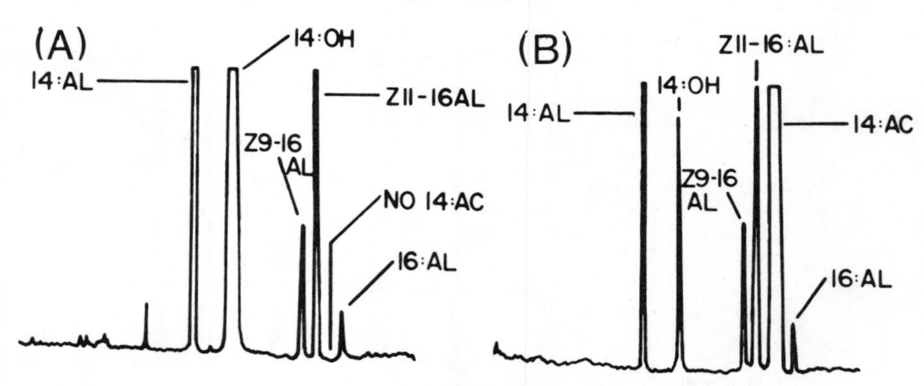

FIGURE 3. Chromatograms of extracts of H. subflexa pheromone gland after treatment with 14:OH (A) and 14:AC (B) and incubation for 1 h prior to extraction. SPB1 column, captions as in text.

Application of 0.5 ug of Z9-14:AC to the surface of the pheromone gland of precalling females of H. virescens resulted in production of 22.9 (± 15.7) ng of Z9-14:AL and 40.32 (± 16.0) ng Z9-14:OH (n = 10) after a 1 h incubation period. Control preparations contained less than 1 ng of Z9-14:AL and no detectable amounts of the corresponding alcohol. Similar experiments were performed using E11-14:AC and in these cases 26.4 (± 17.3) ng of E11-14:AL and 38.3 (± 19.4) ng of E11-14:OH were produced.

DISCUSSION

Male Pheromone Components.

While sex pheromones produced by females of a large number of species of moths have been identified, few studies have been conducted on the pheromones produced by males of these genera (1). To date, the components identified as male pheromones, for example, benzaldehyde and 2-phenylethanol (1), have had no structural similarity to the components released by females. The identification of the acetates, alcohols and acids which correspond to the aldehydic components released by females of H. virescens (9,11) is novel and shows a close relationship between pheromone biosynthesis by males and females of this species.

Various functional roles have been ascribed to the pheromone produced by males of H. virescens. For example, Hendricks and Shaver (22) suggested that the pheromone stopped females from releasing their sex pheromone, while laboratory flight tunnel studies on mating caused Teal et al. (23) to hypothesize that the male pheromone caused females to stop attempting to escape after the male had clasped her genitalea. The description of the bioassays used and indepth discussion of the results obtained in our study are beyond the scope of this paper. However, our results have indicated that the male signals his presence to females using his pheromone which causes females to assume an acceptance posture for mating and that the male pheromone can induce females to begin to release their sex pheromone.

The identification of 16:AC, 16:OH and 16:Acid in a 80.2: 10.0: 9.8 ratio as the major components of the hairpencil pheromone and collectively comprising over 200 ng per male differs from the identification of Z9-14:AL, reported as the male produced pheromone by Jacobson *et al.* (15). In fact, we did not find even trace amounts of this aldehyde when analyzing extracts of up to 10 ME although Jacobson *et al* (15) suggest that Z9-14:AL is present in a concentration of 1 mg per male. While we agree that large amounts of Z9-14:AL will disrupt the pheromone mediated behavior of males of *H. virescens*, as is indicated by results reported by Hendricks (24) and Tingle and Mitchell (25), the results of our chemical and behavioral studies do not support the hypothesis that male pheromone is Z9-14:AL.

An interesting feature of our study was that acetates corresponding to 5 of the 6 aldehydes which comprise pheromone released by females were identified by GC analysis of hairpencil extracts. The remaining acetate, Z9-14:AC was not found. This was not surprising when considering that the ratio of 16:AC to Z11-16:AC was 68:1. Thus, if the same ratio were to exist between 14:AC and Z9-14:AC each male would only maintain a concentration of .013 ng of Z9-14:AC which was not detectable. The ratio of the acetates to the corresponding alcohols was relatively constant being about 8:1 for 16:AC and 16:OH, 9:1 for Z11-16:AC: Z11-16:OH and 6:1 for 14:AC: 14:OH. Therefore, although we hypothesize that the alcohols corresponding to Z7-16:AC, Z9-16:AC and very probably Z9-14:AC are present in the male pheromone, and that the small amounts present have precluded their identification. The presence of these acetates is supported by our studies on biosynthesis which have proven that the acetates are converted to the corresponding alcohols.

While strong evidence that the pheromone produced by males of *H. virescens* is composed of the acetates, alcohols and acids which correspond to the aldehydic components released by females of this species, is presented, the ratio of the components is different from females. For example, the ratio of 16:AC to Z11-16:AC present in the hairpencil pheromone is approximately 68:1 while that of hexadecanal to (*Z*)-11-hexadecenal given off by females is 1:8. This suggests that both sexes possess the same enzymes involved in biosynthesis of the pheromone precursors but that the enzyme activities are regulated differently by each sex.

Pheromone Biosynthesis.

Both sexes of *H. virescens* and females of *H. subflexa* converted the topically applied acetates to the corresponding alcohols and showed little specificity for either chain length or number and geometry of double bonds. The reaction has been inhibited by the addition of 3-octylthio-1,1,1-trifluoro-2-propanone in *in vitro* studies (16) which indicates that an acetate esterase controls this step. In addition, we have demonstrated that females of these species as well as *H. zea* (12) regulate the production of the aldehydic pheromone components via an alcohol oxidase specific for primary alcohols. *In vitro* studies (16) have also indicated that males of *H. virescens* possess the same oxidase and that both sexes of this species utilize an aldehyde dehydrogenase to convert the aldehydes to the corresponding acids. This indicates a close phylogenetic relationship between the members of this genus.

Our studies have indicated that pheromone-mediated reproductive isolation within the *Heliothis* genus can be explained by the biosynthetic activities of the enzymes involved. Females of *H. zea* use the same series of 16 carbon aldehydes that the other 2 species use but do not use the 14 carbon compounds although they are capable of converting the alcohol precursors to the aldehydes. In fact, in the laboratory we have successfully overcome pheromone-mediated isolation between *H. zea* females and *H. virescens* males by topically applying appropriate amounts of the two 14 carbon alcohols (12). This coupled with the work of Bjostad and Roelofs (4) suggest that *H. zea* do not employ enzymes capable of chain shortening long chain compounds for pheromone production. Therefore, species divergence between *H. zea* and the *H. virescens*/*H. subflexa* complex may have been the result of the evolution of chain shortening enzymes by the progenitor of the latter 2 species. The sex pheromone used by females of *H. subflexa* differs from those of *H. zea* and *H. virescens* in that in addition to the alcohols and aldehydes the gland also contains the acetates in a ratio that approximates that of the aldehydes (13,14). The acetates are important for maximizing the behavioral response of males to pheromone lures (13) but, as demonstrated, are also the precursors of the alcohols and aldehydes. Our hypothesis is that the major portion of the acetate is converted to the analogous alcohol within the pheromone gland cells and that both acetates and alcohols are secreted into the cuticle. The alcohol oxidase which has been demonstrated to be present in the cuticle of *H. virescens* females (12) then converts the alcohols to aldehydes leaving the acetates intact. The presence of the same acetate esterase in the gland of *H. virescens* females is an enigma because none of the acetates corresponding to the aldehydic pheromone components are present in the glands (12). One possible explanation is that the acetates are converted to the alcohols as rapidly as they are produced but this hypothesis is not supported by our data and warrants further study. The esterase does function in the production of the male pheromone however, and because the acetates are present in very large quantities compared to the alcohol and acid components indicates sex specific differences in the biochemistry of pheromone production by this species. The presence of the acids in the pheromone produced by males of *H. virescens* was of interest because such acids are not commonly used for pheromone communication by moths (1). However, behavioral and morphological examination of the mechanism involved in hairpencil protrusion indicate that the force with which males expose their hairpencils would be sufficient to disseminate the acids into the air. Biosynthesis of the acids can be explained by conversion of the alcohols to their aldehyde analogs via the alcohol oxidase common to females, with immediate and complete conversion of the aldehydes to the corresponding acids. Enzymes capable of performing these steps have been identified in tissue homogenates of the male pheromone gland (16) and in some instances we have found a peak corresponding to 16:AL in GC analyses of extracts of whole hairpencil glands. However, this peak is not present in all extracts and the small amounts (0.32 $\pm$ 0.18 ng 5 of 15 glands) preclude identification at present. Thus, within the numbers of the *Heliothis* genus we see a common biosynthetic theme of pheromone production. Permutations in the theme which result in the production of specific blends can be attributed to the evolution of genes responsible for specific enzymes and to genetically controlled species specific enzyme activities.

REFERENCES

1. Tamaki Y (1985). In Kert GA, Gilbert LI (eds): "Comprehensive Insect Physiology Biochemistry and Pharmacology. Volume 9," New York: Pergamon, p 145.

2. Jones IF, Berger RS (1978). Incorporation of (1-^{14}C) acetate into cis-7-dodecen-1-ol acetate, a sex pheromone in the cabbage looper (Trichoplusia ni). Environ Entomol 7:666.
3. Bjostad LB, Roelofs WL (1981). Sex pheromone biosynthesis from radiolabeled fatty acids in the redbanded leafroller moth. J Biol Chem. 256:7936.
4. Bjostad LB, Roelofs WL (1983). Sex pheromone biosynthesis in Trichoplusia ni: key steps involve delta-11 desaturation and chain shortening. Science 220:1387.
5. Bjostad LB, Roelofs WL (1984). Biosynthesis of sex pheromone components and glycerolipid precursors from sodium (1-^{14}C) acetate in redbanded leafroller moth. J Chem Ecol 10:681.
6. Morse D, Meighen E (1984). Aldehyde pheromones in Lepidoptera: Evidence for an acetate ester precursor in Choristoneura fumiferana. Science 226:1434.
7. Wolf WA, Roelofs WL (1983). A chain-shortening reaction in orange tortrix moth sex pheromone biosynthesis. Insect Biochem 13:375.
8. Tumlinson JH, Teal PEA (in press). Relationship of structure and function to insect pheromone systems. In Prestwich GD, Blomquist GJ (eds): "Pheromone Biochemistry," Orlando: Academic, p 1.
9. Teal PEA, Tumlinson JH, Heath RR (1986). Chemical and behavioral analysis of volatile sex pheromone components released by calling Heliothis virescens (F.) females (Lepidoptera: Noctuidae). J Chem Ecol 12:107.
10. Teal PEA, Tumlinson JH, McLaughlin JR, Heath RR, Rush, RA (1984). (Z)-11-hexadecen-1-ol: A behavioral modifying chemical present in the pheromone gland of female Heliothis zea (Lepidoptera: Noctuidae). Can Entomol 116:777.
11. Vetter RS, Baker TC (1983). Behavioral responses of Heliothis virescens in a sustained-flight tunnel to combinations of the 7 components identified from the female sex pheromone gland. J Chem Ecol 9:747.
12. Teal PEA, Tumlinson JH (1986). Terminal steps in pheromone biosynthesis by Heliothis virescens and H. zea. J Chem Ecol 12:353.

13. Teal PEA, Heath RR, Tumlinson JH, McLaughlin JR (1981). Identification of a sex pheromone of Heliothis subflexa (Gn.) and field trapping studies using different blends of components. J Chem Ecol 7:104.
14. Klun JA, Leonhardt BA, Lopez JD, LeChance LE (1982). Female Heliothis subflexa (Lepidoptera: Noctuidae) sex pheromone: Chemistry and congeneric comparisons. Environ Entomol 11:1084.
15. Jacobson M, Adler VE, Baumhover AH (1984). A male tobacco budworm pheromone inhibitory to courtship. J Environ Sci Health A19:469.
16. Ding YS, Prestwich GD (1986). Metabolic transformations of tritium-labeled pheromone by tissues of Heliothis virescens moths. J Chem Ecol 12:
17. Mangold HK (1969). Aliphatic lipids. In Stahl E (ed): "Thin-Layer Chromatography: A Laboratory Handbook," New York: Springer, p 363.
18. Morrison WR, Smith LM (1964). Preparation of fatty acid methyl esters and dimethylacetals from lipids with boron fluoride-methanol. J Lipid Res 5:600.
19. Bjostad LB, Roelofs WL (1984). Sex pheromone biosynthetic precursors in Bombyx mori. Insect Biochem 14:275.
20. Heath RR, Sonnet PE (1980). Techniques for in situ coating of Ag^+ onto silica gel in HPLC columns for the separation of geometrical isomers. J Liq Chromatogr 3:1129.
21. Tumlinson JH, Heath RR, Teal PEA (1982). Analysis of chemical communication systems of Lepidoptera. In Leonhardt BA, Beroza M (eds): "Insect Pheromone Technology," Amer Chem Soc Symp 190:1.
22. Hendricks DE, Shaver TN (1975). Tobacco budworm: Male pheromone surpressed emission of sex pheromone by the female. Environ Entomol 4:555.
23. Teal PEA, McLaughlin RR, Tumlinson JH (1981). Analysis of the reproductive behavior of Heliothis virescens (F.) under laboratory conditions. Ann Entomol Soc Am 74:324.
24. Hendricks DE (1976). Tobacco budworm: Disruption of courtship behavior with a component of the synthetic sex pheromone. Environ Entomol 5:978.
25. Tingle FC, Mitchell ER (1978). Response of Heliothis virescens to pheromonal components and an inhibitor in olfactometers. Experientia 34:153.

Molecular Entomology, pages 95–105

IDENTIFICATION, RECONSTITUTION AND EXPRESSION OF NEURONAL ACETYLCHOLINE RECEPTOR POLYPEPTIDES FROM INSECTS[1]

H. Breer, D. Benke, W. Hanke, R. Kleene[2], M. Knipper and L. Wieczorek

Fachbereich Biologie/Chemie, Universität Osnabrück D-4500 Osnabrück, West Germany.

ABSTRACT The nicotinic acetylcholine receptor from locust nervous tissue has been identified as a membrane protein composed of identical subunits, which produces acetylcholine-activated cation channels when reconstituted in lipid bilayers. Immunological and sequence data demonstrate a certain relatedness to the peripheral vertebrate receptor. PolyA$^+$-RNA from insect nervous tissue microinjected into Xenopus oocytes induced the expression of functional receptors. Receptor specific cDNA has been cloned in λgt11.

INTRODUCTION

Acetylcholine (ACh) is one of the most frequent excitatory neurotransmitter in the nervous tissue of arthropods (1,2) and cholinergic synapses apparently play a major role in signal transmission in the nervous system of insects. The task to unravel the complex neurochemical mechanisms and the molecular structure of functional elements of cholinergic synaptic sites is therefore of fundamental importance towards a molecular neurobiology of insects. Specific receptors for acetylcholine are key element at all cholinergic synapses: the receptor transmitter interaction triggers the transduction of a chemical signal in a postsynaptic response, usually ACh binding is converted into a change of transmembrane potential by opening specific ion channels.

[1] This work was supported by the Deutsche Forschungsgemeinschaft, SFB 171.

[2] Present address: Dept. Physiol. Chem. Univ. München

Ligand Binding

Corresponding to the very high concentrations of ACh, rather high levels of AChRs have been detected in the nervous tissue of various insect species; much higher than in the vertebrate brain (Table 1).

TABLE 1
NICOTINIC AND MUSCARINIC BINDING SITES IN THE NERVOUS TISSUE

Species	Tissue	nAChR	mAChR	Reference
Locusta	ganglia	1775	116	4
Drosophila	heads	800	65	5
Cockroach	nerve cords	910	138	6
Mouse	brain	180	570	7
Rat	hippocampus	60	1000	5

Values are given as fmol/mg protein

Electrophysiological research (3) and ligand binding studies have revealed that the receptor for ACh in insects as in vertebrates can be distinguished in nicotinic and muscarinic receptor types; however a different receptor type predominates in each animal group: in the vertebrate brain mostly muscarinic receptors, in insect ganglia very high concentration of putative nicotinic receptors (Table 1). The observation, that the nervous tissue of insects and vertebrates both contain nicotinic and muscarinic binding sites, suggests that the classical dichotomy of cholinergic receptors (nAChR, mAChR) developed very early in evolution.

Biochemical Characterization

Due to its high concentration and its high affinity for α-toxins, a biochemical identification of nicotinic AChR-proteins from insect nervous tissue seemed an amenable task; such an approach not only would allow to explore the molecular properties of a nicotinic AChR from nerve cells but furthermore might provide some interesting comparative data which together with the wealth of information on the AChR from vertebrate electrocytes and muscle cells may offer important clues for unravelling the course of receptor evolution and furthermore may provide valuable information to solve the central questions concerning the molecular basis of channel functions like chemical gating, ion selectivity, desensitization etc.
The putative nAChR, as probed by α-bungarotoxin binding, were solubilized from locust neuronal membranes using mild detergents, which allowed to preserve the toxin binding and its pharmacology. Approaches to estimate the size of the solubilized binding sites, using analytical linear sucrose density gradient centrifugation and polyacrylamide gradient gel electrophoresis of native, toxin binding, receptor complexes revealed that it represents a macromolecule with a sedimentation coefficient of 10 S and an apparent M_r of 250-300.000 (Fig.1).

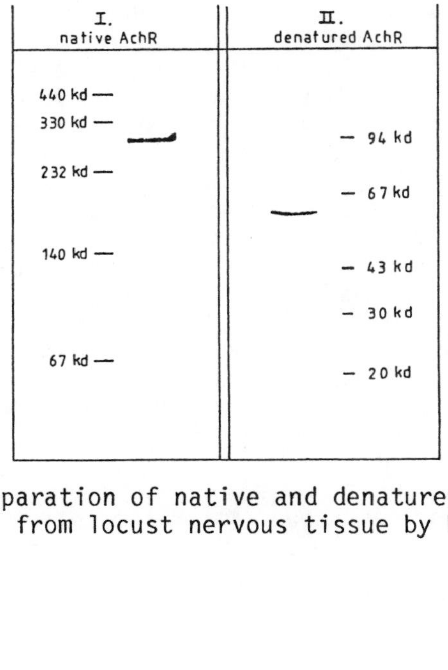

FIGURE 1. Separation of native and denatured AChR polypeptides from locust nervous tissue by PAGE.

The binding entity was further purified by affinity chromatography on Sepharose 4 B gel covalently coupled with α-bungarotoxin; the adsorped receptor proteins were recovered by biospecific eulution using nicotinic ligands. Analysis of the recovered receptor proteins on SDS-polyacrylamid electrophoresis under denaturing conditions, a treatment which dissociates the vertebrate receptor in its subunits, resulted in a single polypeptide band (Fig.1), suggesting that the macromolecular receptor complex obviously dissociates into subunits of identical or very similar size (M_r = 65.000) (8); a similar result was obtained on cockroach (9) suggesting that the neuronal AChR from insects represents an oligomer of 65 Kd subunits.

Immunochemistry of the Receptor

Further evidence that the α-toxin binding protein in insect membranes indeed represents a constituent of an AChR was achieved when immunological approaches were employed. A library of monoclonal antibodies raised against the acetylcholine receptor from _Torpedo marmorata_ have been analyzed for crossreactivity with neuronal membrane preparations from locust ganglia in ELISA assay system (10). It was shown that a few monoclonals, which are mostly known to interact with the ligand binding site of the receptor, showed a considerable cross-reactivity with the toxin binding sites in insect membranes. Immunoblotting techniques were used to demonstrate that anti(Torpedo AChR)-antibodies recognize the purified putative receptor subunit from insect as well. The significant antigenic cross-reactivity between α-toxin binding proteins from _Locusta_ and the _Torpedo_ acetylcholine receptor represents independent evidence that the isolated 65 Kd polypeptide is a major constituent of the insect AChR and implies that some structural homology between both receptor types exist. A localization of the identified putative receptor polypeptides might be another independent criterion for establishing its receptor nature. Antisera against affinity purified receptor proteins were raised in rabbits. Purified monospecific antisera were applied to tissue sections from the locust thoracic (9) and cockroach 6th abdominal ganglia (11). The specifically bound antibodies were subsequently visualized using the sensitive peroxidase-antiperoxidase technique. After completing the protocol, labelled antigenic sites were largely confined to very distinct areas in the neuropil of both ganglia. The specifically labelled areas in both ganglia are known to be very rich in synaptic contacts and in the cockroach terminal ganglion very high labelling was found in the posterior neuropil (11), in areas of the cercal glomerulus, a region where the cercal sensory fibers synapse onto dendritic processes of the giant interneurons (3). Furthermore, the pericarya of these neurons, the site of synthesis and processing of the receptor protein showed significant antigenic reactivity. Thus the histochemical and cytochemical localization of the receptor antigenic site in the insect nervous system supports the view that the purified polypeptides are in fact constituents of the nicotinic AChR.

Reconstitution of the Purified Receptor Protein

The ultimate proof however that the identified polypeptides represent in fact a functional chemically gated ion translocating system can only be achieved by reconstituting the purified protein in artificial membranes; the potential of planar lipid bilayers for studying acetylcholine receptor function is based on the possibility to record the current flowing through individual channels (12); a successful reconstitution process quantitatively restores the ion permeability control of AChR.

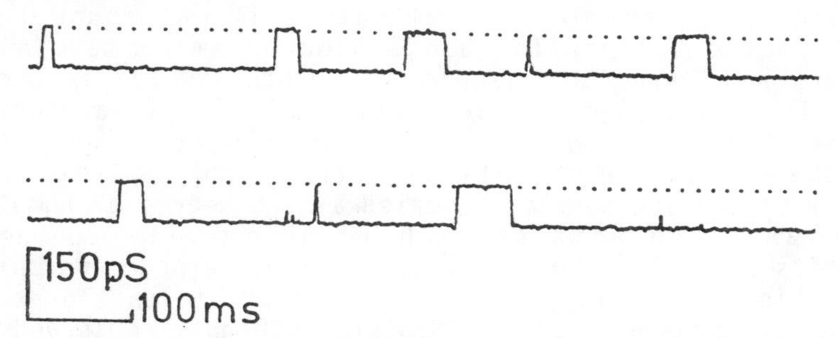

FIGURE 2. Fluctuations of a AChR channel from locust nervous tissue reconstituted in planar lipid bilayer.

The affinity purified insect receptor proteins were first incorporated in liposomes by dialysis techniques and the resulting proteoliposomes subsequently fused with the preformed bilayer thus introducing receptor molecules into the bilayer membrane. The bilayers containing the affinity purified AChR protein from locust responded to the application of cholinergic agonists like carbamylcholine and suberyldicholine, by a marked increase of conductance. The agonist effect was saturable and was blocked by the competitive antagonist d-tubocurarine. At low receptor concentration and high time resolution fluctuations between discrete states in conductance which resemble single channel recordings could be observed. In physiological saline, a conductance of 75 pS and a lifetime of a few msec for the individual channel was estimated (13). These results demonstrate that most if not all of the known electrophysiological criteria for functional acetylcholine activated channels appear satisfied. The reconstitution experiments have thus unambiguously shown that the affinity purified membrane protein contains all the structural elements for the regulation of ion permeability by acetylcholine.

Partial Amino Acid Sequence of the Receptor Protein

Based on the observation that there is some immunological crossreactivity between the peripheral vertebrate AChR and the neuronal AChR of locust, the relatedness of both receptor types at the molecular level is of particular interest not only concerning structure-function relationships but also with regards to the molecular evolution of the ACh-receptor. The affinity purified AChR was subjected to gasphase-microsequencing; unfortunately it was found that the isolated polypeptide had a blocked amino terminus. However, a fragment of the receptor protein which is occasionally present in the preparation (8), gave a readily identifiable single sequence (14). Comparing this sequence with the primary structure of the receptor polypeptides from various vertebrate sources revealed a high degree of homology to an amino acid stretch found in all sequenced α-subunits as well as in the δ-subunits of Torpedo and chick (15). This result emphasizes the view that at least part of the acetylcholine receptor protein is highly conserved in evolution, as has already been proposed on the vertebrate data. Furthermore it is of great interest that the identified peptide sequence alignes to an amino acid stretch in the transmembrane region II/III of the vertebrate polypeptides; this has recently been identified as most striking examples of regional conservation, typical for proteins with multiple transmembrane segments (15). This consideration points to the possibility that the insect neuronal receptor polypeptides display a similar transmembrane organisation as the vertebrate receptor subunits.

The sequence information obtained may be useful to construct an appropriate oligonucleotide probe for screening cDNA- and genomic librabries and will be helpful for an unequovocal identification of the receptor specific DNA-clones.

Expression of Receptor-specific mRNA

An essential early step before the powerful approaches of molecular genetics can be applied is the recognition of the mRNA coding for the receptor polypeptides. The translation of isolated mRNA in a cell-free system in combination with immunoprecipitation using anti-(AChR)-antibodies represents a suitable essay for identifying receptor specific mRNA. PolyA$^+$-RNA was isolated from nervous tissue of locust and translated in a nuclease-treated, i.e. mRNA-dependent, rabbit reticulocyte lysate in the presence of ^{35}S-labelled methionine. Precipitation experiments followed by electrophoretic analysis indicated that the RNA-preparation from locust ganglia contained intact mRNA, capable to code even for high molecular weight polypeptides. To evaluate whether receptor protein have been produced as well, monospecific antiserum against the native AChR have been used for immunoprecipitation of specific polypeptides; it was found that 0.1-0.5% of the polypeptides could be separated, showing that antigenic sites of the native receptor already exist in the non-processed polypeptides and that the polyA$^+$-RNA fraction in fact contained receptor-specific mRNA (unpublished results). The *in vitro* translation system is however incapable to produce intact ligand binding, functional receptors. As a next step in the analysis, an *in vivo* translation system, the *Xenopus* oocyte, was used. The oocyte technique introduced by Gurdon (16) as an efficient translation system for exogenous mRNA, capable to perform all post-translational modifications, has been adapted by Barnard and his colleagues for studying the expression of mRNA for receptors and ion channels (17). When polyA$^+$-RNA isolated from the nervous tissue of young locusts were microinjected into oocytes after 1 day α-bungarotoxin binding activity could be detected (Fig.3), depending linearly on the amount of applied RNA in the range of 0-15 ng per oocyte (18). Binding sites for α-toxin were found in the surface membrane of the oocytes, indicating that some of the AChR molecules are inserted in the memrane. Immunoprecipitation of receptor polypeptides synthesized in the presence of ^{35}S methionine, followed by electrophoresis and autoradiography revealed, that obviously a homogenous population of polypeptides was isolated (M_r = 65.000) indicating that the in ovo synthesized

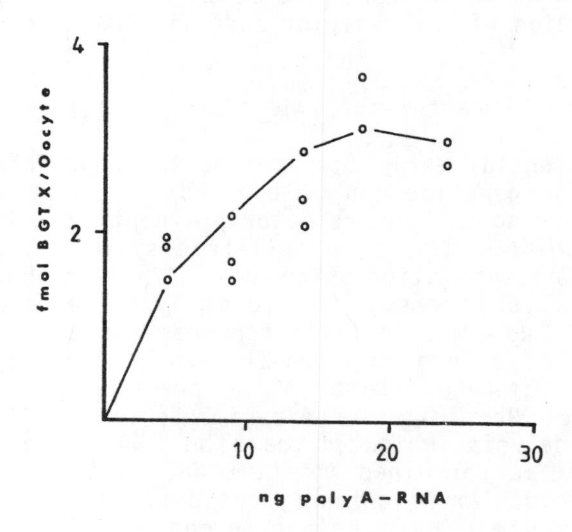

FIGURE 3. Expression of α-BGTX binding sites in Xenopus oocytes after microinjection of locust mRNA.

constituents of the receptor display the same or a similar size as the native subunits. However to verify that the expressed binding proteins represent in fact functional AChRs it was necessary to provide evidence for ACh-gated ion channel. This question was addressed using ion flux studies. When oocytes, previously injected with insect mRNA were treated with cholinergic agonists a significant influx of ^{86}Rb isotops was induced, which was blocked by d-tubocurarine. Thus it can be concluded, that after microinjection of mRNA from the nervous tissue of insects, Xenopus oocytes produce ACh-controlled ion channels (19); thus represents an excellent system to analyze the translation of specific insect nucleic acid samples and possibly the expression of interspecies receptor hybrids (20) in forthcoming molecular-genetic studies on the insect receptor.

Cloning of a Receptor-specific cDNA

A full length cDNA library was constructed by the procedure of Gubler and Hoffman (21) using polyA$^+$-RNA preparations from young locusts, probed for receptor specific mRNA in oocytes. The cDNA was cloned into the λ-gt11 expression vector which promotes synthesis of fusion proteins. An amplified λ-gt11 library was plated, the expression induced with IPTG and the fusion proteins transferred onto NC-filters. Approximately 10^5 recombinant phage plaques were assayed for the production of a β-galactosidase AChR fusion protein that was reactive with monospecific anti(locust AChR)-antiserum. Eight plaques reacted with the antibodies and two were further plaque purified. Lysogens of the AChR-2 clone were made in E.coli Y1089 and grown under conditions to induce the synthesis of β-galactosidase fusion protein; a fusion protein synthesized by the AChR-2 lysogen was identified on Western blots using anti-β-galactosidase and anti-receptor antibodies. The cloned cDNA insert has been isolated and subcloned into a M13 vector for sequence analysis to elucidate the primary structure of the locust neuronal AChR-polypeptides.

Conclusion

Properties of a neuronal acetylcholine receptor from locusts have been explored by an application of biochemical, biophysical and immunological approaches. The neuronal insect receptor appeared to be a homooligomeric protein as was predicted for an ancestral receptor type. Unravelling the complete primary structure of the receptor polypeptides using recombinant DNA techniques will evaluate the relatedness of the neuronal insect receptor and the peripheral vertebrate receptor.

ACKNOWLEDGEMENTS

We thank Prof. Lueken for encouragement during the course of this work and G. Moehrke for typing the manuscript. The excellent technical assistance of G. Hinz and U. Mädler is greatly acknowledged.

REFERENCES

1. Florey E (1963). Acetylcholine in invertebrate nervous system. Can. J. Biochem. Physiol. 41: 2619.
2. Pitman RM (1971). Transmitter substances in insects: A review. Comp. Gen. Pharmacol. 2: 347.
3. Sattelle DB, Harrow D, Hue B, Gepner J, Hall LM (1983). α-bungarotoxin blocks excitatory synaptic transmission between cercal sensory neurons and giant interneurone 2 of the cockroach, Periplaneta americana. J. Exp. Biol. 107: 473.
4. Breer H, (1981). Properties of putative nicotinic and muscarinic cholinergic receptors in the central nervous system of Locusta migratoria. Neurochem. Int. 3: 43-52.
5. Dudai Y (1979). Cholingeric receptors in insects. Trends Biochem. Sci. 4: 40.
6. Lummis SCR, Sattelle DB (1985). Binding of N- [propionyl-^{3}H] propionylated α-bungarotoxin and L- [benzilic-4,4]-^{3}H quinuclidinyl. Comp. Biochem. Physiol. 80C: 75.
7. Breer H (1981). Comparative studies on cholinergic activities in the central nervous system of Locusta migratoria. J. Comp. Physiol. 141: 271.
8. Breer H, Kleene R, Hinz G (1985). Molecular forms and subunit structure of the acetylcholine receptor in the nervous system of insects. J. Neurosci 5: 3386.
9. Sattelle DB, Breer H (1985). Purification by affinity-chromatography of a nicotinic acetylcholine receptor from the CNS of the cockroach Periplaneta americana. Comp. Biochem. Physiol. 82C: 349.
10. Fels G, Breer H, Maelicke A (1983). Are there nicotinic acetylcholine receptors in invertebrate ganglianic tissue? In: Hucho F, Ovchinnikov YA (eds): "Toxins as Toolsin Neurochemistry", Berlin: de Gruyter p. 127.
11. Mädler U, Heilgenberg H, Sattelle DB, Breer H (1986). Localization of acetylcholine-receptor antibodies and α-bungarotoxin in the 6th abdominal ganglion of cockroach, Periplaneta americana (in preparation).
12. McNamee MG, Ochoa ELM (1982). Reconstitution of acetylcholine receptor function in model membranes. Neurosci. 10: 2305.
13. Hanke W, Breer H (1986). Channel properties of a neuronal acetylcholine receptor isolated from the nervous system of insects reconstituted in planar lipid bilayer. Nature (in press).

14. Zenssen, Hinz G, Beyreuther K, Breer H (1986). Sequence homology between the peripheral vertebrate receptor and the neuronal insect receptor. FEBS Lett.
15. Boulter J, Luyten W, Evans K, Mason P, Ballivet M, Goldman D, Stengelin S, Martin G, Heinemann S, Patrick J (1985). Isolation of a clone coding for the α-subunit of a mouse acetylcholine receptor. J. Neurosci. 5: 2545.
16. Gurdon JB, Lane C, Woodland HR, Marbaix G (1971). Use of frog eggs and oocytes for the study of mRNA and its translation in living cells. Nature 233: 177.
17. Barnard EA, Beeson D, Bilbe G, Brown DA, Constanti A, Houamed K, Smart TG (1984). A system for the translation of receptor mRNA and the study of the assembly of the functional receptors. J. Recept. Res. 4: 681.
18. Breer H, Benke D (1985). Synthesis of acetylcholine receptors in Xenopus oocytes induced by poly(A)$^+$-mRNA from locust nervous tissue. Naturwissenschaft 72: 213.
19. Breer H, Benke D (1986). Messenger RNA from insect nervous tissue induces expression of neuronal acetylcholine receptors in Xenopus oocytes. Molecular Brain Research.
20. Sakmann B Metfessel C, Mishina M, Takahashi T, Kurasaki M, Fukuda K, Numa S (1986). Role of acetylcholine subunits in gating of the channel. Nature 318: 583.
21. Gubler U, Hoffmann BJ (1983). A simple and very efficient method for generating cDNA libraries. Gene 25: 263.

Molecular Entomology, pages 107–116

CHEMICAL DETERRENCE OF PLANTS

E.A. Bernays and R.F. Chapman

Division of Biological Control and Department of Entomological Sciences, University of California, Berkeley, California 94720

ABSTRACT Most plants are not eaten by most insects in part because of the secondary compounds which they contain. These compounds act by stimulating deterrent receptors or by interaction with the receptors responding to phagostimulatory compounds. Deterrent receptors generally respond to a wide spectrum of compounds, matching the great diversity of secondary compounds in plants. It is suggested that they have evolved from receptors with a very wide spectrum of sensitivity and that specificity arises from loss of sensitivity to compounds in acceptable host plants rather than acquisition of sensitivity as an adaptation to secondary compounds in non-hosts. This thesis therefore questions the link between deterrence and toxicity of plant secondary compounds.

INTRODUCTION

A majority of phytophagous insects are restricted in the range of plants which they will eat and even polyphagous species are selective, totally rejecting some plants and eating others only in relatively small quantities. Although the more specific feeders may respond positively to chemicals characteristic of their host plants, in all the cases so far studied they are also inhibited from feeding by features of the non-hosts. In other words, the failure to feed on a plant is not, in general, only due to the lack of a characteristic feeding cue. Among less specific feeders, the range of plants eaten may be determined wholly by the inhibiting effects of unacceptable plants.

This inhibition of feeding or oviposition may be due to various features of the plant. The potential herbivore may be repelled from a distance by volatile compounds so that it never reaches the plant or it may be prevented from feeding by physical characteristics of the leaf. Commonly, however, feeding or oviposition is inhibited by chemicals on the surface of, or in the plant. This type of inhibition following physical contact with the plant is called deterrence. It is a sensory phenomenon; rejection is not due to any adverse physiological effects which eating the plant might cause (antibiosis), though the two phenomena may be related, evolutionarily or through associative learning.

Probably all green plants, in at least some stages of the life history, contain one or more chemicals which are deterrent to at least some insects. We do not know how many different deterrents exist, but they are usually plant secondary compounds and it is usual to assume that all these are potential deterrents. Swain (1) estimates that there may be as many as 400,000 different compounds.

Until recently our knowledge of the chemistry of secondary compounds has been largely governed by non-entomological considerations. As a consequence, the alkaloids stood out as probably the most important group until the 1960s. With the recognition of the importance of secondary compounds in host plant selection by insects, following the pioneer work of Fraenkel (2), and the demand for increased food production to meet the needs of the population explosion, much work in natural products chemistry has been specifically insect-related. This is tending to suggest that terpenoids are particularly important, but is also revealing increasing numbers of secondary compounds with novel types of structures (3,4,5). The diversity is compounded by the presence in most plants of an array of different secondary compounds, often including series of related structures.

It is apparent from the diversity of phytophagous insects and their ranges of feeding habits that the impact of deterrent compounds on insects and their capacity to deter varies with the species. Experimental studies substantiate this observation which is clearly demonstrated by examples of compounds tested at naturally occurring concentrations causing rejection behavior in some species, but being phagostimulants for others (Table 1). Moreover, if it is true that all, or at least most, non-host plants are deterrent to an insect species, as the work of Jermy (6) and Bernays and Chapman (7) indicates, it follows that every insect must be capable of perceiving and responding to an enormous range of

TABLE 1
VARIATION IN THE RANGE OF FEEDING RESPONSES TO SOME SELECTED PLANT SECONDARY COMPOUNDS

Compound	Chemical class	Insects stimulated	Insects deterred
Nordihydro-guaiaretic acid	lignan	*Bootettix argentatus*	*Ligurotettix coquilletti*
Tannic acid	tannin	*Lymantria dispar*	*Heliothis zea*
Sinigrin	glucosinolate	*Pieris rapae*	*Manduca sexta*
Cucurbitacin	triterpenoid	*Diabrotica* spp.	*Epilachna tredecemnota*
Ipolamiide	iridoid	*Euphydryas editha*	*Locusta migratoria*
Salicin	phenolic glycoside	*Chrysomela aenicollis*	*Pieris brassicae*

different deterrents. Few species have been studied sufficiently extensively for this thesis to have been adequately tested, but *Locusta migratoria* is known to be deterred from feeding by over 70 different compounds in 11 chemical classes and even the polyphagous *Schistocerca gregaria* is affected by about 40 of the same compounds (7).

These two facts, the variability of the response between species and the broad spectrum of response within a species, focus on the important biological questions: how are deterrent compounds perceived by insects and how did this ability evolve? These are the questions we address in this paper.

PERCEPTION OF DETERRENTS

Sensory Responses

The deterrent properties of any plant are perceived by an insect's contact chemoreceptors, particularly by those on the mouthparts. Contact chemoreceptors are characterized by a terminal pore and a small number, often four or six, chemosensory neurons (8). In caterpillars each neuron in a sensillum is maximally sensitive to different classes of chemical compounds and in some sensilla one cell responds to chemicals which cause rejection. It is inferred that the information transmitted by this cell to the central nervous system leads to rejection behavior and it is consequently called the deterrent cell.

In practice, deterrent compounds do not occur in isolation, but in complex mixtures. In some species of insects, the individual cells in a sensillum continue to fire as expected from their responses to individual compounds when the sensillum is stimulated by a simple mixture of chemicals. For example, the sugar and deterrent receptors of _Spodoptera exempta_ respond to a mixture of sucrose and azadirachtin in an entirely predictable way (Fig. 1). This is also true in _Pieris brassicae_ so that the input from different receptor cells in the sensilla on the mouthparts can be weighted in proportion to the effect produced on food intake (9).

In other insects, however, the responses of individual sensory cells are affected by the presence of chemicals in the mixture other than those that directly stimulate them. This effect is not produced only by deterrent compounds and the activity of the deterrent cell may be affected by compounds other than deterrents. Either inhibition or synergism may occur (10,11). For example, interaction of sucrose and azadirachtin with the receptor cells occurs in the medial maxillary sensillum of _Spodoptera littoralis_. As a result

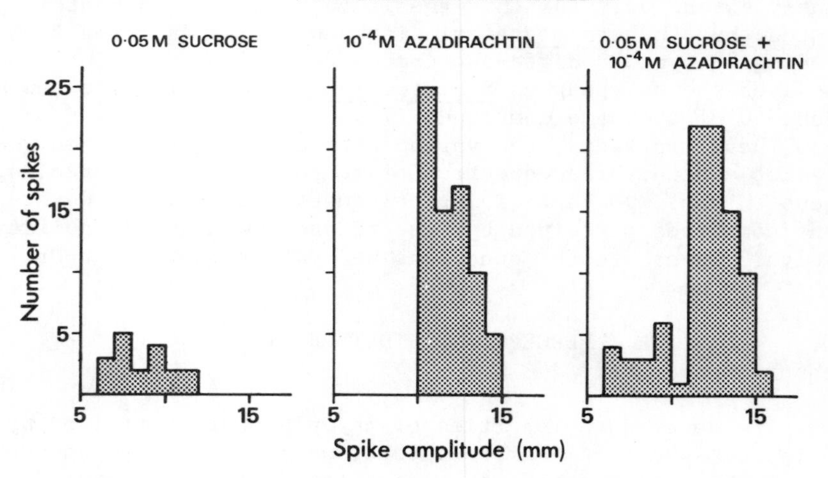

FIGURE 1. Number of action potentials of different amplitude produced in the second 500ms of stimulation by sucrose and azadiractin alone and in a mixture in the medial maxillary sensillum of _Spodoptera exempta_(11).

the overall rate of firing declines as sucrose concentration is increased in the presence of azadirachtin although without azadirachtin the rate of firing increases (Fig. 2) and the response to the mixture is far less than the summed rates of firing when the chemicals are applied separately at the same concentrations. In such cases reduced food intake in the presence of a deterrent may be partly due to the input from the deterrent cell, but this is coupled with a reduction in the rate of firing of the sugar receptor, which alone would normally lead to reduced food intake.

The evidence currently available suggests that in oligophagous caterpillars little or no interaction occurs between deterrents and sucrose and their receptor cells whereas in polyphagous species interaction regularly occurs (11). Consequently in these species it is not possible to predict the effects of a mixture of chemicals on the sensillum from a knowledge of the effects of the individual compounds.

In a few cases deterrent compounds appear to act solely by the suppression of activity of cells responding to phagostimulants. The caterpillars of *Porthetria dispar* and *Malacosoma americana* lack receptors which respond to tannic acid, but it suppresses the activity of the sugar receptor (12). Inhibition of the sugar sensitive cell by alkaloids occurs in the beetle *Entomoscelis americana*. In this case the alkaloids also stimulate another receptor cell, but this apparently does not have a deterrent function (13).

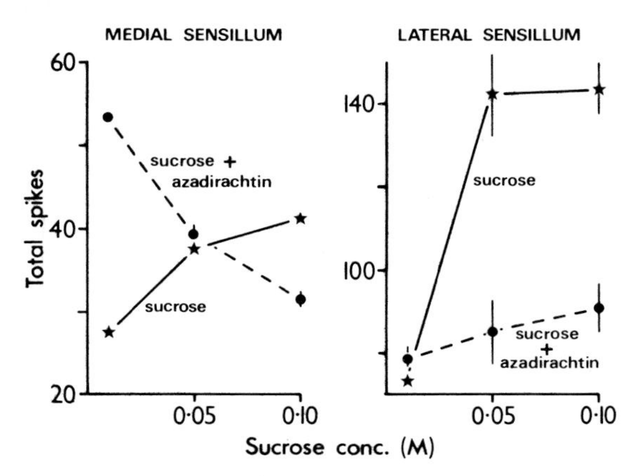

FIGURE 2. Total number of action potentials produced in the first second of stimulation by different concentrations of sucrose, alone and mixed with 10^{-8}M azadirachtin in the maxillary sensilla of *Spodoptera littoralis* (11).

How such interaction is effected is not known, but the alkaloids affecting *E. americana* are known to affect ion conductance in molluscan neurons (13).

Some compounds which cause rejection do not stimulate receptor cells in the normal way, but cause irregular firing in them. This is presumed to be a damage response. Quinine is the most widely investigated of such substances. In *Danaus plexippus* it causes all four cells in the medial and lateral sensilla to fire (14) and the responsiveness of these cells to subsequent stimulation may be reduced (15). Similar effects of other alkaloids and of the saponin digitonin have been recorded in other insects (12,13).

An extreme example of the interaction of a deterrent compound with receptor cells is that of warburganal, a sesquiterpene. 1mM warburganal at first elicits a weak, but normal response from cells in the maxillary sensilla of *Spodoptera exempta*. However, if stimulation continues for more than a minute irregular bursting discharges occur, fading away after 2-5 minutes. Subsequently the response to sucrose is reduced to a very low level, but the cells recover in about an hour (16). It is not certain, however, that this is the primary mode of perception of the compound since behavioral rejection is observed at concentrations one thousandth of that causing these physiological effects (17).

No specific deterrent cell has been identified in Acrididae and most of the substances that have been tested electrophysiologically elicit responses from several cells within a sensillum. The range of amplitudes of the action potentials is generally similar for different compounds (18). This does not necessarily mean that cells equivalent to lepidopteran deterrent cells do not exist, but they have not been recognized. Blaney's (19) data on *Schistocerca gregaria* show that sucrose-impregnated glass-fiber disks treated with azadirachtin were completely rejected irrespective of the concentration of azadirachtin and independently of the calculated sensory input from a group of sensilla on the tips of the maxillary palps. These results indicate the existence of a neuron signalling deterrence for this compound at least. For other substances, including deterrents, discrimination is believed to be essentially a central nervous phenomenon dependent on the total input from a number of sensilla (19).

Spectrum of Sensitivity

The deterrent cell of caterpillars is often sensitive to chemicals from a number of different chemical classes, but a response does not occur to all chemicals in a class. Marked differences also occur between species in the range of compounds eliciting a response from the deterrent cell.

The mechanism by which this cell is able to perceive a wide range of different compounds, and at the same time exhibit various degrees of selectivity, is unknown. The broad range of sensitivity argues against the possession of specific receptor sites. A possible method of perception involves an interaction between the stimulating compound and the phospholipids in the cell membrane of the neuron. It is suggested that salts associate with polar regions of the lipid molecules, while bitter substances (deterrents) associate with the non-polar parts. As a result, changes in the membrane potential occur which lead to the development of the receptor potential (20,21).

EVOLUTION OF DETERRENT RECEPTORS

Dethier (14) concludes that the deterrent receptor of insects has evolved from receptors that were sensitive to a very wide range of compounds. The apparent lack of distinct deterrent cells in Acrididae may be a reflection of this, and it may be hypothesized that the discrete deterrent receptors of hemipteroid and endopterygote insects are derived from this generalized condition.

This implies that the sensitivity to plant secondary compounds is not an adaptive response, and it does seem inconceivable that sufficient selection pressure can have been maintained to lead to the adaptive development and subsequent maintenance of sensitivity to each of the hundreds, and perhaps thousands, of compounds to which the deterrent cell is capable of responding. In this respect it is also significant that the deterrent cells of many insects are known to be sensitive to chemicals which the species cannot have experienced in recent evolutionary history. For example, many Palearctic and Neotropical Lepidoptera exhibit a response to azadirachtin (11). This compound comes from the neem tree, which is native to north west India so that the geographic separation will have prevented contact.

However, if it supposed that the deterrent cell is sensitive to a wide spectrum of compounds by virtue of the properties of its receptor membrane, why are all plants not deterrent? The implication must be that the association with a particular host plant involves a loss of sensitivity by the deterrent cell to those potential deterrents that the host contains. There is one example in support of this thesis. Van Drongelen (22) examined the sensitivity patterns of the receptors in a range of closely related *Yponomeuta* species living on different host plants. These are believed to be derived from an ancestor feeding on Celastraceae and in the species currently living on Celastraceae the deterrent cells in the lateral and medial contact chemoreceptors on the maxilla are sensitive to the compounds phloridzin and salicin among others. Sensitivity to phloridzin also exists in other *Yponomeuta* species, but is absent from *Y. mallinellus* which feeds on *Malus* having high concentrations of phloridzin. In *Y. rorellus*, which feeds on *Salix*, where salicin is present, the deterrent cell in the lateral galeal sensillum is insensitive to salicin, but that in the medial sensillum is sensitive to it; in other species, which do not normally encounter salicin, both receptors are sensitive to the compound. Van Drongelen argues that the larval switch must almost certainly have followed a change in adult behavior since larval distribution is largely governed by adult oviposition. The sensitivity change is thus an adaptive one occurring *after* the behavioral change and involves the loss of sensitivity to key compounds of the hosts rather than any gain in sensitivity to non-hosts.

Thus we consider that the possession of a neuron sensitive to a wide range of plant secondary chemicals and signalling deterrence is a primitive feature of phytophagous insects and it is no surprise that novel foods are commonly rejected by phytophagous insects. This rejection is a consequence of the sensory perception of the insect; it does not reflect any ecological adaptation.

In those instances where rejection behavior involves interference with the perception of phagostimulants, Mitchell and Sutcliffe (13) also argue that this does not necessitate a co-evolutionary relationship between the insect and the plant. The deterrent compound need only have the fortuitous effect of interfering with normal processes in the receptor cell membrane.

In either of these cases, stimulation of a deterrent cell or interference with other receptors, the question now arises "how is specificity achieved?" It will only be possible to answer this question when we have a much fuller understanding of the basic receptor mechanisms.

CONCLUSION

The conclusion that the perception of deterrent

compounds is a primitive feature of insect chemoreceptors or is a fortuitous consequence of their physical properties is at variance with the commonly held view that deterrence arises as an adaptation to the toxic properties of plant secondary compounds. We have argued that there is little experimental evidence to support the link between deterrence and toxicity, while there are good reasons, other than the avoidance of toxins, for the host specificity to which deterrence contributes (23). This is not to say that links between deterrence and toxicity do not sometimes occur, and in some cases deterrence may be a direct consequence of toxicity. But a consideration of the mechanisms by which deterrent compounds are perceived strongly suggests that the evolution of deterrence is not closely linked to the evolution of plant secondary compounds.

REFERENCES

1. Swain T (1977). Secondary compounds as protective agents. Ann Rev Plant Physiol 28:479.
2. Fraenkel G (1969). Evaluation of our thoughts on secondary plant substances. Entomol Exp Appl 12:473.
3. Munakata K (1977). Insect feeding deterrents in plants. In Shorey HH, McKelvey JJ (eds): "Chemical Control of Insect Behavior," New York: Wiley & Sons, p 93.
4. Kubo I, Matsumoto T (1985). Potent insect antifeedants from the African medicinal plant Bersama abyssinica. In Hedin PA (ed): "Bioregulators for Pest Control," ACS Symp.
5. Mabry TJ, Gill JE (1979). Sesquiterpene lactones and other terpenoids. In Rosenthal GA, Janzen DH (eds): "Herbivores. Their Interaction with Secondary Plant Metabolites," New York: Academic Press, p 502.
6. Jermy T (1966). Feeding inhibitors and food preference in chewing phytophagous insects. Entomol Exp Appl 9:1.
7. Bernays EA, Chapman RF (1978). Plant chemistry and acridoid feeding behaviour. In Harborne JB (ed): "Biochemical Aspects of Plant and Animal Coevolution," London: Academic Press, p 99.
8. Altner H, Prillinger L (1980). Ultrastructure of invertebrate chemo-, thermo-, and hygroreceptors and its functional significance. Int Rev Cytol 67:69.
9. Schoonhoven LM (in press). What makes a caterpillar eat? The sensory code underlying feeding behavior. In Chapman RF, Stoffolano JG, Bernays EA (eds): "Perspectives in Chemoreception and Behavior," New York: Springer Verlag.
10. Dethier VG, Kuch JH (1971). Electrophysiological studies

of gustation in lepidopterous larvae. I. Comparative sensitivity to sugars, amino acids and glycosides. Z Vergl Physiol 72:343.

11. Simmonds MSJ, Blaney WM (1984). Some neurophysiological effects of azadirachtin on lepidopterous larvae and their feeding response. In Schmutterer H, Ascher KRS (eds): "Natural Pesticides from the Neem Tree and other Tropical Plants," Proc 2nd Int Neem Conf Ravisch-Holzenhausen Germany, p 163.
12. Dethier VG (1982) Mechanism of host-plant recognition. Entomol Exp Appl 32:49.
13. Mitchell BK, Sutcliffe JF (1984). Sensory inhibition as a mechanism of feeding deterrence: effects of three alkaloids on leaf beetle feeding. Physiol Entomol 9:57.
14. Dethier VG (1980). Evaluation of receptor sensitivity to secondary plant substances with special reference to deterrents. Amer Nat 115:45.
15. Blaney WM, Simmonds MSJ (1983). Electrophysiological activity in insects in response to antifeedants. Unpub. report. Birkbeck College, University of London.
16. Ma W-C (1977). Alterations of chemoreceptor function in armyworm larvae (Spodoptera exempta) by a plant-derived sesquiterpenoid and by sulfhydryl reagents. Physiol Entomol 2:199.
17. Schoonhoven LM (1982) Biological effects of antifeedants. Entomol Exp Appl 31: 57.
18. Blaney WM (1975). Behavioural and electrophysiological studies of taste discrimination by the maxillary palps of Locusta migratoria (L.). J Exp Biol 62:555.
19. Blaney WM (1980). Chemoreception and food selection by locusts. In van der Starre H (ed): "Olfaction and Taste VII," London: Information Retrieval Limited, p 127.
20. Kurihara K, Kame N, Ueda T, Kobatake Y (1975). Origin of receptor potential in chemoreception. In Denton DA, Coghlan JP (eds): "Olfaction and Taste V," NY: AP, p77.
21. Yoshii K, Kurihara K (1983). Mechanism of the water response in carp gustatory receptors: Independent generation of the water response from the salt response. Brain Res 279:191.
22. van Drongelen W (1979). Contact chemoreception of host-plant specific chemicals in larvae of various Yponomeuta species (Lepidoptera). J Comp Physiol 134:265.
23. Bernays EA, Chapman RF (in press). The evolution of deterrent behavior in plant-feeding insects. In Chapman RF, Stoffolano JG, Bernays EA (eds): "Perspectives in Chemoreception and Behavior," New York: Springer Verlag.

II. MOLECULAR ENDOCRINOLOGY AND DEVELOPMENT IN INSECTS

Molecular Entomology, pages 119–128

PROTHORACICOTROPIC HORMONE(PTTH) OF THE SILKMOTH, *BOMBYX MORI*: 4K-PTTH[1]

Hironori Ishizaki, Akira Mizoguchi, Masaaki Hatta, *Akinori Suzuki, *Hiromichi Nagasawa, *Hiroshi Kataoka, *Akira Isogai, *Saburo Tamura, **Masahiko Fujino, and **Chieko Kitada

Biological Institute, Faculty of Science, Nagoya University, Chikusa-ku, Nagoya, *Department of Agricultural Chemistry, Faculty of Agriculture, The University of Tokyo, Tokyo, and **Takeda Chemical Industry, Juso-Honmachi, Osaka, Japan

ABSTRACT The *Bombyx* brain contains 22K-PTTH and 4K-PTTH. Complete amino acid sequence of 4K-PTTH has been determined. Sequence homology exists between 4K-PTTH and insulin-family peptides. The 4K-PTTH-producing cells have been identified immunohistochemically.

INTRODUCTION

Moulting and metamorphosis of insects are dictated by ecdysone, a steroid hormone, which is secreted by a paired thoracic organ, prothoracic glands. Synthesis and release of ecdysone by these glands are controlled by a cerebral neuropeptide, prothoracicotropic hormone(PTTH). Because of its pivotal role in the endocrine network controlling insect development and because of its importance as an neuroendocrine transducer that mediates environmental signals(e.g. photoperiod) to internal milieu to effect modulation of development (e.g. diapause), PTTH has extensively been studied for its biological facets(reviews, 1,2,3). The chemical nature of this hormone, however, has long remained unclarified. We started to purify PTTH of the silkmoth, *Bombyx mori*, in 1959(4) and

[1]This work was partly supported by Grant-in-Aid for Scientific Research(Nos. 58470106, 59360012, 60560133, and 6105005) from the Ministry of Education, Science and Culture of Japan.

purified 4K-PTTH, one molecular form of PTTH, to hmomgeneity in 1982(5,6). Many reports appeared to purify PTTH from various insects which include *Bombyx*, *Periplaneta americana*, and *Manduca sexta*(see 6) but none of these succeeded in obtaining this hormone in a pure form.

RESULTS

PTTH in *Bombyx* brain: 4K-PTTH and 22K-PTTH

As a source of PTTH for purification, we used *Bombyx*, first its brains and later its heads. *Bombyx* can be supplied in a large enough number on the commercial basis from sericultural industry.

When we started purification study, two methods to assay PTTH were available. Kobayashi(7) had developed the *Bombyx* assay: PTTH can induce adult development when injected into *Bombyx* dormant pupae from which the brain has surgically been removed shortly after the pupal moult. On the other hand, Ichikawa(4) had established a similar assay using brain-removed dormant pupae of the saturniid moth *Samia cynthia ricini*. Since *Bombyx* brain extracts were active on both of these assays, we assumed that the PTTH molecule of *Bombyx* was active in a species non-specific manner on *Bombyx* and *Samia*(4). Because of several technical advantages, we employed the brainless *Samia*-pupa assay to purify PTTH derived from *Bombyx*.

Long after the start of this purification work, we came to notice that the *Bombyx* brain contained two distinct molecular classes of PTTH(8,9), one with a molecular weight of ca. 4,400 which activates *Samia* pupae but not *Bombyx* pupae (4K-PTTH or PTTH-S) and the other with a molecular weight of ca. 22,000 which specifically activates *Bombyx* pupae(22K-PTTH or PTTH-B). 4K-PTTH and 22K-PTTH could be readily separated from each other by Sephadex gel-filtration, DEAE-Sepharose chromatography, or acetone precipitation. Clearly, the molecule that we had been purifying up to this time was 4K-PTTH.

The fact that the *Bombyx* brain possesses a PTTH which is not concerned with activation of its own prothoracic glands but activates the glands of other species is puzzling. All experimental data on the action of 4K-PTTH on *Samia* pupae, however, satisfy the criteria for this molecule to be PTTH: the dose-response relationship, the effective concentration *in vivo*, and the activation of the prothracic glands *in vitro* were all physiological(6). Though we find no way at present to explain the physiological significance of 4K-PTTH in Bombyx, it seems improbable that such a physiological molecule plays no role in Bombyx. The possibility remains, for example, that 4K-PTTH might act on the prothoracic glands of Bombyx at the developmental stages other than pupa-adult development. Yet, strictly speaking, 4K-PTTH must be referred to as "PTTH-like peptide", though we call it PTTH simply for brevity, until its physiological and phylogenetical significance is fully understood. On the other hand, 22K-PTTH is obviously a genuine PTTH of Bombyx as far as pupa-adult development is concerned. Accordingly, we immediately began another purification work on 22K-PTTH with assay using Bombyx brainless pupae.

The above historical background accounts for the present status of our PTTH study in which the structural analyses of 4K-PTTH is much advanced compared to those of 22K-PTTH. Amino acid sequencing of 22K-PTTH has just started.

Purification of 4K-PTTH

After a 20 year struggle and by the use of about 20 millions of Bombyx heads, we have established the purification scheme with which 4K-PTTH was purified to homogeneity, with a 10^6-fold purification. Details of purification have been described(1,6,9,10,11). Three molecular forms, namely 4K-PTTH -I, -II, and -III, were resolved by a reversed-phase HPLC on Develosil ODS-5, the final step of purification. Their retention times differed slightly but definitely from one another. The yield of these three PTTHs in a representative run were 50, 36, and 63μg from 0.65 million heads, respectively. As little as ca. 0.1 ng of these PTTHs elicited adult development when injected into a debrained Samia pupa and they enhanced ecdysone release by a prothoracic gland in vitro at a concentration of 10^{-11}M.

Primary Structure of 4K-PTTH-II

Edman degradation of 4K-PTTH-I, -II, and -III by automated gas-phase sequencer(Applied Biosystems, model 470A), directly and after generation of S-carboxamide methylated cysteine, redily disclosed a 19 amino acid sequence from the N-terminus(11). The sequence of these three molecules differed from one another by substitution of 3-6 amino acid residues.

Further sequence analyses were performed for 4K-PTTH-II only. When an intact material was subjected to Edman degradation, no phenylthiohydantoin derivatives of amino acid were formed beyond the 19th residue. After reductive alkylation of disulfide bonds, on the other hand, 4K-PTTH-II was aplit into two non-identical peptides which were separable from each other by HPLC. Edman degradation and carboxypeptidase A digestion followed by determination of released amino acids readily determined the complete sequence of A-chain, one of the two peptides, which consisted of 20 amino acid residues (Fig.1B). By contrast, Edman degradation of B-chain, the other peptide, produced no phenylthiohydantoin derivatives of amino acid. Edman degradation was effective when B-chain had been digested with pyroglutamate aminopeptidase, indicating that the N-terminus of B-chain was blocked by pyroglutamic acid.

The B-chain sequence analysis was not a simple task because it consisted further of minutely heterogeneous peptides. Edman degradation after pyroglutamate aminopeptidase digestion together with sequence analyses of peptide fragments after trypsin or chymtrypsin digestion led us to finally construct the sequences of four B-chains(Fig.1A). B-chain consisted of two groups of peptides, one with 28 amino acid residues and the other with 26 residues. Furthermore, the 28-residue group was a mixture of two peptides with alanine or glycine at position 5, while the 26-residue group consisted of two peptides with alanine or glycine at position 3. Thus, 4K-PTTH-II was a mixture of four molecules which differed from one another only in the N-terminal region of B-chain, as shown in Fig.1. We suspect that similar microheterogeneities may exist in PTTH-I and -III and if so, a very high degree of heterogeneity is extected to exist in 4K-PTTH.

Sequence Homology of 4K-PTTH with Insulin-Family Peptides

The sequence homology of N-terminal portion(now known as A-chain) of 4K-PTTH with insulin-family peptides of vertebrates has been reported previously(11). Elucidation of the complete sequence of 4K-PTTH-II has further confirmed this conclusion (Fig.1B). A 47% homology is observed in A-chains between 4K-PTTH-II and human insulin, while a 30% homology is seen for B-chains. 4K-PTTH-II shares another important feature with porcine relaxin, another member of insulin family, in having pyroglutamic acid at N-terminus. The resemblance at A-chain C-terminus between 4K-PTTH-II and porcine relaxin of terminating by cysteine also seems important regarding the functional

```
1                 5
pGlu-Gln-Pro-Gln-Ala-Val- - - - -
pGlu-Gln-Pro-Gln-Gly-Val- - - - -
        pGlu-Gln-Ala-Val- - - - -
        pGlu-Gln-Gly-Val- - - - -
```

Fig.1A. N-terminal region of B-chain of four different 4K-PTTH-IIs.

A chain

```
                                     1               5                  10                 15                20
Human insulin                        H-Gly-Ile-Val-Glu-Gln-Cys-Cys-Thr-Ser-Ile-Cys-Ser-Leu-Tyr-Glu-Leu-Glu-Asn-Tyr-Cys-Asn-OH
4K-PTTH-II                           H-Gly-Ile-Val-Asp-Glu-Cys-Cys-Leu-Arg-Pro-Cys-Ser-Val-Asp-Val-Leu-Leu-Ser-Tyr-Cys-OH
Porcine relaxin   H-Arg-Met-Thr-Leu-Ser-Glu-Lys-Cys-Cys-Glu-Val-Gly-Cys-Ile-Arg-Lys-Asp-Ile-Ala-Arg-Leu-Cys-OH
```

B chain

```
                   1                   5                        10              15              20
Human insulin                   H-Phe-Val-Asn-Gln-His-Leu-Cys-Gly-Ser-His-Leu-Val-Glu-Ala-Leu-Tyr-Leu-Val-Cys
4K-PTTH-II        pGlu-Gln-Pro-Gln-Ala-Val-His-Thr-Tyr-Cys-Gly-Arg-His-Leu-Ala-Arg-Thr-Leu-Ala-Asp-Leu-Cys
Porcine relaxin   pGlu-Ser-Thr-Asn-Asp-Phe-Ile-Lys-Ala-Cys-Gly-Arg-Glu-Leu-Val-Arg-Leu-Trp-Val-Glu-Ile-Cys

                  23      25
Human insulin     Gly-Glu-Arg-Gly-Phe-Phe-Tyr-Thr-Pro-Lys-Thr-OH
4K-PTTH-II        Trp-Glu-Ala-Gly-Val-Asp-OH
Porcine relaxin   Gly-Val-Trp-Ser-OH
```

Fig.1B. Complete sequence of 48-residue 4K-PTTH-II with alanine at position 5 of B-chain. Human insulin and porcine relaxin are aligned to show sequence homology(boxed).

significance. The actual presence of disulfide bonds has not been proved yet but the presence of six cysteine residues in PTTH at exactly the same positions as in insulin makes it highly probable that A- and B-chains of PTTH are connected to each other in the same manner as in insulin.

From these data, we conclude that *Bombyx* 4K-PTTH and vertebrate insulin-family peptides have derived from the common ancestral molecule. Though a considerable information has accumulated which detected chemically or immunohistochemically the insulin-like immunoreactive material in insect nervous tissues(12), our PTTH study is the first to demonstrate an insulin-related peptide in insects by amino acid sequencing of pure material.

Monoclonal Antibody against Synthetic Peptides

All trials so far made to generate antibody against the native preparations of PTTH has been unsuccessful. Knowledge of PTTH sequence has now enabled us to prepare synthetic peptides corresponding to various parts of PTTH and generate monoclonal antibodies against these synthetic peptides. This approach has been very successful and one example of the results is described below.

We synthesized a decapeptide corresponding to the N-terminal region of 4K-PTTH A-chain[PTTH(1-10)]. 85μg of PTTH(1-10)-BSA complex(5:1 molar ratio) were injected into BALB/c mice three times. A month later, all the mice produced the antiserum that reacted with PTTH(1-10) and a partially purified 4K-PTTH named "highly purified PTTH"(10). Antibody detection was made by the enzyme-linked immunosorbent assay (ELISA). Spleen cells were fused with the mouse myeloma(NS-1) cells and more than 10 among 1,000 hybridoma clones secreted the antibody reacting both the synthetic and native PTTH. Among these only a few clones produced the antibody that was specific to PTTH without cross-reacting with BSA, and only one of these clones with the highest specificity and sensitivity to PTTH was selected for detailed studies.

Quantitative analyses with competitive ELISA which utilizes the inhibition of antibody binding to the solid-phase antigen by the antigen added to the antibody solution showed that this antibody was bound to PTTH(1-10) with a dose-dependent manner at the concentrations between 0.02 and 2μM, and to "highly purified PTTH" between 40 and 10,000 head equivalents/ml(for head equivalent, see 10).

Interestingly, we found later that a lot L329 of "highly

purified PTTH" did not react with this antibody at all, in contrast to a lot L323 which had been used for the initial screening of hybridoma clones and for the above competitive ELISA. The lot L329 retained a high biologic activity while L323 was as one third active as L329. Since PTTH possesses disulfide bonds which are essential for its biologic activity (10), we tested the immunoreactivity of L329 before and after disulfide reduction with dithiothreitol. A clear-cut immunoreactivity appeared after disulfide reduction. We concluded, therefore, that the antigenic site for this antibody is burried within an intact PTTH molecule whereas it becomes exposed after disulfide reduction.

Immunohistochemistry

Immunohistochemical study was performed to localize 4K-PTTH in the cephalic endocrine organs of Bombyx using PTTH (1-10) antibody. As shown in Fig.2 & 3, four pairs of dorsomedial neurosecretory cells of pars intercerebralis were immunoreactive. Nerve fibers in the periphery of corpora allata also reacted strongly(Fig.2), in accord with the previous suggestion that this organ is the neurohaemal site for PTTH(13). In the previous section, we concluded that this antibody could react with 4K-PTTH after disulfide reduction. Therefore, we suspect that the antigenic site was exposed through denaturation of PTTH during histochemical procedures.

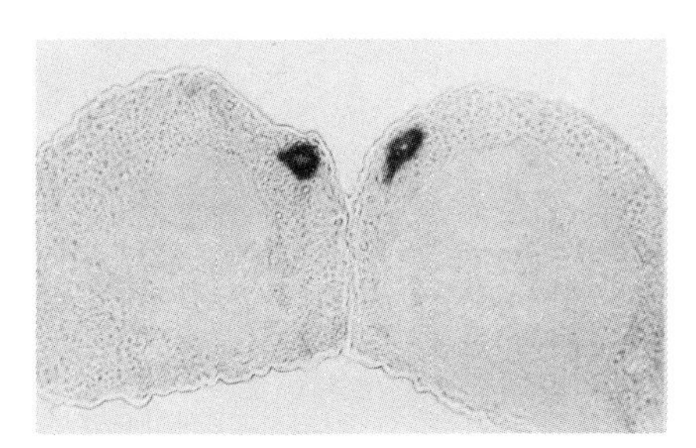

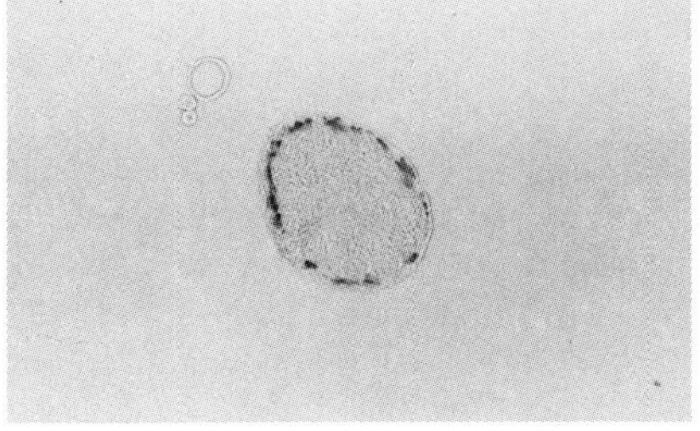

Fig.2. PTTH(1-10) immunoreactive material in day-0 5th-instar larva of Bombyx mori. Left, brain. This transverse section contains two immunoreactive neurosecretory cells out of altogether eight immunoreactive cells. X 120. Right, corpus allatum. Peripheral nerve fibers are immunoreactive. X 140.

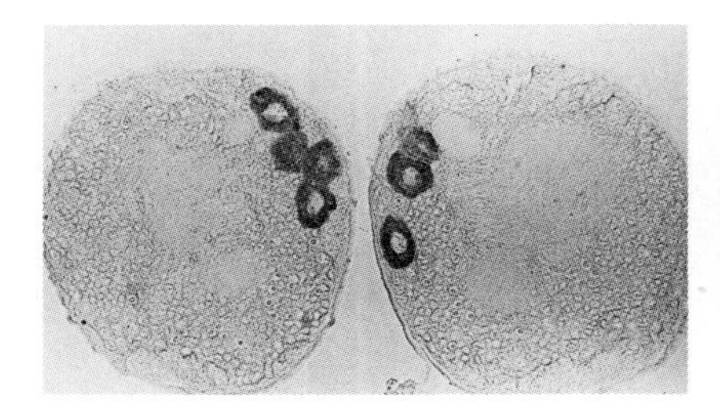
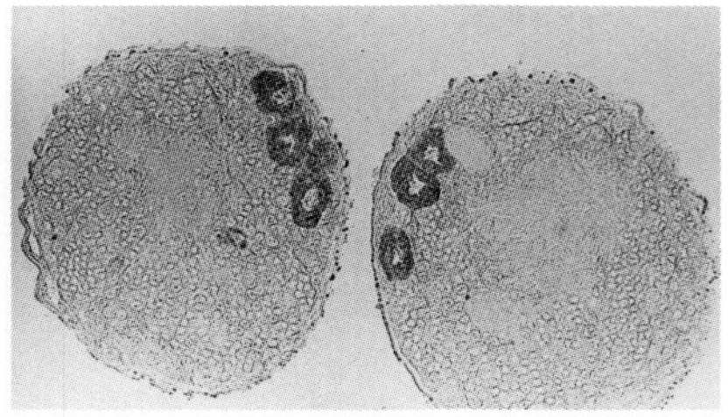

Fig.3. Consecutive sections of a brain of day-0 5th-instar larva of Bombyx mori treated with PTTH(1-10) antibody (left) and insulin antibody(right). The same cells are immunoreactive to both antibodies. Horizontally sectioned so that all of the four immunoreactive cells in a cerebral hemisphere were contained in a single section. X 120.

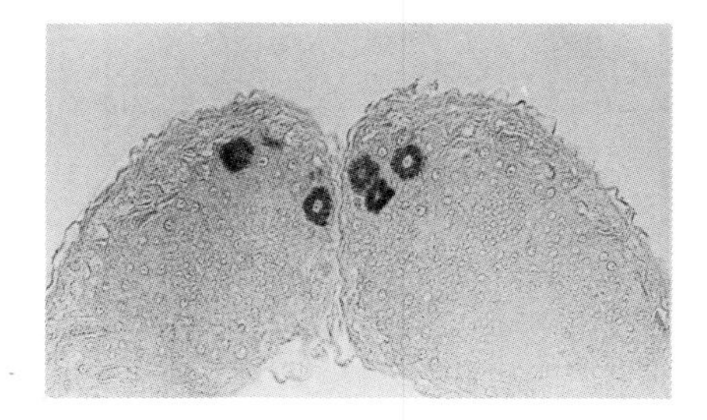

Fig.4. Transverse section of a brain of day-0 5th-instar larva of Samia cynthia ricini treated with PTTH(1-10) antibody. Serial sections(not shown) revealed altogether six pairs of immunoreactive neurosecretory cells. X 120.

Neurosecretory cells and corpora allata of Bombyx have been shown to be insulin immunoreactive(14). Consecutive sections treated with PTTH(1-10) antibody and bovine insulin antibody(Cooper Biomedical) showed that the identical cells were reactive to these antibodies(Fig.3).

The brain neurosecretory cells(six pairs) and corpora allata of Samia were also immunoreactive to PTTH(1-10) antibody(Fig.4), suggesting that the PTTH molecule of Samia is probably immunologically related to Bombyx 4K-PTTH.

CONCLUDING REMARKS

Considerable progress has been made towards the molecular understanding of PTTH. An extremely high degree of molecular heterogeneity and the presence of 22K-PTTH and 4K-PTTH in the *Bombyx* brain seem unique. Unexpected findings often open a new area, and so evidence unexplainable by existing knowledge is particularly important. We are now cloning PTTH genes. This approach will provide us with much information which is extremely tedious or even impossible to gain by purification and sequencing of peptides.

REFERENCES

1. Ishizaki H, Suzuki A (1980). Prothoracicotropic hormone. In Miller TA (ed): "Neurohormonal Techniques in Insects." New York: Springer, p 244.
2. Raabe M (1982). "Insect Neurohormones." New York: Plenum, p 57, 90.
3. Bollenbacher WE, Granger NA (1985). Endocrinology of the prothoracicotropic hormone. In Kerkut GA, Gilbert LI (eds): "Comprehensive Insect Physiology, Biochemistry, and Pharmacology." Oxford, New York: Pergamon, vol 7, p 109.
4. Ichikawa M, Ishizaki H (1961). Brain hormone of the silkworm, *Bombyx mori*. Nature 191:933.
5. Suzuki A, Nagasawa H, Kataoka H, Hori Y, Isogai A, Tamura S, Guo F, Zhong X, Ishizaki H, Fujishita M, Mizoguchi A (1982). Isolation and characterization of prothoracicotropic hormone from silkworm, *Bombyx mori*. Agr Biol Chem 46:1107.
6. Nagasawa H, Kataoka H, Hori Y, Isogai A, Tamura S, Suzuki A, Guo F, Zhong X, Mizoguchi A, Fujishita M, Takahashi SY, Ohnishi E, Ishizaki H (1984). Isolation and some characterization of the prothracicotropic hormone from *Bombyx mori*. Gen Comp Endocr 53:143.
7. Yamazaki M, Kobayashi M (1969). Purification of the proteininc brain hormone of the silkworm, *Bombyx mori*. J Insect Physiol 15:1981.
8. Ishizaki H, Mizoguchi A, Fujishita M, Suzuki A, Moriya I, O'oka H, Kataoka H, Isogai A, Nagasawa H, Tamura S, Suzuki A (1983). Species specificity of the insect prothoracicotropic hormone(PTTH): the presence of *Bombyx*- and *Samia*-specific PTTHs in the brain of the silkworm, *Bombyx mori*. Devel Growth Differ 25:593.

9. Ishizaki H, Suzuki A (1984). Prothoracicotropic hormone of Bombyx mori. In Hoffmann J, Porchet M (eds):"Biosynthesis, Metabolism and Mode of Action of Invertebrate Hormones." Berlin Heidelberg: Springer, p 63.
10. Nagasawa H, Isogai A, Suzuki A, Tamura S, Ishizaki H (1979). Purification and properties of the prothoracicotropic hormone of the silkworm, Bombyx mori. Devel Growth Differ 21"29.
11. Nagasawa H, Kataoka H, Isogai A, Tamura S, Suzuki A, Ishizaki H, Mizoguchi A, Fujishita M, Suzuki A.(1984). Amino-terminal amino acid sequence of the silkworm prothoracicotropic hormone: homology with insulin. Science 226:1344.
12. Thorpe A, Duve H (1984). Insulin- and glucagon-like peptides in insects and molluscs. Mol Physiol 5:235.
13. Agui N, Bollenbacher WE, Granger NA, Gilbert LI (1980). Corpus allatum is release site for insect prothoracicotropic hormone. Nature 285:669.
14. Yui R, Fujita T, Ito S (1980). Insulin-, gastrin-, pancreatic polypeptide-like immunoreactive neurones in the brain of the silkworm, Bombyx mori. Biomed Res 1:42.

Molecular Entomology, pages 129–139

NEUROPEPTIDE CONTROL OF ECDYSONE BIOSYNTHESIS[1]

Wendy A. Smith[2], Wendell L. Combest, Dorothy B. Rountree, and Lawrence I. Gilbert

Department of Biology, Wilson Ø46A, University of North Carolina, Chapel Hill, North Carolina 27514

ABSTRACT The cellular mechanism of action of the cerebral neuropeptide, prothoracicotropic hormone (PTTH), was investigated *in vitro* using prothoracic glands (PG) from the tobacco hornworm, *Manduca sexta*. PTTH stimulates the PG to synthesize the steroid hormone, ecdysone. The peptide appears to act via a Ca^{2+} -dependent increase in cyclic AMP (cAMP) formation, with the cyclic nucleotide in turn enhancing glandular ecdysone synthesis. PTTH-stimulated ecdysone synthesis is accompanied by the activation of cAMP-dependent protein kinase (cAMP-PK). In day 3 fifth instar larval glands, brief exposure to PTTH results in >9Ø% activation of soluble cAMP-PK. Activation of soluble cAMP-PK in day Ø pupal glands, however, requires the additional presence of a phosphodiesterase inhibitor. In cell-free PG homogenates, protein phosphorylation is enhanced by the addition of cAMP. Further, incubation of intact PG in the presence of PTTH reveals the selective phosphorylation of a single protein (M_r= 34,ØØØ). It is postulated that this phosphoprotein controls a rate-limiting step in ecdysone biosynthesis.

[1]This work was supported by the following: NIH grant AM-3Ø118 and training grant #AM-Ø7271, NSF grant DCB-85Ø2194 and graduate fellowship to D.B.R., and the Monsanto Company.
[2]Present address: Department of Biology, Northeastern University, Boston, MA Ø2115.

INTRODUCTION

Prothoracicotropic hormone (PTTH) is a cerebral neuropeptide which stimulates the prothoracic glands (PG) to synthesize increased quantities of the steroid prohormone ecdysone. Ecdysone is then hydroxylated peripherally to the major insect molting hormone, 20-hydroxyecdysone. We have attempted to define the biochemical processes underlying peptide-stimulated ecdysone synthesis, using the PG of the tobacco hornworm, *Manduca sexta*, as an *in vitro* model system. Both cAMP and Ca^{2+} have been implicated in PTTH action, with current evidence favoring cAMP as the messenger directly regulating ecdysone synthesis (1). Steroidogenesis is enhanced by agents that stimulate an increase in intracellular levels of cAMP (1,2). Further, PTTH stimulates an increase in cAMP formation at doses and times consistent with the steroidogenic effects of the peptide (1). In glands removed from day 3 fifth instar larvae, PTTH-stimulated cAMP synthesis is accompanied by a marked accumulation of intracellular cAMP. By contrast, in glands removed from day 0 pupae, PTTH alone does not stimulate significant accumulation of cAMP. Rather, the effects of the hormone on cAMP synthesis are seen only in the presence of a phosphodiesterase inhibitor (1-methyl-3-isobutylxanthine, MIX).

Extracellular Ca^{2+} is required for PTTH action in both larval and pupal PG (3,4). The cation appears to exert its effects prior to the action of cAMP. Specifically, PTTH-stimulated ecdysone synthesis and cAMP formation are absolutely dependent upon extracellular Ca^{2+}, and both of these effects are mimicked by the Ca^{2+} ionophore, A23187. Pharmacological agents that enhance intracellular cAMP, however, stimulate ecdysone synthesis in the absence of extracellular Ca^{2+}. Thus, current evidence suggests that Ca^{2+} facilitates the activation of adenylate cyclase by PTTH, and that the resulting increase in intracellular cAMP enhances ecdysone synthesis. It is likely that cAMP-dependent protein kinase, the only known intracellular receptor for cAMP, plays an essential role in glandular response to PTTH. Evidence in favor of this hypothesis is described in the present report.

RESULTS

Activation of cyclic AMP-dependent protein kinase by PTTH.

Cyclic AMP-dependent protein kinase (cAMP-PK) may be activated by a subtle increase in cellular levels of cAMP, as appears to occur in vertebrate steroidogenic tissues (5,6). Thus, cAMP-PK may be activated to a similar extent in both larval and pupal PG, despite marked differences in PTTH-stimulated accumulation of cAMP.

To rule out gross differences in the nature of cAMP-PK present in larval and pupal PG, the enzyme was first characterized using homogenates of glands removed at each stage of development (methods described in refs. 7,8). Two isozymes of cAMP-PK (type I and type II) have been characterized in mammalian tissues, distinguishable on the basis of elution profiles from DEAE cellulose columns (9). Two comparable isozymes were found in the PG of *Manduca*, as determined by elution profiles, sensitivity to cAMP, and inhibition by a mammalian protein kinase inhibitor (PKI) (8). The type II isozyme was the predominant form in both day 3 fifth instar larval and day 0 pupal PG (>95% of total cAMP-PK).

Activation of cAMP-PK in response to PTTH was tested in larval and pupal PG. Following incubation in medium containing PTTH or in control medium, glands were homogenized, soluble fractions prepared, and cAMP-PK activity determined by incorporation of the terminal phosphate of [^{32}P]ATP into histone (methods described in ref. 7,8). Each sample was assayed under three conditions: a) in the presence of saturating amounts of cAMP, b) in the absence of added cAMP, and c) in the presence of PKI. This procedure permitted calculation of the percent of total cAMP-PK in each sample that was activated by endogenous cAMP ("activity ratio", 9). Table 1 reveals that 10 min of exposure to PTTH led to enhanced activity of cAMP-PK, from a basal level of approximately 30% to a stimulated level of over 90%. Similar results were obtained when charcoal was added during glandular homogenization (data not shown), indicating that the enzyme was not activated by cAMP released during preparation of tissue samples for enzyme assay (9). Further experiments revealed that the larval enzyme was activated within 3 min of exposure to PTTH, clearly preceding a detectable enhancement of ecdysone synthesis, and was elicited by doses of hormone that stimulated steroidogenesis (8).

In contrast to results obtained with larval PG, there was no detectable activation of soluble cAMP-PK in pupal glands by PTTH (10 min, Table 1; 1, 5, and 30 min, not shown). However, exposure to PTTH in the presence of a phosphodiesterase inhibitor (MIX) led to a significant activation of cAMP-PK over basal levels, and over levels of activation seen in the presence of

MIX alone. This pattern was strikingly similar to that seen in previous studies of cAMP accumulation. The results indicated that PTTH was capable of activating cAMP-PK under experimental conditions in which the accumulation of cAMP was enhanced artificially, but raised doubts regarding the normal involve-

TABLE 1
EFFECTS OF PTTH ON CYCLIC AMP-DEPENDENT PROTEIN KINASE ACTIVITY[a]

Stage	Treatment (n)	Protein Kinase Activity (pmole ^{32}P incorp./mg/min) -cAMP	+cAMP	Activity Ratio -cAMP/+cAMP
Larva	Control (5)	170 ± 15	538 ± 55	.33 ± .05
	PTTH (3)	505 ± 35	542 ± 22	.92 ± .03
Pupa	Control (8)	212 ± 20	505 ± 32	.43 ± .05
	PTTH (3)	170 ± 12	492 ± 98	.36 ± .04
	MIX (3)	260 ± 28	430 ± 33	.60 ± .04
	PTTH + MIX (5)	485 ± 60	508 ± 38	.88 ± .04

[a]Enzyme activity was measured as incorporation of $[^{32}P]O_4$ into histone (7,8). Each sample consisted of 12 larval (day 3 fifth instar) or pupal (day 0) glands, pre-incubated for 10 min in medium containing no hormone (Control); PTTH (0.1 U of semi-purified hormone, ref. 4,8); MIX (50 μM), or PTTH + MIX. Assays were conducted using supernatant fractions of homogenized samples (8). Values represents mean + S.E.M. of 3 to 8 samples, as indicated in parentheses.

ment of this enzyme in the response of pupal PG to PTTH. Several alternate lines of inquiry, described below, provided indirect evidence in favor of a role for cAMP-PK at this stage of development.

PTTH refractoriness during diapause.

When exposed to short day lengths (LD 12:12) as larvae, <u>Manduca</u> enters a period of diapause during the pupal stage. During this time, the PG are refractory to PTTH, i.e., the peptide does not enhance basal rates of steroidogenesis (2, 1Ø). Previous studies revealed that glands from diapausing pupae retained functional PTTH receptors, as indicated by their ability to synthesize relatively large amounts of cAMP in response to PTTH (2). However such glands were incapable of responding to cAMP analogs or other agents that pharmacologically enhanced intracellular levels of cAMP. This result indicated that PTTH refractoriness during diapause resulted from a lesion in the steroidogenic pathway beyond the level of cAMP formation (2). A comparison of cAMP-PK in glands from diapausing, as opposed to non-diapausing pupae, revealed a significant decrease in total levels of this enzyme during diapause (Table 2).

TABLE 2
CYCLIC AMP-DEPENDENT PROTEIN KINASE ACTIVITY IN PROTHORACIC GLANDS FROM DIAPAUSING AND NON-DIAPAUSING PUPAE

Stage	Protein Kinase Activity ($pmole\ ^{32}P$ incorporated/mg/min) -cAMP	+cAMP	Activation ratio
ND_0	258 ± 5 (3)	585 ± 38 (7)	.44 ± .Ø4 (3)
D_{10}	193 ± 8 (3)	382 ± 28 (7)	.43 ± .Ø1 (3)

Enzyme activity was measured as described in Table 1. Each sample consisted of 16 glands removed from day Ø non-diapausing pupae (ND_0) or from day 1Ø diapausing pupae (D_{10}). Values represent mean ± S.E.M. of 3 to 7 samples, as indicated in parentheses.

Total protein content of the PG did not differ significantly between the two stages of pupae examined (2), nor were differences apparent in the number of proteins separable by SDS-PAGE. In fact, a decrease in cAMP-PK has been the only difference so far detected in the protein composition of PG from diapausing vs. non-diapausing pupae. This finding suggests that cAMP-PK may be an essential element in the maintenance of cellular responsiveness to PTTH.

Presence of endogenous substrates for cAMP-PK in pupal PG.

In order for cAMP-PK to have a role in the action of PTTH during the pupal stage, glands at this stage must possess endogenous substrates for the enzyme. The existence of such substrates was assayed by incubating soluble and particulate fractions from pupal gland homogenates in the presence of [^{32}P]ATP, with or without cAMP or PKI. Phosphorylated proteins were then separated by SDS-PAGE, and visualized by autoradiography. Results of one such experiment (Figure 1, repeated 2 times with identical results) revealed that

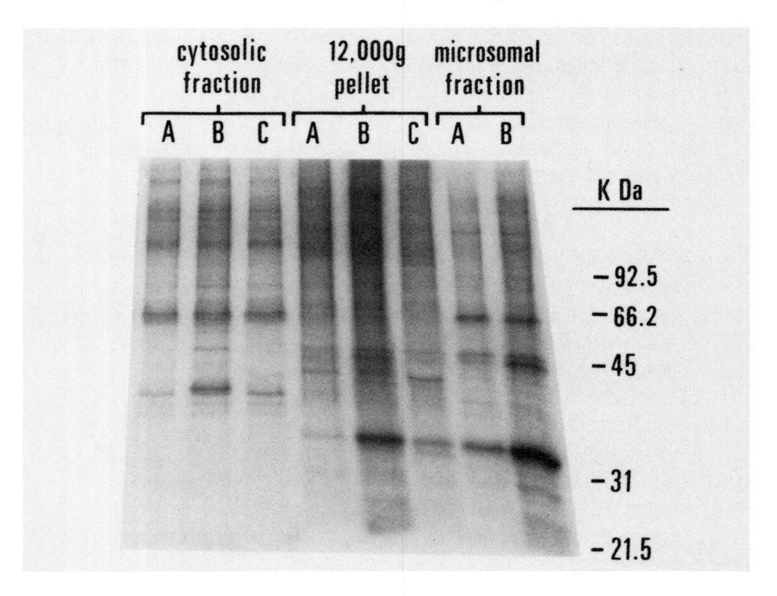

FIGURE 1. Autoradiograph of phosphorylated proteins in subcellular fractions of day Ø pupal PG. Frozen glands were homogenized in 5Ø mM Hepes buffer (pH 6.8) containing Ø.15 M NaCl, 6.Ø mM MgCl , and a protease inhibitor and separated into cytosol, 12,ØØØ x g particulate and 18Ø,ØØØ x g microsomal fractions by centrifugation. Membrane fractions were washed twice by rehomogenization. Each fraction was incubated for 2 min at 3Ø°C under phosphorylating conditions with 5.Ø μCi [γ-^{32}P]ATP; 2Ø μM. Lanes: A) basal; B) +1Ø μM cAMP; C) +1Ø μM cAMP and 1 mg/ml PKI. Phosphorylated proteins were separated by SDS-PAGE on 13 1/2% acrylamide gels and visualized by autoradiography.

the incorporation of radiolabelled phosphate into a peptide of 34 kDa present in the 12,000 x g particulate fraction was enhanced markedly by cAMP and inhibited by PKI, suggesting that this protein was capable of serving as a substrate for endogenous cAMP-PK. A microsomal fraction (180,000 x g pellet) also contained two prominent cAMP-dependent bands, corresponding to M_r 's of 34,000 and 44-45,000. A densitometric scan of the 34,000 band revealed a 4-fold increase in the incorporation of [^{32}P] in the presence of 10 to 100 uM cAMP (Ka = 2.0 μM) , with no corresponding increase in the presence of cGMP. While phosphorylation was inhibited by the addition of PKI, it was not affected by low Ca^{2+} (addition of EGTA) or by the addition of the Ca^{2+} -calmodulin inhibitor, trifluoperazine (not shown). Preliminary experiments conducted with day 3 fifth instar larval glands revealed a similar pattern of basal and cAMP-stimulated phosphorylation of cellular proteins. Thus, it appears that both larval and pupal PG possess potential substrates for cAMP-PK.

Effects of PTTH on protein phosphorylation.

Proteins labelled under cell-free conditions such as those described above may not necessarily serve as substrates in intact cells. Such a discrepancy can result from the artifactual phosphorylation in cellular homogenates of proteins normally isolated from the effect of protein kinase (e.g. via subcellular compartmentalization). Thus, an intact cell approach was critical to establish that the labelling pattern observed using tissue homogenates was physiologically relevant. To this end, intact pupal glands were incubated in medium containing [^{32}P]O_4 which was subsequently incorporated into cellular ATP. The glands were then challenged with PTTH or control medium, and phosphorylated proteins were separated and visualized as described above. Results of a representative experiment (repeated 3 times) are shown in Figure 2.

As seen in Figure 2, when intact pupal PG were exposed for 45 min to PTTH alone, the phosphorylation of a 34,000 dalton protein was clearly enhanced (Rountree, Combest, and Gilbert, unpublished). Phosphorylation of this protein was also increased following exposure of intact glands to dibutyryl cAMP plus MIX. The Ca^{2+} ionophore, A23187, was not as effective as PTTH or dibutyryl cAMP in stimulating protein phosphorylation. Similar results were obtained using glands removed from day 3 fifth instar larvae (not shown).

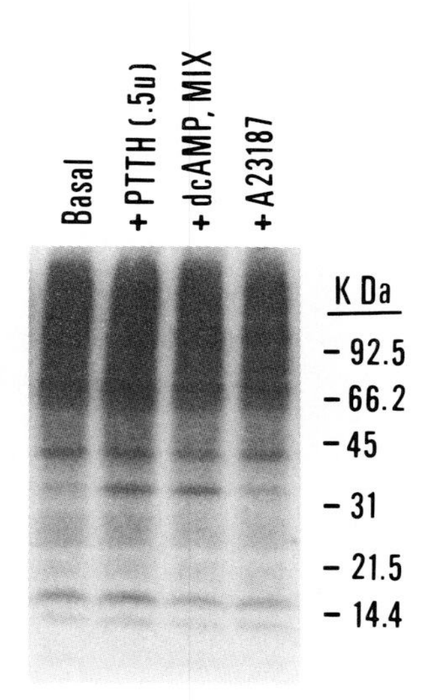

FIGURE 2. Autoradiograph of phosphorylated proteins in intact day 0 pupal PG. Groups of 4 glands were preincubated for 1 hr in 150 μl phosphate free Grace's medium containing 20 μCi $[^{32}P]O_4$ to label intracellular ATP. Glands were then challenged for 45 min with PTTH (0.5U) lane 2, dibutyryl cAMP (1 mM) and MIX (0.1 mM) lane 3, calcium ionophore A23187 (5.0 μg) and calcium (1.0 mM) lane 4. Glands were solubilized with SDS and subjected to SDS-PAGE on 13 1/2% acrylamide gels and visualized by autoradiography.

The intact cell technique does not permit direct identification of specific protein kinases responsible for a hormone-stimulated increase in protein phosphorylation. Thus, the possibility exists that the effects of PTTH were mediated by a non-cAMP-dependent PK (e.g. protein kinase C). However, the simplest conclusion to be drawn from the above results is that, in response to PTTH, a 34,000 protein is phosphorylated by a cAMP-dependent enzyme.

DISCUSSION

Evidence to date regarding the mechanism of action of PTTH, particularly in early fifth instar larval PG, is in keeping with the following model: 1) the peptide stimulates a Ca^{2+} -dependent increase in cAMP synthesis, 2) cAMP enhances the activity of cAMP-PK, and 3) this enzyme in turn phosphorylates cellular proteins involved in ecdysone synthesis. These elements of the response pathway, including synergistic effects of cAMP and Ca^{2+} , are typical of many cAMP-mediated responses in vertebrates, including peptide-stimulated steroidogenesis (see 11).

Indirect evidence described in the present report suggests that the model also applies to pupal PG. However, it remains to be demonstrated that cAMP-PK is activated by PTTH at this stage in the absence of phosphodiesterase inhibitors. This situation is similar to that found in the adrenal cortex following stimulation by low, but physiologically effective doses of ACTH. Activation of cAMP-PK by low doses of ACTH has been reported in some studies, particularly those employing adrenal tumor cells (12, 13). However, in cells derived from normal glands, activation of cAMP-PK requires the combined presence of ACTH and MIX (5). Several explanations could account for the results obtained with pupal PG, including: a) steroidogenesis can be stimulated by more than one second messenger (e.g., cAMP and diacylglcyerol, the latter acting via protein kinase C), or b) hormone-sensitive cAMP-PK is associated with a particulate cell fraction (e.g., plasma membrane or endoplasmic reticulum). The applicability of these hypotheses to the pupal PG response is currently being explored. It should be noted that the mechanism of action of PTTH may ultimately prove to be the same for larval and pupal PG. That is, rather than asking how pupal glands can synthesize ecdysone in the absence of dramatic changes in intracellular levels of cAMP, we should perhaps address the possibility that the large amounts of cAMP produced by larval PG serve a function in addition to the acute stimulation of ecdysone synthesis.

Results described in the present report indicate that cAMP-dependent protein phosphorylation likely plays a role in PTTH-stimulated ecdysone synthesis. The data raise obvious questions regarding the nature and function of the single band phosphorylated in intact cells upon exposure to PTTH. The band conceivably consists of more than one protein, which could be resolved, for example, by the use of two-dimensional gels. A logical hypothesis is that the phosphoprotein(s) controls a rate-limiting step in ecdysone synthesis. Unfortunately, a direct test of this hypothesis is hampered by the fact that details of the biosynthetic pathway for ecdysone in the PG are presently conjectural (14).

Additional elements of the hormonal regulation of ecdysone synthesis remain to be clarified. For example, in the present model, an assumption is made that cAMP is directly involved in ecdysone synthesis. However, the nucleotide may serve to stimulate a kinase involved in the release of bound Ca^{2+} from intracellular sites. A Ca^{2+} -mobilizing effect of cAMP has been observed in such diverse cells as vertebrate kidney, pancreas, and blowfly salivary glands (see 11). Further, several Ca^{2+} -dependent hormones have been observed to enhance intracellular levels of inositol phosphates, via hydrolysis of membrane phosphatidylinositides, which in turn appear to mobilize intracellular calcium stores (15). The potential importance of intracellular Ca^{2+} in PTTH-stimulated ecdysone synthesis clearly requires further attention.

ACKNOWLEDGMENTS

We thank Ms. Susan Whitfield for assistance with graphics, and Ms. Shelia King for preparation of the manuscript. Purified PTTH was provided by Dr. Walter Bollenbacher, and portions of the diapause studies were conducted in collaboration with Drs. M. F. Bowen and W. E. Bollenbacher.

REFERENCES

1. Smith WA, Gilbert LI, Bollenbacher WE (1984). The role of cyclic AMP in ecdysone synthesis. Molec Cell Endocrinol 37:285.
2. Smith WA, Bowen MF, Bollenbacher WE, Gilbert LI (1986). Cellular changes in the prothoracic glands of diapausing pupae of Manduca sexta. J Exp Biol (In press).
3. Smith WA, Gilbert LI, Bollenbacher WE (1985). Calcium-cyclic AMP interactions in prothoracicotropic hormone stimulation of ecdysone synthesis. Molec Cell Endocrinol 39:71.
4. Smith WA, Gilbert LI (1986). Cellular regulation of ecdysone synthesis by the prothoracic glands of Manduca sexta. Insect Biochem 16:143.
5. Sala GB, Hayashi K, Catt KJ, Dufau ML (1979). Adrenocorticotropin action in isolated adrenal cells. J Biol Chem 254:3861.

6. Sala GB, Dufau ML, Catt KJ (1979). Gonadotropin action in isolated ovarian luteal cells. J Biol Chem 254:2077.
7. Combest WL, Gilbert LI (1986). Characterization of cyclic AMP-dependent protein kinase activity in the larval brain of Manduca sexta. Insect Biochem (In press).
8. Smith WA, Combest WL, Gilbert LI (1986). Involvement of cyclic AMP-dependent protein kinase in prothoracicotropic hormone-stimulated ecdysone synthesis. (Submitted).
9. Corbin JD, Keely SL, Soderling TR, Park CR (1975). Hormonal regulation of adenosine 3',5'-monophosphate-dependent protein kinase. Adv Cyclic Nucleotide Res 5:265.
10. Bowen MF, Bollenbacher WE, Gilbert LI (1984). In vitro studies on the role of the brain and prothoracic glands in the pupal diapause of Manduca sexta. J Exp Biol 108:9.
11. Rasmussen H (1981). "Calcium and cAMP as Synarchic Messengers." New York: John Wiley and Sons, p 178.
12. Schimmer BP (1980). Cyclic nucleotides in hormone regulation of adrenocortical function. Adv Cyclic Nucleotide Res 13:181.
13. Murray SA, Byus CV, Fletcher WH (1985). Intracellular kinetics of free catalytic units dissociated from adenosine 3',5'-monophosphate-dependent protein kinase in adrenocortical tumor cells (Y-1). Endocrinology 116:364.
14. Rees H (1985). Biosynthesis of ecdysone. In Kerkut G, Gilbert LI (eds): "Comprehensive Insect Physiology, Biochemistry, and Pharmacology," Oxford: Pergamon, 7:249.
15. Berridge MJ, Irvine RF (1984). Inositol trisphosphate, a novel second messenger in cellular signal transduction. Nature (Lond) 312:315.

Molecular Entomology, pages 141–153

THE PEPTIDES, mRNA, AND GENES OF THE AKH FAMILY[1]

Martin H. Schaffer and Barbara E. Noyes

Department of Psychiatry, University of Chicago
Chicago, IL 60637

ABSTRACT Over the last 2 years several groups have between them isolated and sequenced 5 different peptides from the corpora cardiaca (C.C.) of various insects. All have proven to be members of the structurally related peptide family which includes adipokinetic hormone (AKH). The isolation of several of these hormones is described. Progress in the study of the biosynthesis of several of the peptides, and in the identification, sequencing and cloning of the AKH mRNA and gene are also discussed.

INTRODUCTION

The RPCH-AKH family of neuropeptides is a large group with strikingly similar amino acid sequences and similar bioactivities in a variety of arthropod preparations. They occur in a wide array of arthropods with typical neuropeptide localization in neuroendocrine tissues and neurons. Several of the earliest recognized bioactivities are of uncertain physiological significance and considering their wide distribution both in individual animals and across species it is quite possible that most of these peptides' functions await recognition. Nonetheless the regular presence of multiple family members in each insect studied suggests that these compounds play consistently important roles in insects' endocrine and neural systems. Further because these peptides often are found in settings that are favorable for biochemical and physiological studies they are likely to provide important model systems with implications reaching even beyond the large class Insecta. Thus my collaborators and I have begun to study the biochemistry of some of these peptides with an eye toward the charac-

[1]This work was supported in part by a grant from The Brain Research Foundation of the University of Chicago.

terization of preparations in which biochemical and physiological investigations can be combined into a vivid picture of neuropeptide function. This paper will survey our work on the peptides from several insects and our initial studies on the AKH mRNA and gene with emphasis on the potential of the systems in which these peptides occur for further studies.

RESULTS

The Peptides

The progress in the study of this peptide family has in large measure been linked to the isolation and sequencing of the peptides, and we think this will continue to be true. But after the characterization of AKH (1), the study of other insect peptides in this family went rather slowly. This was so despite the fact that a second peptide was clearly present in locust corporus cardiacum (C.C.) (2), and a similar bioactivity occurred in P. americana C.C. (3,4). Our first entry into this field was a rather naive search for the Periplaneta activity based on the idea that it might be AKH itself. Technical advances in peptide chemistry, notably high performance liquid chromatography (HPLC) and high sensitivity detectors greatly facilitated the search, and so AKH could readily be purified from a methanolic extract of Schistocerca C.C. using a C-18 reverse phase column monitored by a flow fluorimetric detector (Figure 1A). The procedure is not only a testimony to the power of HPLC but also to the richness of C.C. as a source for peptide. This has repeatedly proved important in the study of the AKH family biochemistry. When the same purification procedure was applied to the Periplaneta americana C.C. it was clear that the reported bioactivity could not be due to AKH. But there were two very prominent fluorescent peaks isolated from cockroach C.C. (Figure 1B) by this procedure. At just this time Dr. Michael O'Shea and Dr. Jane Witten, then neighbors at the University of Chicago, began a search for myoactive peptides in P. americana C.C. and they encountered the same two compounds. Thus we pooled our efforts and were able to show that these peptides, which we call MI and MII, are released in a calcium dependent manner from C.C., have myoactivity and lipid mobilizing activity in locust, and are structurally related to AKH as shown by purification and amino acid analysis (5). In collaboration with Dr. Kenneth Rinehart, Jr. and his colleagues we used fast atom bombardment mass spectrometry (FAB-MS) to sequence these peptides (6) confirming their close structural relationship to AKH and to the first of these arthropod

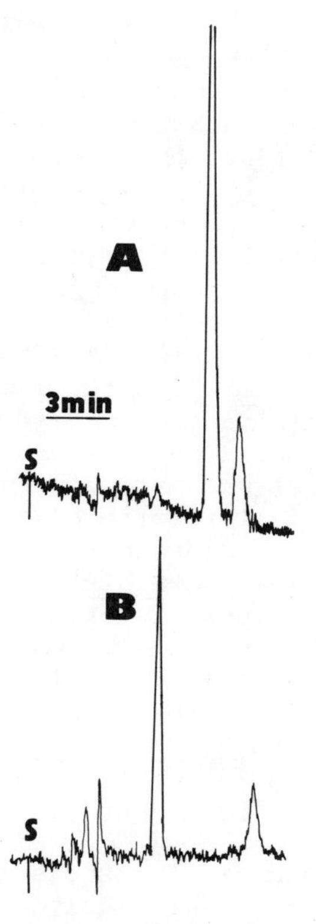

FIGURE 1. HPLC of methanolic extract of S. nitans (A) and P. americana (B) C.C.'s. The C-18 column was eluted at 1.5 ml/min with 25 mM ammonium acetate pH4.5 - 27% acetonitrile, and monitored by fluorescence λ_E=276.

peptides to be sequenced, prawn red pigment concentrating hormone (RPCH). At the same time Scarborough et al. isolated and sequenced the same two peptides from P. americana C.C. using stimulation of isolated cockroach heart as an initial bioassay (7). This activity confirmed that the peptide MI is the same as neurohormone D described by Bauman and Gersch (8). Further Scarborough et al. demonstrated that each peptide has the hyperglycemic activity first observed by Steele (4,7). Thus the identification of these two peptides accounted for many of the bioactivities described in Periplaneta C.C. over the last 40 years.

AKH II was the next peptide in this family to be sequenced. As seen in Figure 1A C.C. extracts from grasshoppers chromatographed on reverse phase HPLC show two major peaks. The larger is AKH and the smaller peak has appropriate mass and bioactivities to fit the initial description of AKH II (2). This led several groups to apply the same techniques and to obtain sequence information. Gäde and Goldsworthy (9) and Siegert *et al.* (10) isolated AKH II from *Schistocerca gregaria* and *Locusta migratoria* and showed by FAB-MS and gas phase sequencing respectively that the peptide does have a similar structure to other family members, especially the crustacean peptide RPCH. While AKH has the identical structure in both these locust species there is a single amino acid substitution, gly to ala, between AKH II in *S. gregaria* and *L. migratoria*. At the same time our laboratory had accumulated a significant amount of a guanidinium thiocyanate extract of *S. nitans* C.C. as a side product from our studies on AKH mRNA. This material was used to isolate AKH II by minor modifications of the usual reverse phase techniques, and again in collaboration with Dr. Kenneth Rinehart (9) the peptide was sequenced by FAB-MS. The two *Schistocerca* peptides have identical sequences, and fortunately Yamashiro *et al.* (11) had previously synthesized this peptide based on the reported amino acid composition ([6-thr]RPCH). Their generous gift of some of this material allowed us to demonstrate co-chromatography of the synthetic and natural *Schistocerca* peptides under a variety of conditions and to confirm that AKH II has lipid mobilizing activity in locust, but can only achieve 70% of the maximal response (9).

Gäde surveyed a number of insects (12) and identified by lipid mobilization assay other likely members of the AKH family in each. He purified, and in collaboration with Dr. Kenneth Rinehart, sequenced one of these peptides found in *Carausius morosus* (13). Table 1 lists the peptides in the RPCH-AKH family with known sequences. This list will surely grow as other species are investigated and perhaps even as more work is done on the species already studied.

TABLE 1
THE RPCH - AKH FAMILY

Peptide	Sequence
RPCH	pGlu Leu Asn Phe Ser Pro Gly Trp NH_2
AKH	pGlu Leu Asn Phe Thr Pro Asn Trp Gly Thr NH_2
M I	pGlu Val Asn Phe Ser Pro Asn Trp NH_2
M II	pGlu Leu Thr Phe Thr Pro Asn Trp NH_2
AKH II	pGlu Leu Asn Phe Ser Thr Gly/Ala Trp NH_2
C. morusus	pGlu Leu Thr Phe Thr Pro Asn Trp Gly Thr NH_2

While we are currently studying the mRNA and gene for S. nitans AKH we hope to eventually extend our investigations at the nucleic acid level to Drosophila melanogaster to use the power of genetic approaches to study neuropeptide function. Thus it is important to establish which of these peptides are present in Drosophila. Thanks to Dr. Brian Charlesworth who generously gave us some flies we have begun to investigate this by making acidic methanolic extracts of either whole thoraxes or of the corpus allatum - C.C. complex along with a short segment of gut, following the dissection described by King et al. (14). These studies are just beginning and are hampered by the rather slow dissection and the very small size of the insect. Nonetheless it is encouraging that both extracts give rather simple patterns of fluorescent peaks when chromatographed on a C-18 HPLC column (Figure 2). In fact the "gland"

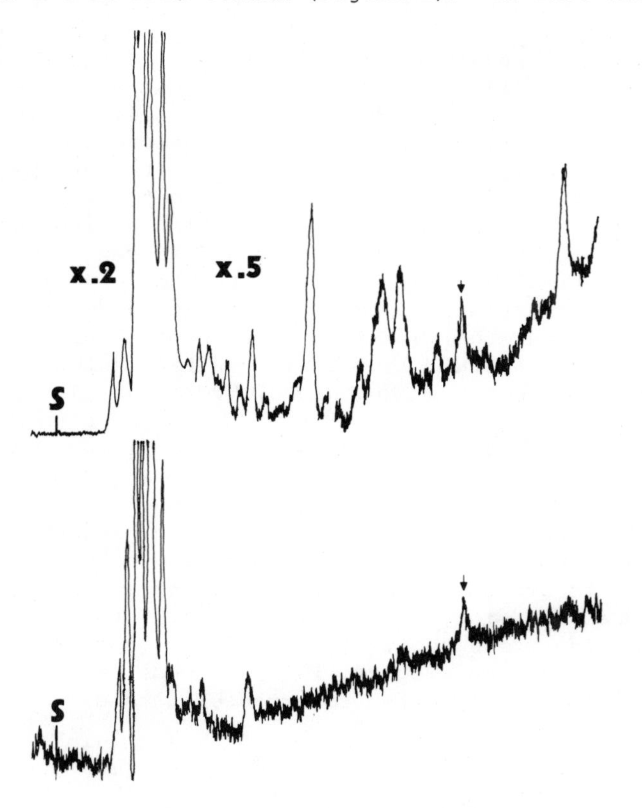

FIGURE 2. HPLC of Drosophila thorax (above) or C.C. (below) under conditions as in Figure 1 except an acetonitrile gradient 20-38% over 30 mins. was used.

preparation gives a well isolated peak (arrow) which runs just after AKH. Preliminary experiments suggest that this peak is destroyed by treatment with pGlu aminopeptidase. Thus it is quite likely to be another peptide in the AKH family. The purification and sequence determination of this peptide along with the characterization of its biological activities are actively being pursued.

The functional roles of these peptides in insects are likely to be quite varied. Immunohistochemical and HPLC studies in locust and cockroach demonstrate the presence of related compounds in neurons of the CNS and nerve endings in gut (5,12,16). Biological activities suggest a role in the regulation of heart beat and skeletal muscle function (5,7,8,17). The clearest example of an important endocrine function is AKH's role in the modulation of energy supply during locust flight (18). The hyperglycemic effect of MI and MII in *Periplaneta* is robust but its slow time course makes its relation to energy metabolism less obvious and suggests other functions such as a role in chitin synthesis (19). Nonetheless, because of the evidence for MI and, or MII release during feeding (17) and their effects on heart rate, skeletal muscles, and trehalose metabolism, along with their localization in foregut, it is tempting to speculate that one or both of these peptides may function as humorally transmitted messengers modulating eating behavior.

The variation in structure of the different family members is an interesting combination of careful conservation and seemingly random variation. Both structure-activity studies (11, 20,21) and analysis of the bioactivities of natural peptides suggest that the overall length and general design of the molecule (terminally blocked, uncharged, hydrophobic) are most important in determining its activity, although certain residues do appear to be important for specific activities. Given the rather similar bioactivities of all family members it is not obvious why insects synthesize more than one in their C.C.'s particularly since these peptides are often present and releasable in large amounts. Nor is it clear why each genus seems to require its own subtle variations of the basic sequence. Presumably as the study of the biochemistry of the receptors which recognize these peptides progresses, and our understanding of their diverse physiological roles deepens, we will be in a better position to judge the importance of this.

Biosynthesis in the C.C.

No other aspect of neuropeptide biochemistry is receiving as much attention as is their biosynthesis. There are several reasons for this including the popular idea that this may be an important locus of control both in cellular differentiation and in function, and the fact that advances in molecular biology have made this an eminently addressable area. The C.C. is particularly useful in the study of AKH family peptide biosynthesis because it is such a rich source. It should not be taken for granted, however, that just because the C.C. is a rich storehouse of peptides that it is also a major site of synthesis. Fortunately synthesis does occur in the C.C., and this can be easily demonstrated by incubating C.C.'s in organ culture with [^{3}H]- amino acids (Figure 3). This has been demonstrated both for AKH in locust and for MI and MII in cockroach (5). So far this observation has been utilized to begin an investigation of the structure of the AKH mRNA (see below) and to search for protein precursors of the AKH peptide (22). In addition it is reasonable to suppose that the C.C. will prove a very favorable tissue for the study of synthetic intermediates and processing and modification enzymes. Thus while the study of synthesis of neuropeptides in neurons is routinely very difficult because of the low abundance of compounds of interest in a very complex tissue, the availability of the C.C. means that the biosynthesis of the AKH family neuropeptides may be studied much more easily. Of course the results of those studies including synthetic schemes, nucleic acid probes, and antibodies to particular

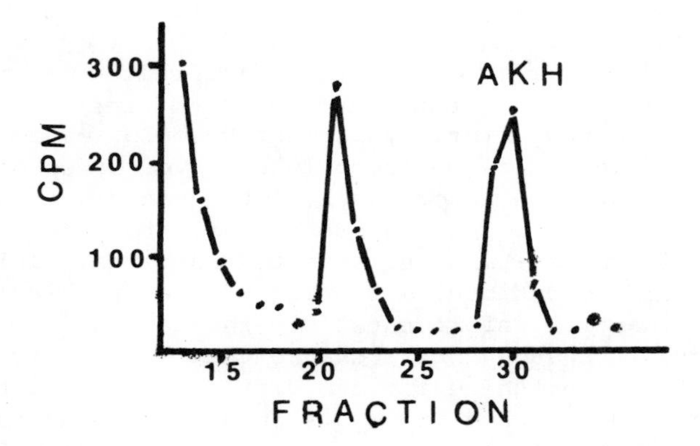

FIGURE 3. Methanolic extract of S. nitans C.C. after incubation with [^{3}H]-leu in vitro for 2 hrs., and chromatography on a reverse phase column as in Figure 1. Fractions were collected over 0.3 min. intervals and counted. The large number of counts in early fractions is unincorporated leu. The second peak is unidentified. The third is AKH.

protein intermediates can be applied to these same peptides in other locations such as identified neurons.

The AKH mRNA.

An interest in neuropeptide biosynthesis naturally leads one to the powerful new recombinant DNA techniques. In particular these approaches allow one to gain both detailed structural information about the nucleic acids and protein precursors which give rise to the neuropeptide and recombinant DNA molecules which are essential tools in the investigation of genetic diversity and gene expression. Therefore we began a study of the AKH mRNA and gene.

For all their promise, recombinant DNA techniques pose certain problems particularly in their application to neuropeptides. In general the problems involve constructing the recombinant DNA molecules of interest and selecting these molecules from the complex mixtures in which they are found in cDNA and gene libraries. Usually cDNA (mRNA derived) libraries provide a good starting point, even though they are more difficult to construct than gene libraries. This is true because their complexity is so much less than gene libraries that the task of selecting recombinants of interest is much simpler. Because the C.C. is such a rich source, a C.C. cDNA library would seem a particularly favorable starting point. Unfortunatel the C.C. is a very small organ, and given the number of grasshoppers our colony can provide, it is difficult to accumulate sufficient RNA to be confident of a good result.

Even given a cDNA library, screening would remain something of a problem. There are two commonly employed modes of cDNA library screening. The newer involves the construction of an expression library in which the protein of interest is made as part of a fusion protein in the infected bacteria. This protein is then selected by antibody screening of the proteins produced by the recombinants. Unfortunately in the case of small neuropeptides there is a substantial chance that any particular antibody will fail to recognize the unmodified version of the short neuropeptide sequence when it occurs in a large precursor which in turn is part of a fusion protein. While this task is by no means impossible and in fact has been applied successfully (23), it requires antibodies with particular selectivities and a favorable mRNA to provide the template for the cDNA to be cloned. In the case of AKH neither of these critical resources was readily available. All other selection techniques exploit, in one way or another, the ability of complementary strands of nucleic acids to select each other out and specifically form a double helix even in the presence of very complex mixtures of nucleic acids. The key to any of these techniques then is a suitable complementary nucleic acid (cDNA), and the problem to be solved is how to obtain such a cDNA. More and more, investigators are doing this by using known protein sequences to predict mRNA sequences and then chemically synthesizing a relatively short presumptive cDNA. This strategy, often called an oligo probe approach, is complicated by the degeneracy of the genetic code. Unfortunately this problem is well illustrated by the protein sequence of AKH.

TABLE 2
POSSIBLE AKH mRNA SEQUENCES

AKH Peptide		pGlu	Leu	Asn	Phe	Thr	Pro	Asn	Trp	Gly	Thr NH_2
Possible AKH mRNA Sequences	5'	CA^{A}_{G}	CUX UU^{A}_{G}	AA^{U}_{C}	UU^{U}_{C}	ACX	CCX	AA^{U}_{C}	UGG	GGX	ACX[a]
Oligo Mix I	3'							TT^{A}_{G}	ACC	CC^{T}_{G}	TG

[a]X denotes any of the four bases is possible.

As shown in Table 2 for any 12 nucleotide cDNA there are at least 32 possible sequences, and for a 15 nucleotide probe there are at least 128 possible sequences. Although there are some impressive examples of gene library screening in just such situations (24), this approach seemed an unattractive alternative. While a cDNA library seemed much more workable, problems remained. Perhaps most troubling was the very limited amounts of RNA. There were also uncertainties about producing cDNAs which contained the AKH sequence, since in the worst case the AKH sequence might occur just once in a message of hundreds or even thousands of nucleotides, and the enzymes used to produce such cDNAs usually do not make full length copies of the RNA template. Finally there was the unattractive choice of screening with either a very short or a highly complex mixed probe. In light of all this we chose to synthesize a mixture of 4 probes, 11 nucleotides in length, which covered half the possible sequences (Table 2, Oligo Mix I), and we began our study by using primer extension (25) rather than a cloning technique. In this approach the probe is radiolabelled at its 5' hydroxyl group and mixed with the RNA, deoxynucleotide triphosphates, and the enzyme reverse transcriptase. The probe then serves as a primer for the enzymatic synthesis of an extended, radioactive cDNA which can be purified by polyacrylamide gel electrophoresis and used for Maxam-Gilbert sequencing. This approach tolerates short probes even with certain base pair mismatches, and it requires small amounts of RNA. It is also attractive in difficult cases because it leads in short order to sequence information allowing one to judge whether the probe is selecting the correct RNA or not. Unfortunately when this technique was applied to total C.C. RNA prepared by extraction in gaunidinium thiocyanate followed by centrifugation through a CsCl cushion (26), the resulting cDNA pattern was very complex, and an AKH-related cDNA was not identified. We are not yet sure of the cause of this result. Our experience to date suggests that AKH mRNA is difficult to prepare and we may not have provided a good template for the enzyme. In addition two problems inherent in the technique tend to make the pattern complex. First priming may occur at multiple sites since short primers, even with mismatches are acceptable. Second the enzyme reverse transcriptase does not uniformly produce full length cDNAs but rather tends to stop short of the 5' end of the RNA. Unfortunately these premature terminations do not occur at random but rather at relatively descrete sites along the template. In this way a single priming site can give rise to a series of bands, and the complexity of the cDNA pattern increases geometrically.

In view of the complex cDNA pattern, a modified approach was necessary to simplify the product mixture. Inspection of the possible AKH mRNA sequences suggested that repeating the priming experiment in the absence of one of the four nucleotide triphosphate substrates might accomplish this. The pattern should be simplified since most priming sites will terminate after being extended by just a few nucleotides. For example, only about 3% of all priming sites will be extended by 12 nucleotides or more when one of the triphosphates is excluded from the reaction. Further the product of interest is concentrated in just one gel band rather than being divided between premature termination products, and one can predict based on the possible nucleotide sequences what size bands should be investigated. In this way, a 23 nucleotide cDNA, enzymatically synthesized without dCTP, was purified and shown by sequence analysis to be AKH derived. Despite its short length the identification of this product as AKH cDNA is unambiguous. Not even allowing for the specificity of the primer, the probability of obtaining an AKH compatible sequence by chance is 0.0000038. This sequence data allowed us to synthesize a 22 nucleotide DNA probe which exactly complements the AKH mRNA sequence. This probe seems sufficient to screen a gene library.

The AKH Gene.

S. nitans. DNA was prepared from testes following the tissue disruption protocol of Klett and Smith (27) and the DNA purification procedure of Dillela and Woo (28). The DNA was partially digested using the restriction enzyme MboI and a 15 to 20 kb fraction was prepared as described by Maniatis *et al*. (29). This material was ligated with purified arms of lambda phage Charon 35 (29,30) and packaged using a "Gigapack" packaging extract (Vector Cloning Systems). More than 47 million original recombinants were produced with a background (ligated Charon 35 arms) of less than 1%. This gene library was used without amplification. In the initial screen one-half million plaque replicas were hybridized with the 5'-[^{32}P] labelled 22 nucleotide probe. Hybridization was carried out at 37°C as described by Maniatis *et al*. (29). The nitrocellulose filters were washed extensively with 6 X SSC at 4°C and briefly at 55-60°C. This procedure identified 4 positive plaques. On plaque purification and rescreening two of the plaques gave consistently positive signals under these conditions and exposure of the filters overnight at -70°C with an intensifying screen. These recombinant phage which contain approximately 15 and 20 kb inserts are currently being characterized. We are hopeful that their isolation marks the beginning of a long series of interesting studies on the AKH gene.

REFERENCES

1. Stone JV, Mordue W, Batley KE, Morris HR (1976). Structure of locust adipokinetic hormone, a neurohormone that regulates lipid utilisation during flight. Nature 263:207.
2. Carlsen J, Herman WS, Christensen M, Josefsson L (1979). Characterization of a second pigment concentrating activity from the locust copora cardiaca. Insect Biochem. 9:497.
3. Brown FA Jr., Meglitsch A (1940). Comparison of the chromatophorotropic activity of the insect copora cardiaca with that of crustacean sinus glands. Biol. Bull. 79:409.
4. Steele JE (1961). Occurrence of a hyperglycaemic factor in the coporus cardiacum of an insect. Nature 192:680.
5. O'Shea M, Witten J, Schaffer M (1984). Isolation and characterization of two myoactive neuropeptides: further evidence of an invertebrate neuropeptide family. J Neuroscience 4:521.
6. Witten JL, Schaffer MH, O'Shea M, Cook JC, Hemling ME, Rinehart KL Jr. (1984). Structures of two cockroach neuro-

peptides assigned by fast atom bombardment mass spectrometry. Biochem Biophys Res Comm 124:350.

7. Scarborough RM, Jamieson GC, Kalish F, Kramer SJ, McEnroe GA, Miller CA, Schooley DA (1984). Isolation and primary structure of two peptides with cardioacceleratory and hyperglycemic activity from the copora cardiaca of Periplaneta americana. Proc Natl Acad Sci USA 81:5575.
8. Bauman E, Gersch M (1982). Purification and identification of neurohormone D, a heart accelerating peptide from the corpora cardiaca of the cockroach Periplaneta americana. Insect Biochem 12:7.
9. Gade G, Goldsworthy GJ, Schaffer MH, Cook JC, Rinehart KL Jr. (1986). Sequence analyses of adipokinetic hormones II from Corpora Cardiaca of Schitocerca nitans, Schistocerca gregaria, and Locusta migratoria by fast atom bombardment mass spectrometry. Biochem Biophys Res Comm 134:723.
10. Siegert K, Morgan P, Mordue W (1985). Primary structures of locust adipokinetic hormones II. Biol Chem Hoppe-Seyler 366:723.
11. Yamashiro D, Applebaum SW, Li CH (1984). Synthesis of shrimp red pigment-concentrating hormone analogs and their biological activity in locusts. Int J Peptide Protein Res 23:39.
12. Gäde G (1984). Adipokinetic and hyperglycaemic factors of different insect species: separation with high performance liquid chromatography. J Insect Physiol 9:729.
13. Gäde G, Rinehart KL Jr. (1985). Unpublished observation.
14. King RC, Aggarwal SK, Bodenstein D (1966). The comparative submicroscopic cytology of the corpus allatum-corpus cardiacum complex of wild type and fes adult female Drosophila melanogastes. J Exp Zool 161:151.
15. Schooneveld H, Tesser GI, Veenstra JA, Romberg-Privee HM (1983). Adipokinetic hormone and AKH-like peptide demonstrated in the corpora cardiaca and nervous system of Locusta migratoria by immunocytochemistry. Cell Tissue Res 230:67.
16. Witten J, Worden MK, Schaffer MH, O'Shea M (1984). New classification of insect motoneurons: expression of different peptide transmitters. Society for Neuroscience 14th Annual Meeting. Anaheim. Abstract 46.4.
17. Davey KG (1962). The release by feeding of a pharmacologically active factor from the corpus cardiacum of Periplaneta americana. J Ins Physiol 8:205.
18. Mayer RJ, Candy DJ (1968). Control of haemolymph lipid concentration during locust flight: an adipokinetic hormone from the corpora cardiaca. J Insect Physiol 15:611.

19. Steele JE (1963). The site of action of insect hyperglycemic hormone. Gen Comp Endocrinology 3:46.
20. Stone JV, Mordue W, Broomfield CE, Hardy PM (1978). Structure-activity relationships for the lipid mobilising action of locust adipokinetic hormone. Eur J Biochem 89:195.
21. Josefsson L (1983). Invertebrate neuropeptide hormones. Int J Peptide Protein Res 21:459.
22. Hekimi S, O'Shea M (1985). Adipokinetic hormone in locusts: synthesis and developmental regulation. Society for Neuroscience 15th Annual Meeting. Dallas. Abstract 282.12, p 959.
23. Lechman RM, Wu P, Jackson IMD, Wolf H, Cooperman S, Mandel G, Goodman RH (1986). Thyrotropin-releasing hormone precursor: characterization in rat brain. Science 231:159.
24. Seeburg PH, Adelman JP (1984). Characterization of cDNA for precursor of human luteinizing hormone releasing hormone. Nature 311:666.
25. Noyes BE, Mevarech M, Stein R, Agarwal KL (1979). Detection and partial sequence analysis of gastrin mRNA by using an oligodeoxynucleotide probe. Proc Natl Acad Sci USA 76:1770.
26. Fiddes JC, Goodman HM (1979). Isolation, cloning and sequence analysis of the cDNA for the α-subunit of human chorionic gonadotropin. Nature 281:351.
27. Klett RP, Smith M (1968). Isolation of deoxyribonucleate from invertebrates. Methods in Enzymology 12B:112.
28. Dillela A, Woo SLC (1985). Cosmid cloning of genomic DNA. Focus 7.2:1.
29. Maniatas T, Fritsch EF, Sambrook J (1982). "Molecular Cloning a Laboratory Manual." Cold Spring Harbor: Cold Spring Harbor Laboratory Publications, p 270.
30. Loenen WAM, Blattner FR (1983). Lambda Charon Vectors (Ch 32, 33, 34, and 35) adapted for DNA cloning in recombination-deficient hosts. Gene 26:171.

Molecular Entomology, pages 155–164

PROPERTIES OF GENES EXPRESSED DURING ECDYSONE-INDUCED IMAGINAL DISC MORPHOGENESIS IN DROSOPHILA[1]

Jeanette E. Natzle[2], James W. Fristrom, Dianne K. Fristrom, John Moore, David Osterbur[3], Stephenie Paine-Saunders, and Donald Withers

Department of Genetics, University of California
Berkeley, California 94720

ABSTRACT Expression of specific classes of genes is precisely regulated during imaginal disc prepupal morphogenesis by the steroid hormone 20-hydroxyecdysone. Some "early" gene sequences (eg. 2B5) may regulate the expression of other genes. In contrast, restructuring of the disc epithelium (evagination) might be mediated, at least in part, by 20-HOE responsive genes that encode membrane or secreted proteins. We have isolated and characterized genes of this second type. The temporal and spatial expression of these genes is consistent with predictions for genes that mediate imaginal disc morphogenesis. The regulatory classes of ecdysone-responsive genes in imaginal discs correspond formally to those in another ecdysone target tissue, the larval salivary gland.

INTRODUCTION

Variation in the titer of the steroid hormone, 20-hydroxyecdysone (20-HOE), co-ordinates a number of important transitions in Drosophila development. One transition, metamorphosis, involves histolysis of juvenile

[1]Supported in part by a Leukemia Society postdoctoral fellowship to JEN, and by USPHS Grant GM 19937 to JWF.
[2]Present address:Department of Zoology, University of California, Davis, CA 95616.
[3]Present address:Department of Biology, Indiana University, Bloomington, IN 47405.

tissues and morphogenesis and differentiation of embryonic epithelia in imaginal discs into adult forms. The complex cellular responses to 20-HOE probably reflect modulation of expression of specific gene sets. Some gene products might secondarily alter expression of other genes while other gene products might directly cause morphogenesis and/or differentiation of adult tissues.

The hormone-dependent morphogenesis of discs (evagination) proceeds mainly by short distance rearrangements of cells in the epithelium (1). Changes in cell-cell relationships may require changes in cell surface properties that result from differences in the composition or topology of cell surface and/or extracellular matrix components(2). Indeed, recent evidence from vertebrates (3, 4) underscores the importance of cell adhesion molecules (CAMs) and substrate adhesion molecules (SAMs) in cell-cell interactions during development.

While genetic approaches in *Drosophila* may identify genes involved in disc morphogenesis (5) our work offers a molecular approach specifically designed to investigate genes that influence morphogenesis by alteration of the cell surface. This approach is possible because of mass isolation and *in vitro* culture techniques (6) that make imaginal discs biochemically accessible for the study of epithelial morphogenesis distinct from later differentiation (7). Our analysis identifies some putative regulatory (2B5) and structural (IMP-E1-3, IMP-L1-3) loci that probably comprise part of a gene set devoted to "prepupal morphogenesis" in imaginal discs. Moreover, the spatial expression of these genes suggests morphogenetic functions for the genes in discs during metamorphosis.

METHODS

Isolation of DNA and RNA (8), construction of subclones, radioactive labeling of nucleic acids with ^{32}P and filter hybridization studies were done as described (9). Hybridization signals on autoradiograms were quantitated with a Zeineh laser densitometer and integrator. Synthesis of single-stranded, single polarity ^{3}H-RNA probes for tissue hybridization studies utilized DNA subclones constructed in the cloning vector pIBI-76 (International Biotechnology, Inc.) using either SP6 or T7 RNA polymerase according to the manufacturer's specifications. Frozen tissue sections were processed as described (10) except that sections were treated with protease K rather than with pronase.

RESULTS

Our interest in gene products that mediate cell rearrangement during imaginal disc prepupal morphogenesis led to isolation of ecdysone regulated genes that encode membrane or secreted proteins. In _in vitro_ cultures we observed a set of specific new, 20-HOE dependent mRNA transcripts associated with membrane-bound polysomes, the site of synthesis of secreted and many membrane proteins. We used this induced RNA fraction in a differential hybridization screen of a _Drosophila_ recombinant genomic lambda library to isolate genes that encode membrane or secreted proteins (9). The differential screen yielded six gene sequences (Table 1, IMP E1-3, IMP L1-3) that encode _i_nducible _m_embrane-bound _p_olysomal transcripts. These six non-homologous genes have unique cytogenetic locations on the polytene chromosome map (Table 1).

TABLE 1
PROPERTIES OF INDUCIBLE GENES

Gene	Location	Expression in: Imaginal Discs	Salivary Glands	Regulatory Class	Sensitive to Cycloheximide
2B5	2B5	+	+	Early	No
IMP-E1	66C7-10	+	-	Early	No
IMP-E2	63E	+	-	Early	No
IMP-E3	84E	+	n.d.	Early	n.d.
IMP-L1	70A	+	n.d.	Late	Yes
IMP-L2	64B	+	-	Late	No
IMP-L3	65B	+	n.d.	Late	n.d.

n.d.:Not determined.

A 20-HOE-dependent gene product used for morphogenesis must be produced at the time and under the experimental conditions that cause evagination. The genes we have isolated (Fig. 1) are active in imaginal discs _in vitro_ within the first 10 hours following hormone exposure, a period when dramatic morphogenesis occurs. The IMP genes were isolated using mRNA from discs cultured continuously with 1 ug/ml 20-HOE, conditions that block synthesis of the pupal cuticle and its components (7). In those cases examined (Table 1), expression of IMP sequences was not found in other major hormone target tissues such as salivary glands. This provides additional evidence that these genes are utilized for morphogenetic processes in imaginal discs.

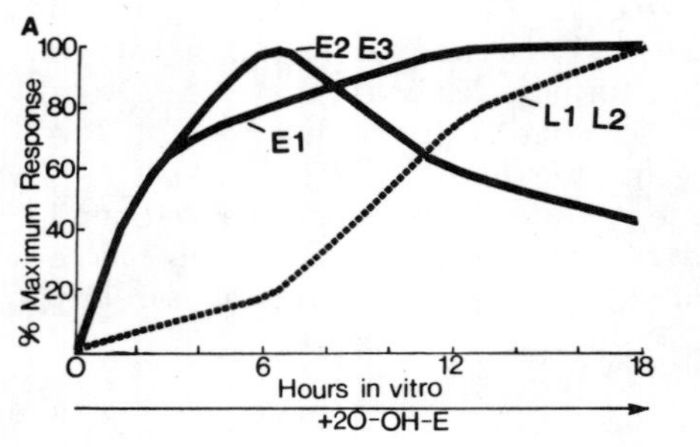

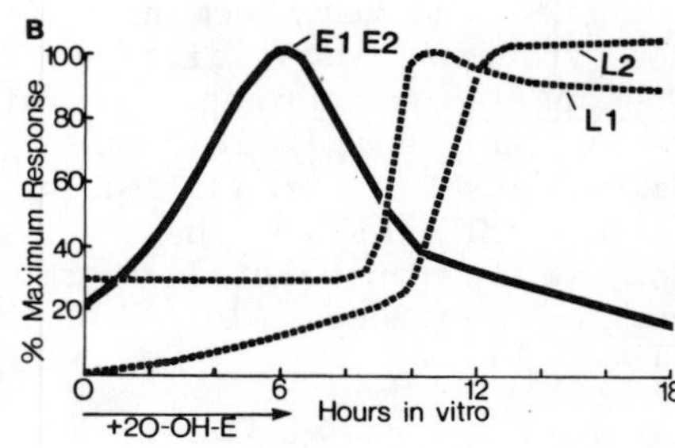

Figure 1. Induction of 20-HOE responsive genes in imaginal discs _in vitro_. Transcript levels quantitated by densitometry of autoradiograms are expressed as per cent of maximal induction versus time in culture. A, 20-HOE at 1 ug/ml present continuously in the culture. B, 20-HOE at 1 ug/ml present only for the first six hours of culture.

An additional type of hormone-responsive gene may be exemplified by the 2B5 locus on the X chromosome. This locus controls the developmental processes that occur at metamorphosis, including imaginal disc morphogenesis and histolysis of larval tissues (5,11). It is unlikely that the locus encodes a single product that directly mediates both morphogenesis and histolysis. Rather, genetic and molecular data (12,13), indicate that 2B5 encodes regulatory product/s/ that control other ecdysone-responsive genes. Genomic DNA within and near the genetically defined 2B5 locus has been cloned by Chao and Guild (14) and by Galceran et al (15) and expression in some tissues and developmental periods has been analyzed. In imaginal discs, we find that 2B5 hormone-responsive sequences are expressed within one hour of hormone addition (Table 1). Ultimately, a link between regulatory sequences, like 2B5, and possible regulated sequences like the IMP's may be found in imaginal discs.

Based on analyses of specific transcript levels in discs cultured in vitro with 20-HOE, the genes can be grouped into two basic regulatory classes (Table 1 and Fig. 1). IMP-E1, E2, E3, and 2B5 fall into the "early" class. Transcripts of these genes accumulate to maximum levels within 4 hours of addition of 20-HOE to our in vitro disc cultures (Fig. 1A). Removal of the hormone from the medium causes a decline in the induced level of IMP-E1, E2, and E3 transcripts in the discs (Fig. 1B). Accumulation of IMP-E1, IMP-E2, and 2B5 transcripts can be induced in the absence of protein synthesis (Table 1). These data, taken together, imply that these "early" genes are primary response loci that interact directly with the steroid hormone receptor complex.

The "late" IMP genes (IMP-L1, L2, and L3) do not reach their maximal induction until 8-10 hours after hormone exposure. In contrast to the "early" genes, accumulation of the "late" gene transcripts continues in hormone-pulse cultures after hormone is removed (Fig. 1B). A pulse of hormone is, however, not required for induction of the transcripts (Fig. 1A). The simplest interpretation of these results is that these late genes are secondary response loci; ie. their induction indirectly depends on 20-HOE via a hormone-dependent secondary inducer. This is probably the case for IMP-L1 since the transcript is not induced if protein synthesis is inhibited (Table 1). IMP-L2 constitutes a more complex case. Even though the highest induced level of expression occurs 10-12 hours after hormone exposure, consistent with a "late" pattern (Fig. 1), transcripts from this gene begin to accumulate at low levels very soon after addition of 20-HOE. Therefore, IMP-L2 may not exemplify a typical secondary hormone response. The transcript accumulation time course indicates, however, that the gene product may be used in a later phase of imaginal disc prepupal morphogenesis.

The sizes of the transcripts expressed in imaginal discs are summarized in Table 2. With the exception of 2B5 transcripts, a single size class of inducible transcript homologous to each gene is found. The multiple transcripts from 2B5 detected in imaginal disc mRNA (10 kb, 8.5 kb, 6.7 kb) as well as the transcript complexity reported for other tissues and developmental stages (14, 15) may reflect functional complexity of the locus.

TABLE 2
DEVELOPMENTAL EXPRESSION

Gene	Transcript size: Imaginal Disc	Whole Animal	Developmental Time 0-12h embryo	12-24h embryo	1st	2nd	3rd	Prepupa	Pupa
2B5	10kb	n.d.			n.d.				
	8.7								
	6.7								
IMP-E1	8.5	8.5kb	−	+	+	+	+	+	−?
		3.8	−	−	+	+	+	+	+
		2.95	−	+	+	+	+	+	+
		2.0	−	+	+	+	+	−	−
		1.5	+	+	+	+	+	−	−
		0.8	−	−	+	+	+	−	−
IMP-E2	1.8	1.8	+	+	−	−	+	+	+
IMP-E3	1.4	1.4	−	−	−	−	+	+	+
IMP-L1	1.6	n.d.			n.d.				
IMP-L2	3.0	3.0	−	−	−	−	+	+	+

n.d.:not determined.
kb:kilobases.

Imaginal disc morphogenesis occurs at pupariation in response to the increase in 20-HOE titer. Other peaks of 20-HOE titer occur during Drosophila development (16) and the IMP genes could be expressed at times other than metamorphosis. Poly-A+ mRNA isolated from whole animals at sequential stages of development extending from embryogenesis through pupation has been isolated and analyzed on Northern blots using probes representing coding regions of some of the IMP genes. The transcript sizes and temporal distributions are summarized in Table 2. Some of the genes (IMP-E3 and IMP-L2) appear to be expressed mainly during late time periods surrounding pupariation when imaginal disc morphogenesis occurs. The developmental pattern for IMP-E2 expression is more complex. A similar transcript size class accumulates both during embryonic development and at pupariation. The IMP-E1 transcription pattern is even more complex. The 8.5 kb transcript size class observed in disc cultures is also present at other developmental stages. In addition, a number of other transcript size classes are present in RNA samples from whole animals. This complex IMP-E1 transcription pattern is also observed with a probe made from a relatively short (1.1 kb) imaginal disc cDNA clone representing the 3' end of the gene. We hypothesize that the multiple transcript forms result from differential processing or initiation within the same gene sequence rather than by detection of transcripts from adjacent or interspersed exons of different genes. Since the RNA used in these experiments, in most cases, was isolated from developing animals with a relatively wide range of ages, we cannot precisely correlate expression with specific developmental cues, other than the hormone-dependent expression in imaginal discs at pupariation.

Some of the complexity of the IMP-E1 expression pattern may stem from differences in gene expression in specific tissues. We have used single-stranded ^{3}H-RNA hybridization probes to investigate the spatial distribution of the IMP-E1 transcript in frozen sections of pupariating animals. The probe hybridizes strongly to RNA in all types of discs with the exception of the ommatidial region of the eye disc. This is of interest because the eye disc does not evaginate during metamorphosis. Hybridization is also seen in a cell layer at the periphery of the brain and within the brain at the interface of the cellular cortex and the neuropil. We do not yet understand the significance of the brain expression of IMP-E1, but since the brain undergoes extensive remodeling at metamorphosis, detection of a signal here is not inconsistent with a morphogenetic role for the gene product.

DISCUSSION

The imaginal disc genes that we have isolated are active under conditions selective for a defined morphogenetic process and are expressed, in the cases studied up to this point, in tissues undergoing morphogenesis. Thus, these genes may provide us with insight into both the mechanisms by which particular form and structure are achieved (morphogenesis) as well as the means for precise control of gene activity that contributes to developmental events.

Parallels have been drawn between the sequence of events in imaginal disc development during metamorphosis and the precise progression of salivary gland puffing patterns that occurs over the same time interval in response to the same changes in 20-HOE titer (17, 18). With the isolation of ecdysone-responsive gene sequences expressed in imaginal discs we now have tools formally similar to salivary gland puffs with which to analyze the hormone response in an imaginal, in contrast to a larval, tissue. Indeed, as suggested (17), we have found evidence for multiple, sequentially expressed classes of responsive genes in discs. Some genes belong to an early, apparently primary response class. Within this group are transcriptionally and genetically complex loci like 2B5 that may regulate specific aspects of the subsequent hormone response and that are also active in larval tissues. But we have also identified a class of early genes that probably encode structural rather than regulatory gene products, based on their presumed location either within cellular membrane systems or external to the cell. The imaginal disc hormone response also includes a set of late sequences that are apparently activated by a secondary inducer. In addition, there is a set of pupal cuticle genes expressed only in response to a pulse of 20-HOE (19) whose expression parallels the "prepupal" puffs described by Richards (20). While the overall <u>patterns</u> of regulation are similar in salivary glands and imaginal discs, the precise <u>sets</u> of genes active in each tissue differ. Thus, tissue-specific mechanisms exist for regulating the hormone response. Further analysis of these hormone-responsive genes may help us to understand at a molecular level how a single hormonal stimulus can elicit different responses from two target tissues.

Specific gene activation could contribute to a complex cellular behavior like epithelial morphogenesis in a number of ways. We have focused upon events presumably occuring at the cell surface based on both morphological (1) and biochemical (2) evidence indicating that the cell rearrangement that occurs during evagination is accompanied by cell surface-associated protein changes. Products of genes activated at this time could modify surface interactions directly by providing new cell surface, cell junction, or extracellular matrix proteins, by enzymatically modifying pre-existing surface molecules (eg. proteolytically) or by modifying processing pathways of cell surface or secreted molecules. Alternatively, new gene products might alter cytoskeleton-membrane association, resulting in effects on cell motility or organization of membrane protein domains. In embryonic vertebrate systems it is becoming increasingly clear that changing patterns of cell surface molecules (CAMs and SAMs) correlate closely with changes in cell behavior and tissue organization and may play a causal role in establishing developing biological patterns (3, 4). We propose that the genes we have isolated from imaginal discs encode protein products that play key roles in the remodeling of the insect epithelium during metamorphosis.

ACKNOWLEDGMENTS

We thank members of our laboratory for assistance and G. Guild for 2B5 clones.

REFERENCES

1. Fristrom D (1976). The mechanism of evagination of imaginal discs of Drosophila melanogaster. III. Evidence for cell rearrangement. Dev Biol 54:163.
2. Rickoll WL, Fristrom JW (1983). The effects of 20-hydroxyecdysone on the synthesis of membrane proteins in Drosophila imaginal discs. Dev Biol 95:275.
3. Edelman GM (1984). Cell adhesion and morphogenesis: The regulator hypothesis. Proc Natl Acad Sci USA 81: 1460.
4. Grumet M, Hoffman S, Crossin KL, Edelman GM (1985). Cytotactin, an extracellular matrix protein of neural and non-neural tissues that mediates glia-neuron interaction. Proc Natl Acad Sci USA 81:8075.
5. Kiss I, Beaton A, Tardiff J, Fristrom JW (1986). Manuscript in preparation.
6. Eugene O, Yund MA, Fristrom JW (1979). Preparative isolation and short term culture of imaginal discs of Drosophila melanogaster. Tissue Culture Assoc Man 5:1055.
7. Fristrom JW, Doctor J, Fristrom DK, Logan WR, Silvert DJ (1982). The formation of the pupal cuticle by Drosophila imaginal discs in vitro. Devel Biol 91:337.
8. Natzle JE, McCarthy BJ (1984). Regulation of Drosophila alpha- and beta-tubulin genes during development. Dev Biol 104:187.
9. Natzle JE, Hammonds AS, Fristrom JW (1986). Isolation of

genes active during hormone-induced morphogenesis in Drosophila imaginal discs. In press, J Biol Chem.

10. Hafen E, Levine M, Garber RL, Gehring WJ (1983). An improved method for the detection of cellular RNAs in Drosophila tissue sections and its application for localizing transcripts of the homeotic Antennapedia gene complex. EMBO J 2:617.
11. Fristrom DK, Fekete E, Fristrom JW (1981). Imaginal disc development in a non-pupariating lethal mutant in Drosophila melanogaster. W Roux's Archiv 190:11.
12. Belyaeva ES, Vlassova IE, Biyasheva ZM, Kakpakov VT, Richards G, Zhimulev IF (1981). Cytogenetic analysis of the 2B3-4 - 2B11 region of the X-chromosome of Drosophila melanogaster.II.Changes in the 20-OH ecdysone puffing caused by genetic defects of puff 2B5. Chromosoma 84:207.
13. Crowley TE, Mathers PH, Meyerowitz EM (1984). A trans-acting regulatory product necessary for expression of the Drosophila melanogaster 68C glue gene cluster. Cell 39:149.
14. Chao A, Guild G (1986). In press, EMBO J.
15. Galceran J, Gimenez C, Edstrom JE, Izquierdo M (1986). Micro-cloning and characterization of the early ecdysone puff region 2B of the Drosophila melanogaster chromosome. Insect Biochem 16:249.
16. Richards G (1981). The radioimmune assay of ecdysteroid titres in Drosophila melanogaster. Molec Cell Endocr 21:181.
17. Fristrom JW, Natzle J, Doctor J, Fristrom D (1985). The regulation of a developmental sequence during imaginal disc metamorphosis. In Balls M, Bownes M (eds):"Metamorphosis," Oxford:Clarendon Press, p.162.
18. Ashburner M, Chihara C, Meltzer P, Richards G (1974). Temporal control of puffing activity in polytene chromosomes. Cold Spring Harbor Symp Quant Biol 38:655.
19. Fechtel K, Natzle J, Fristrom JW (1986). Unpublished.
20. Richards G (1981). Sequential gene activation by ecdysone in polytene chromosomes of Drosophila melanogaster. IV. The mid-prepupal period. Dev Biol 54:256.

Molecular Entomology, pages 165–177

THE MECHANISM OF STEROID REGULATION OF PEPTIDE ACTION ON THE INSECT NERVOUS SYSTEM[1]

James W. Truman and David B. Morton

Department of Zoology, University of Washington
Seattle, Washington 98195

ABSTRACT The insect central nervous system (CNS) has alternating periods of responsiveness and non-responsiveness to eclosion hormone (EH). The responsive periods are associated with molts and are caused by the appearance and then withdrawal of ecdysteroids. After steroid priming, the CNS responds to EH by elevating its levels of cGMP, the second messenger in this system. Initially, though, the insects do not show a behavioral response even though the second messenger system is functional. Behavioral responsiveness occurs about 16 hr later with the appearance of two 54kD proteins which are phosphorylated in response to EH and are thought to mediate EH action. Thus, the final step that regulates responsiveness appears to be at the level of phosphoprotein substrates rather than at the level of receptors.

INTRODUCTION

Research over the last decade has shown that peptides play a major role in regulating CNS function in both vertebrates (1) and invertebrates (2). In many cases they act as modulators, by altering how neurons respond to classical transmitters. However, the action of peptides may itself be modulated by other factors. In the present paper we explore a possible mechanism for one such

[1]Unpublished research funded by grants from the McKnight Foundation and NIH (NS 13079).

interaction which involves the steroid regulation of the action of the neuropeptide, eclosion hormone (EH), on the nervous system of the moth *Manduca sexta*.

THE ECLOSION HORMONE SYSTEM OF INSECTS

One of the most striking behaviors shown by insects is ecdysis behavior, which involves a complex sequence of motor patterns used by the insect at the end of a molt to escape from the cuticle of the preceding stage (3). In Lepidoptera, and likely in other insect groups (4), this behavior is triggered by EH, which acts directly on the CNS to evoke the stereotyped ecdysis motor programs (5,6). Once begun the ecdysis behavior dominates all other behavior patterns. At its completion, an array of new behaviors appropriate for the new stage are then permanently "turned-on". This activation of stage-specific behaviors is also caused by EH (7).

Exposure to EH alone is not sufficient to cause an insect to undergo ecdysis and the associated behavioral changes. For example, in *Manduca sexta* the nervous system is responsive to this peptide only for brief periods associated with each molt (Fig. 1; 8). This relationship of EH sensitivity to the end of a molt is scarcely surprising since ecdysis behavior only makes sense when an insect is confined within a cuticle that must be shed. This relationship is not causal, however, since ecdysis behavior occurs in response to EH even when the old cuticle is previously removed (9). Furthermore, the deafferented or totally isolated CNS from appropriately staged insects responds to EH exposure by generating a full ecdysis motor pattern (5,6). Thus, the changes that allow the CNS to become responsive to EH appear to occur centrally, arising in parallel with, but independent of, the peripheral changes which make the behavior appropriate.

The association of EH sensitivity with the period of the molt suggested that ecdysteroids might be involved in rendering the CNS responsive to EH. Slama (10) first showed in *Tenebrio* that exogeneous 20-hydroxyecdysone (20-HE) given during the end of the molt delayed pupal ecdysis. A similar delaying effect of 20-HE treatment was found for *Manduca* larval (11), pupal, and adult (12) molts. During each molt in *Manduca*, the ecdysteroid titer rapidly rises to initiate the processes required for production of a new cuticle and then gradually declines so that ecdysis coincides with a low point in the ecdysteroid titer (13). If the normal ecdysteroid decline is interrupted by an injection of 20-HE, ecdysis is then delayed by a span of time that is proportional to the dosage given (12). Continuous infusion of 20-HE into pharate adult Manduca to sustain an elevated ecdysteroid titer blocked ecdysis for the duration of the infusion (6 days). Thus, the normal withdrawal of ecdysteroids at the end of a molt appears essential for ecdysis to occur.

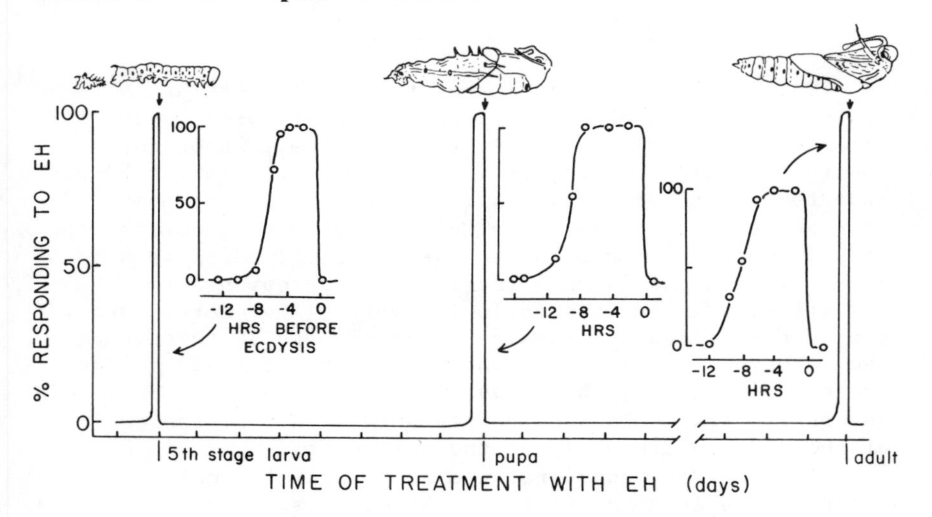

FIGURE 1. The ability of EH to elicit the ecdysis motor pattern when given at various times during the life history of Manduca. Insets provide an expanded time scale for periods prior to each ecdysis (arrows).

The declining ecdysteroid titer appears to influence the ecdysis system in two ways. Firstly, it is a prerequisite for the release of EH by the respective neurosecretory neurons (12,14). Secondly, the steroid withdrawal is necessary for the CNS to become responsive to EH. This latter effect can be seen in the case of pupal ecdysis by injecting prepupae with 20-HE at various times before the predicted ecdysis and then challenging these insects with EH at the time when they would normally be responsive (about 3-4 hr before ecdysis). At 22°C, 20-HE treatments blocked the subsequent development of EH-responsiveness when given up to 14 hr before pupal ecdysis. When injections were given after that time, the insects developed a normal sensitivity to EH, irrespective of the amount of steroid injected (Morton & Truman, unpublished). At this temperature, the ability to respond to EH appears at 7.5 hr before ecdysis in control animals. Based on these observations, we have concluded that the steroid withdrawal sets events in motion at about 14 hr before ecdysis which are steroid-independent and which result in the appearance of EH responsiveness 6.5 hr later.

LEVEL OF ACTION IN THE CNS

The complex organization of the CNS hinders an analysis of the mechanisms by which ecdysteroids could alter CNS responsiveness. Steroids are well known for their organizational action within both vertebrate (15) and invertebrate nervous systems (16). Consequently, one hypothesis is that the neural circuitry which generates the ecdysis motor patterns is at least partially disassembled after ecdysis and is only reassembled in response to ecdysteroids during the following molt. Therefore, even if the EH target neurons remained sensitive to EH throughout the intermolt period, the functional circuit would not be present and, hence, the behavior could not occur. Hormonal responsiveness could only be revealed with the reassembly of the ecdysis circuitry at the next molt.

A number of behavioral and neurophysiological observations argue against this hypothesis. A comparison of larval and pupal ecdysis motor patterns in *Manduca* showed that the insect used the same central motor pattern at both stages, irrespective of the fact that the degeneration of certain muscles in the prepupa rendered the patterned bursting by their motoneurons useless (17). If the ecdysis circuitry were reconstructed at each molt, one might have expected that the useless parts of the larval program would have been discarded when rebuilding the pupal ecdysis pattern generator. A more compelling argument can be made for crickets in which intermolt animals can be induced to show the peristaltic movements characteristic of ecdysis in response to specific stimuli to the legs and abdomen (18). Likewise, removal of the pupal cuticle from developing adult *Antheraea pernyi* moths elicits spontaneous ecdysis movements days before the insect becomes sensitive to EH (7). Thus, the major elements of the ecdysis motor pattern are functional well before the appearance of EH responsiveness. Although the circuitry is apparently functional at these early times, the peptide simply does not seem able to activate it.

An alternative hypothesis is that the EH-sensitive elements are distinct from the pattern-generating circuitry and that these connections must be established before EH can trigger the behavior. Another possibility is that the anatomical connections may be there all of the time and the ecdysteroids act to alter some aspect of the physiology of the target cells which renders them EH-responsive. The experiments in this paper are based on this last hypothesis.

THE MODE OF ACTION OF EH

The changes induced by EH in their target neurons which result in the activation of stereotyped behavior patterns are unknown. They may involve changes in membrane conductances through the alteration of ion channels and/or metabolic changes related to transmitter synthesis or reception. Although the nature of these final steps have not been elucidated, we nevertheless have some insight into the cellular mechanisms that mediate these changes.

In the three systems that have been examined in detail, the adult CNS of _Hyalophora cecropia_ (19), the intersegmental muscles of _Antheraea polyphemus_ (20), and the prepupal CNS of _Manduca_ (21), the action of EH seems to be mediated through increases in cyclic GMP (cGMP). In all systems cGMP or analogues such as 8-bromo-cGMP effectively mimic EH in triggering the appropriate response whereas cAMP is a very poor mimic. The effects of EH are enhanced by pretreatment of the insects with drugs which block cyclic nucleotide degradation. In the case of pupal ecdysis, an increase in cGMP (but not cAMP) occurs in the CNS of _untreated_ animals just prior to the onset of the behavior, at a time coinciding with endogenous EH release. This cGMP increase can be induced prematurely in this and the other two systems by injection of EH. By contrast, cAMP levels are not affected in any of the systems by EH treatment. Importantly, the threshold levels of EH required to induce a cGMP increase are essentially the same as those required to evoke the appropriate physiological or behavioral responses.

Interpretation of changes in levels of cGMP in the CNS is complicated by the fact that this is a heterogenous tissue that undoubtedly contains both EH target cells and non-target cells. Consequently, the observed cGMP increases could be a direct response to EH acting on its target neurons or it could be a secondary response in non-target neurons that are driven synaptically by the target cells. While we cannot exclude the latter hypothesis at present, two pieces of evidence make it unlikely. Firstly, the time course of the cGMP increase suggests a direct response: cyclic nucleotide levels are already significantly elevated within 5 min after EH injection, well in advance of any change in motor output from the CNS (19,21). Secondly, studies on the intersegmental muscles of *A. polyphemus* show that target cells do indeed respond directly to EH with an increase in cGMP (20). These muscles represent a homogenous set of target cells that respond to EH by degeneration. Innervated versus denervated muscles showed comparable increases in cGMP (12-20 x basal) and both degenerate after challenge with EH. Thus, in the case of muscle target cells, the cGMP response is a direct response to EH and not mediated through presynaptic influences. By homology to the case in muscle, we have assumed that the cGMP increase in the CNS is a direct response of neuron targets to EH and that cGMP is the second messenger that mediates the neural response.

Cyclic GMP has provided an avenue by which to examine how ecdysteroids promote EH responsiveness. An intermolt insect, such as a feeding 5th instar caterpillar, when injected with EH shows no behavioral response. When feeding 5th stage larvae were injected with EH and the cGMP levels in their CNS subsequently analyzed, it was found that EH caused no increase in their levels of cGMP. Similarly, wandering stage larvae and prepupae until about 36 hr before pupal ecdysis were also refractory to challenge by EH. By 24 hr before ecdysis, however, the insects responded to EH treatment by showing a cGMP increase in their CNS and they maintained this competence until the normal time of ecdysis (21). After ecdysis, the CNS was again refractory to EH, both in terms of the cGMP increase and the behavior.

This ability of the CNS to elevate cGMP levels in response to EH is controlled by the ecdysteroid titer (21). The transformation from the larval to the pupal stage in *Manduca* is caused by 2 sequential releases of ecdysteroid. An initial, small pulse (the commitment peak) causes the transition to a wandering stage larva. Two days later, a larger surge of ecdysteroid (the prepupal peak) causes the formation of the pupal stage (22). Since the primary source of ecdysteroid is the prothoracic glands in the thorax, isolation of the abdomen before either of the ecdysteroid peaks can deny the abdomen its normal exposure to ecdysteroid. The initial commitment peak of ecdysteroid seems to have no direct role in the acquisition of EH responsiveness; abdomens isolated after the commitment peak but prior to the prepupal peak do not subsequently acquire the competence to elevate their cGMP levels in response to EH challenge (21). By contrast, when such abdomens were infused with levels of 20-HE to mimic the prepupal ecdysteroid peak, the abdominal CNS acquired EH responsiveness by about 48 hr later (21). Thus, the prepupal ecdysteroid peak appears essential for the acquisition of competence by the cGMP system.

One hypothesis for how ecdysteroids might regulate EH responsiveness is through the regulation of EH receptors or the ability of these receptors to affect the synthesis or degradation of cGMP. Since EH can cause a full-scale cGMP increase by 24 hr before ecdysis, we suspect that the receptor-guanylate cyclase system must already be functional at this time (21). It is important to note, however, that although EH can stimulate a cGMP increase by 24 hr before ecdysis, it cannot trigger a behavioral response until about 16 hr later. A similar relationship is also seen in the EH-induced degeneration of the intersegmental muscles (20). These observations suggest that the final changes that confer EH responsiveness reside at a level distal to the receptor-second messenger system.

A similar conclusion is also reached in experiments in which we attempted to by-pass any possible receptor blockade by treatments with cGMP or 8-bromo-cGMP. These compounds were effective in inducing pupal ecdysis behavior only when given at times when the peptide was also effective (21). The same relationship between EH and cGMP analogues was seen for the effects of these substances on the degeneration of moth muscle (20).

Based on the above observations, we have concluded that prior to the prepupal ecdysteroid peak the ability of CNS target cells to respond to EH is blocked at least at 2 levels in the biochemical cascade induced by EH. The first block is possibly at the level of the EH receptor while the second resides distal to the second messenger, perhaps at the level of a cGMP-dependent protein kinase or the proteins phosphorylated by this enzyme. The putative receptor block appears to be repaired relatively early, but it is the unblocking of the latter step that finally results in a fully responsive system.

ROLE OF PHOSPHOPROTEINS IN REGULATING RESPONSIVENESS

In most, if not all, eukaryotic systems, cyclic nucleotides exert their intracellular effects through phosphorylation of endogenous proteins (23). Consequently, we looked for changes in the levels of protein kinase and endogenous substrates found in the ventral CNS of *Manduca* during the acquisition of EH responsiveness. Nervous systems were removed at 24 hr before ecdysis (at a stage when the insects respond to EH with an increase in cGMP but no behavior) and at 4 hr before ecdysis, when both the cGMP and behavioral responses can be induced. The cGMP-dependent protein kinase and the cAMP-dependent protein kinase were partially purified from the 2 stages and their activities measured. No changes were seen in the activity levels of either protein kinase through this period (24).

We then examined the proteins phosphorylated in response to cyclic nucleotides. Extracts of the CNS from -4 hr animals were incubated with ^{32}P-ATP in the presence or absence of cGMP. The labeled proteins were then analysed by 2-dimensional SDS polyacrlamide gel electrophoresis followed by fluorography. Two of the most prominent proteins that were phosphorylated in the presence of cGMP were seen as two elongated spots at 54,000 daltons (the 54kD proteins) (Fig. 2).

Evidence that these proteins are phosphorylated *in vivo* in response to EH was obtained using a "post-hoc" phosphorylation approach. We injected replicate groups of -4 hr prepupae with EH or saline, removed their CNS 90 min later (when the EH-treated insects began ecdysis), prepared cell-free extracts from both groups, and subjected them to *in vitro* phosphorylation in the presence of cGMP and labeled ATP. The only proteins that did not incorporate ^{32}P in the experimental group but that did so in the controls were the 54kD proteins (24). We interpret the lack of label in the EH-treated group as indicating that these proteins had accepted endogenous, unlabeled phosphate *in vivo* in response to EH stimulation and, thus, were unavailable for labeling with ^{32}P phosphate during the subsequent *in vitro* reactions. The 54kD proteins were the only CNS proteins which consistently showed this response.

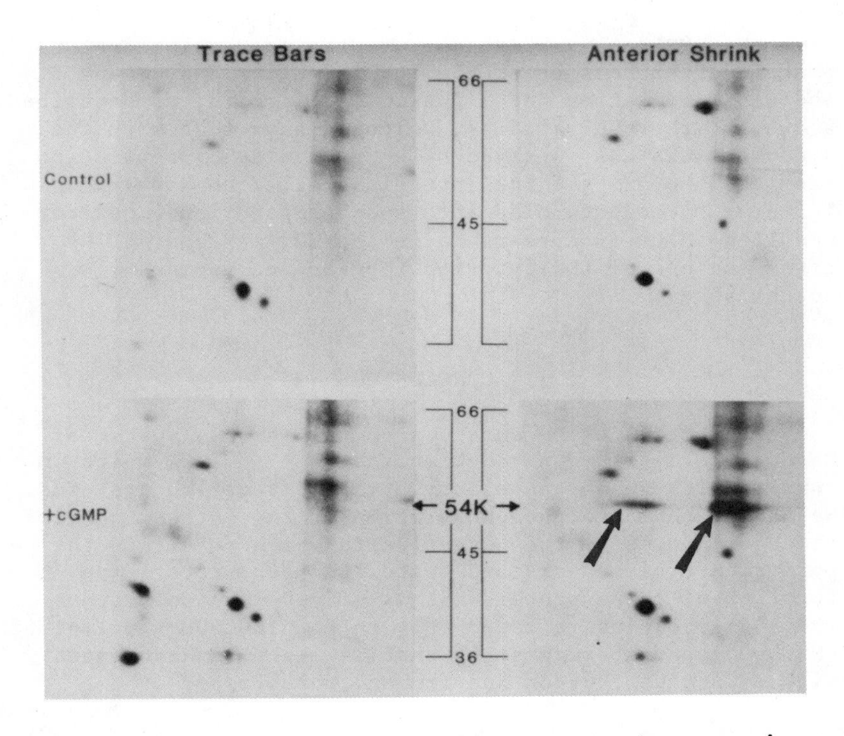

FIGURE 2. Sections of fluorograms showing proteins phosphorylated in cell-free CNS extracts and subjected to 2-D gel electrophoresis. The +cGMP samples were incubated with 10^{-4}M cGMP. Left: the trace bars stage was 24 hr before pupal ecdysis; right: the anterior shrink stage at -4 hr. Large arrows show the 54 kD proteins. The acidic end of each gel is to the left.

At present we can monitor the presence of the 54kD proteins only by their ability to accept labeled phosphate when CNS extracts are subjected to in vitro phosphorylation. Based on this criteria, the 54kD proteins are not found in the CNS of intermolt larvae, nor are they found in the nervous systems of insects at 24 hr before ecdysis when the animal can respond to EH with a cGMP increase but not a behavioral response (Fig. 2). Importantly, they are first detected in the CNS at 8 hr before ecdysis, the same time at which the insects become behaviorally responsive to EH (Morton and Truman, unpublished).

The CNS requires exposure to the prepupal ecdysteroid peak in order to express the 54kD proteins. By means of abdominal isolation and infusion experiments, as described above for the cGMP studies, we could show that when the abdominal CNS was deprived of ecdysteroids by separation from the thorax, the 54kD proteins failed to appear. Infusion of ecdysteroids into such preparations, however, resulted in their appearance in the CNS at about 60 hr after the end of the infusion (Morton and Truman, unpublished).

CONCLUSIONS

The action of EH on the prepupal nervous system of Manduca is mediated through an increase in cGMP which in turn results in the phosphorylation of two 54kD proteins. We assume that the phosphorylation of the latter proteins effects changes in the target cells which result in the performance of ecdysis and associated behaviors. The function of these proteins is as of yet unknown but one of the 54dD proteins is found only in the 100,000 g pellet of CNS homogenates, suggesting that it might be associated with membranes (Morton and Truman, unpublished).

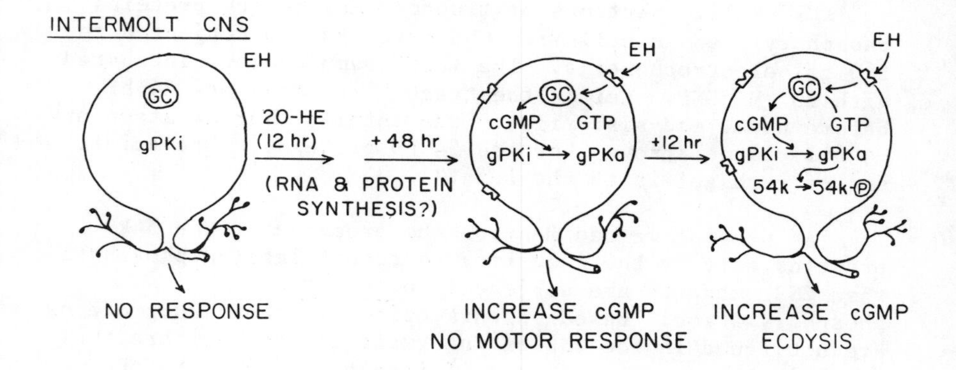

FIGURE 3. A model for the mechanism by which ecdysteroids render target neurons responsive to EH. Notched rectangles: EH receptor; GC: guanylate cyclase; gPK: cGMP-dependent protein kinase either active (a) or inactive (i). See text for details.

Ecydsteroids may act on this system at two levels (Fig. 3). In the intermolt larva, we suggest that both the EH receptors and the 54 kD proteins are absent so that the CNS can show neither a cGMP response nor the appropriate behavior when challenged with EH. EH responsiveness is acquired by the CNS in response to ecdysteroid exposure and then the withdrawal of the steroid. An early result of the ecdysteroid withdrawal is likely the appearance of functional EH receptors. Consequently the nervous system can respond to EH with a cGMP increase but not with the behavior since the 54 kD proteins are not yet present. By 16 to 20 hr later, the appearance of the 54 kD proteins completes the system and the insects show behavioral responsiveness.

Some of the major questions posed by this system have to do with the function of the 54 kD proteins and the manner by which ecdysteroids regulate their expression. Nevertheless, the above results suggest that mediational systems in cells may be regulated at more levels that previously thought, with control involving not only the initial steps in a cascade, the hormone receptor and second messenger systems, but also the presumptive last step, the proteins whose phosphorylation result in the cellular response. This represents a novel mechanism by which steroids might regulate the response of cells to a peptide hormone.

ACKNOWLEDGEMENTS

We thank Prof. L.M. Riddiford for critical comments on the manuscript.

REFERENCES

1. Iversen LL (1983). Neuropeptides - what next? Trends NeuroSci 6:293.
2. O'Shea M, Schaffer M (1985). Neuropeptide function: the invertebrate contribution. Ann Rev Neurosci 8:171.
3. Reynolds SE (1980). Integration of behaviour and physiology in ecdysis. Adv Insect Physiol 15:475.
4. Truman JW, Taghert PH, Copenhaver PF, Tublitz NJ, Schwartz LM (1981). Eclosion hormone may control all ecdyses in insects. Nature 291:70.

5. Truman JW (1978). Hormonal release of stereotyped motor programmes from the isolated nervous system of the Cecropia silkmoth. J Exp Biol 74:151.
6. Weeks JC, Truman JW (1984). Neural organization of the peptide-activated ecdysis behaviors during metamorphosis in Manduca sexta. I. Conservation of the peristalsis motor pattern at the larval-pupal transformation. J Comp Physiol A 155:407.
7. Truman JW (1976). Development and hormonal release of adult behavior patterns in silkmoths. J Comp Physiol A 107:39.
8. Truman JW, Taghert PH, Reynolds SE (1980). Physiology of pupal ecdysis in the tobacco hornworm, Manduca sexta. I. Evidence for control by eclosion hormone. J Exp Biol 88:327.
9. Truman JW (1971). Physiology of insect ecdysis. I. The eclosion behaviour of saturniid moths and its hormonal release. J Exp Biol 54:805.
10. Slama K (1980). Homeostatic function of ecdysteroids in ecdysis and oviposition. Acta Ent Bohemoslovaca 77:145.
11. Curtis AT, Hori M, Green JM, Wolfgang WJ, Hiruma K, Riddiford LM (1984). Ecdysteroid regulation of the onset of cuticular melanization in allatectomized and black mutant Manduca sexta larvae. J Insect Physiol 30:597.
12. Truman JW, Rountree DB, Reiss SE, Schwartz LM (1983). Ecdysteroids regulate the release and action of eclosion hormone in the tobacco hornworm, Manduca sexta (L). J Insect Physiol 29:895.
13. Bollenbacher WE, Smith SL, Goodman W, Gilbert LI (1981). Ecdysteroid titer during larval-pupal-adult development of the tobacco hornworm, Manduca sexta. Gen Comp Endocrin 44:302.
14. Copenhaver PF, Truman JW (1986). Identification of the cerebral neurosecretory cells that produce eclosion hormone in the moth Manduca sexta. J Neurosci, in press.
15. Arnold AP, Gorski RA (1984). Gonadal steroid induction of structural sex differences in the CNS. Ann Rev Neurosci 7:413.
16. Weeks JC, Truman JW (1985). Independent steroid control of the fates of motoneurons and their muscles during insect metamorphosis. J Neurosci 5:2290.
17. Weeks JC, Truman JW (1984). Neural organization of the peptide-activated ecdysis behaviors during

metamorphosis in *Manduca sexta*. II. Retention of the proleg motor pattern despite loss of prolegs at pupation. J Comp Physiol 155:423.

18. Carlson JR (1981). Temporal variation in the availability of an ecdysial motor programme during the last instar and early adult stages of the cricket *Teleogryllus oceanicus*. J Insect Physiol 27:189.
19. Truman JW, Mumby SM, Welch SK (1980). Involvement of cyclic GMP in the release of stereotyped behaviour patterns in moths by a peptide hormone. J Exp Biol 84:201.
20. Schwartz LM, Truman JW (1984). Cyclic GMP may serve as a second messenger in peptide-induced muscle degeneration in an insect. Proc Nat Acad Sci 81:6718.
21. Morton DB, Truman JW (1985). Steroid regulation of the peptide-mediated increase in cyclic GMP in the nervous system of the hawkmoth, *Manduca sexta*. J Comp Physiol A 157:423.
22. Riddiford LM (1981). Hormonal control of epidermal cell development. Amer Zool 21:751.
23. Cohen P (1982). The role of protein phosphorylation in neural and hormonal control of cellular activity. Nature 296:613.
24. Morton DB, Truman JW (1985). Cyclic nucleotide-dependent protein phosphorylation appears to determine behavioral sensitivity to a peptide hormone in the moth *Manduca sexta*. Abstr Soc Neurosci 11:327.

Molecular Entomology, pages 179–188

SELECTION OF METHOPRENE-RESISTANT MUTANTS OF *DROSOPHILA MELANOGASTER*[1]

Thomas G. Wilson and Judy Fabian

Department of Zoology, University of Vermont
Burlington, Vermont 05405

ABSTRACT Methoprene, a chemical analogue of juvenile hormone, is toxic to late-instar larvae of *Drosophila melanogaster*. We have selected high- and low-level resistant mutants following chemical mutagenesis of male parents. One of the high-level mutants (*Met*) has been partially characterized; it is nearly 100-fold more resistant to either methoprene or juvenile hormone III than susceptible wild-type strains. The locus responsible for resistance maps to the X-chromosome. Flies homozygous for *Met* have fecundity and fertility equivalent to wild-type. This finding implies not only that *Met* females have retained their ability for endogenous juvenile hormone regulation but also that such a mutation would be rapidly selected under methoprene pressure in the field.

INTRODUCTION

A common response of insect populations to insecticidal pressure is development of resistance (1). In order to study the mechanism(s) of resistance to a particular insecticide, investigators generally select a resistant population by stepwise selection pressure on a natural population. Selected populations subjected to a genetic analysis have shown in some instances resistance to be due to a single gene (2) and in other instances, resistance to be polygenic (3). Determining the mechanism(s) of resistance

[1]This work was supported by a grant from the National Institutes of Health.

in insects from the latter type of population is often difficult.

We have been interested in the insect growth regulator type of insecticide which mimics juvenile hormone (JH)(4). Soon after the discovery that JH analogues could be potent insecticides, there was hope that insects would have difficulty developing resistance to these chemicals (5). However, investigators working on the house fly, _M. domestica_ (6) and the mosquito, _C. pipiens_ (7) soon demonstrated resistance to methoprene, a JH analogue with potent activity against dipteran insects.

We are interested not only in the genetic basis for resistance but also in understanding the mechanism which can allow these insects to be highly resistant to an analogue of one of their own hormones. To these ends we have induced lesions in the genome of _Drosophila melanogaster_ and screened the F_1 generation for mutants which are resistant to methoprene. In this paper we report the results of an initial screening, and we report a preliminary characterization of a methoprene-resistant mutant.

MATERIALS AND METHODS

The wild-type strain Oregon-RC was obtained from the Bowling Green Stock Center. _FM7_ is a balancer chromosome, and _C(1)M5_ is a stock whose females have attached X-chromosomes (8). JH III was obtained from Sigma Biochemicals, and methoprene was a generous gift from the Zoecon Corporation. It was supplied as either ZR-515 or the more active optically pure isomer, ZR-2008.

Methoprene was administered to _Drosophila_ larvae by incorporation into _Drosophila_ Instant Food (8), by suspending impregnated paper strips above the food (9), or by topical application (8). JH III was administered by topical application (8). The sensitive period for methoprene toxicity is the late larval - early pupal stage, but death is delayed until the late pupal stage (8); therefore, the response of larvae to methoprene was evaluated by determining mortality in the pupal stage. Fecundity and fertility were determined for isolated adults maintained on a standard yeast-agar-cornmeal-molasses _Drosophila_ recipe. This food results in adequate but not maximal oviposition rates.

Two methods were used to mutagenize Oregon-RC males: feeding of ethyl methane sulfonate (EMS)(10), which results in presumed point mutations (11), and irradiation from a Co-60 source (3000 or 5000 rads), which results in more varied DNA damage than EMS (12). Since EMS-induced mutants are usually expressed as F_1 mosaics (11), each F_1 male was individually crossed with an attached-X female to generate F_2 males, which were tested for survival on methoprene-food. In this way a chromosome carrying an induced resistant mutation could be rescued and evaluated without the uncertainty of whether an F_1 male mosaic for that mutation would survive the methoprene food. Since there was no reason to believe γ-rays would result in significant mosaicism in F_1 individuals, progeny from irradiated males were directly tested on methoprene food.

RESULTS

Selection of Resistant Mutants.

We have begun probing the Drosophila genome for mutations which confer methoprene resistance by screening for recessive X-chromosomal and dominant autosomal mutants which survive a toxic dose of methoprene. A standard procedure of crossing mutagenized males with attached-X females (Figure 1) allowed the expression of dominant and X-chromosomal recessive mutations in the F_1 generation. The Oregon-RC strain was selected to be mutagenized because preliminary experiments established this strain as methoprene-sensitive; it was important to select a sensitive strain so that resistant mutants induced in the strain would not be shrouded by a background genome having methoprene resistance.

A total of 2403 F_1 lines were tested following EMS mutagenesis. From these lines 76 showed resistance ($\geq$ 20% pupal survival) on methoprene food, and 21 retested positively. All but two of these lines showed less than 10X more resistance than did Oregon-RC; these low-resistance lines have not been further studied. Two lines showed high (>10X) resistance to methoprene, and some characteristics of one of these mutant lines will be presented. After irradiating Oregon-RC males, we directly tested a total of 35,000 F_1 individuals on methoprene food. Twenty-one F_1 survivors were found, and lines from each fertile survivor were established by crossing with Oregon-RC males or attached-X females. None of the progeny from any of the 21 lines retested positively.

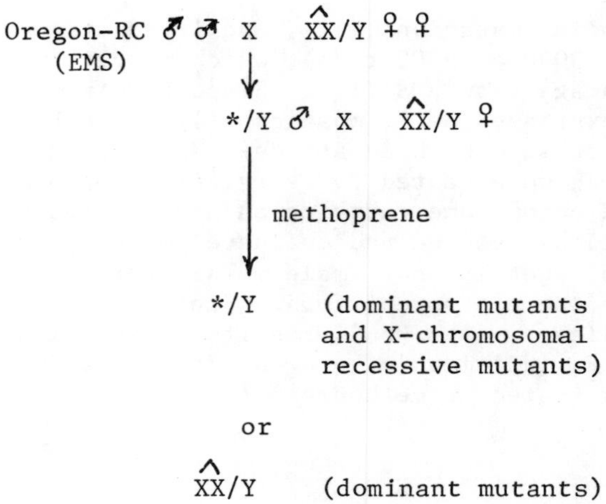

Oregon-RC ♂♂ X $\hat{XX}$/Y ♀♀
(EMS)
↓
*/Y ♂ X $\hat{XX}$/Y ♀
↓ methoprene
*/Y (dominant mutants and X-chromosomal recessive mutants)

or

$\hat{XX}$/Y (dominant mutants)

FIGURE 1. Selection screen for methoprene-resistant mutants of Drosophila. $\hat{XX}$ indicates female with attached X-chromosomes and also carrying a Y-chromosome. * indicates a mutagenized X-chromosome.

Establishment of a Methoprene-Resistant Line.

Two of the lines proved to be >10X more resistant to methoprene than was Oregon-RC. One of the high-resistance EMS lines was inbred on methoprene food for three generations. Simple genetic crosses established that the resistance was associated with the X-chromosome. A homozygous culture of the mutant, termed Met (Methoprene tolerant), was established by crossing Met males with the FM7 balancer chromosome and then selecting non-FM7 flies in the F_2 generation.

Fecundity and Fertility of Met flies. A mutant that can survive a dose of methoprene toxic to susceptible flies might be expected to have difficulty regulating its endogenous JH titer and might express a phenotype reflecting an imbalance of JH. Although no role for JH in Drosophila in preadult development has been established, it is clear that JH is involved in vitellogenic oocyte development (13,14). Therefore, the fecundity of Met females was compared with that of Oregon-RC (Figure 2). No difference was apparent between the two strains, suggesting that Met females have little difficulty regulating the titer of JH required for oogenesis.

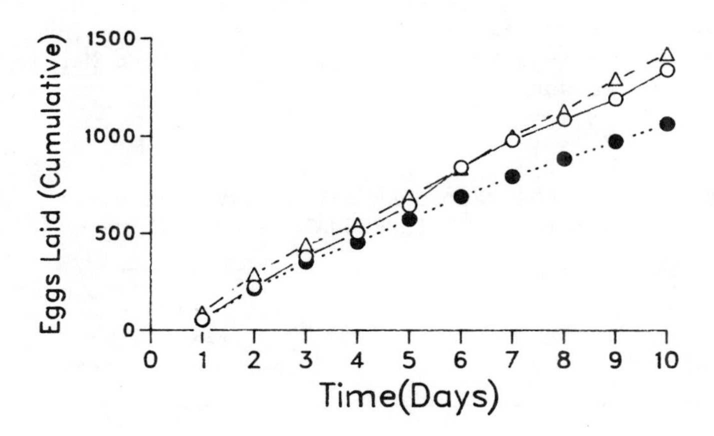

FIGURE 2. Fecundity of Met and FM7 females. Δ, FM7; O, Met raised on non-methoprene food; ●, Met raised on methoprene food. Each point represents cumulative number of eggs laid by 25 females as a function of time after eclosion.

The fecundity of Met females raised on a dose of methoprene lethal to Oregon-RC was somewhat lower than that of Met females raised on non-methoprene food (Figure 2). Nevertheless, Met females raised on methoprene food are far from sterile. The fertility of Met females (Figure 3) and Met males (unpublished results) also compares favorably with wild-type.

Resistance of Met Larvae to Methoprene. The resistance of Met larvae to methoprene was evaluated by measuring pupal survival as a function of the amount of methoprene incorporated into food. The susceptible strain chosen for comparison was FM7 since it had been used to construct a homozygous stock of Met and therefore would share the background autosomal genome with Met. The susceptibility of FM7 to methoprene was found to be similar to Oregon-RC and other wild-type strains tested (unpublished results). As shown in Table 1, Met larvae are 100X more resistant to methoprene than FM7 larvae. Females heterozygous for Met have an intermediate LD_{50} value, indicating that Met is a semidominant mutant.

TABLE 1

LD_{50} VALUES FOR Met PUPAE AFTER LARVAL FEEDING ON METHOPRENE-IMPREGNATED FOOD.[a]

Genotype	LD_{50}	95% confidence interval lower	upper
Met/Y	0.68	0.42	1.16
Met/Met	0.37	0.24	0.59
Met/FM7	0.022	0.014	0.034
FM7/Y	0.0071	0.0045	0.011
FM7/FM7	0.0089	0.0056	0.013

[a] LD_{50} values are given in μl of ZR-515 vial of food (3.6 ml of Instant Drosophila medium). Y refers to Y-chromosome. Slash separates homologous chromosomes.

Met larvae are resistant not only to the lethal effects of methoprene but also to the morphogenetic effects, since Met adults raised on methoprene food do not have the abnormal male genitalia or aberrant abdominal bristle patterns characteristic of susceptible strains treated with JH or JH analogue (15).

It is possible that Met larvae are resistant due to enhanced gut degradation or malabsorption of methoprene. If so, then topical application of methoprene should produce results different from incorporation into diet. To test this possibility, the LD_{50} values for methoprene topically applied to Met or FM7 larvae were determined (Table 2). Met larvae were found to be highly resistant to topically applied methoprene.

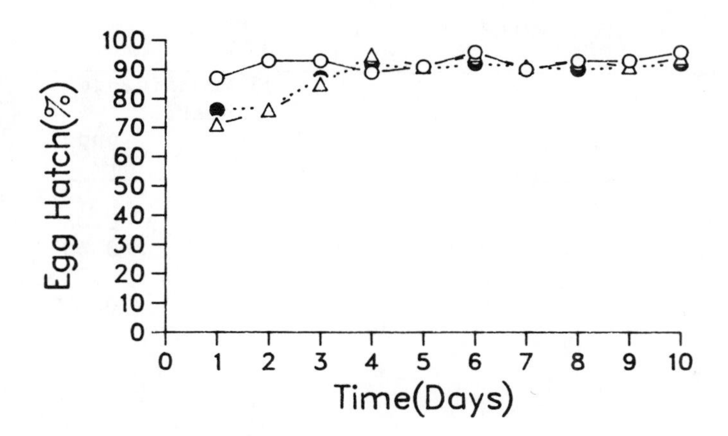

FIGURE 3. Fertility of Met and FM7 females. Symbols same as Figure 2. Each point represents the % hatch of eggs laid by 25 females as a function of time following eclosion.

We next examined the resistance of Met larvae to topically applied JH III, which appears to be the natural hormone in dipteran insects (16). As expected (17), JH III was less toxic than methoprene to FM7 larvae (Table 2). Clearly, Met larvae were more resistant to JH III than FM7 larvae, and Met again showed semidominant expression of the resistance. Thus, the resistant phenotype of Met is not specific for methoprene, but also includes resistance to an amount of the natural hormone toxic to susceptible flies.

Mapping of the Met mutation. In order to determine the location of the Met mutation on the X-chromosome, Met males were crossed to females homozygous for the white, miniature, and forked mutations. F_1 females were crossed with Oregon-RC males, and recombinant F_2 males crossed with attached-X females to establish a line of iso-X-chromosomal males. Each of these lines was grown on methoprene food to test for the presence of Met. We found Met and miniature to be closely linked; 9/416 recombinants between these two genes places the map location of Met to 35.4 on the X-chromosome.

TABLE 2
LD_{50} VALUES FOR TOPICAL APPLICATION OF METHOPRENE OR JH III TO LARVAE

Genotype	LD_{50}[a]	95% confidence interval lower	upper
Met/Y	0.52	0.29	0.88
Met/Met	0.24	0.14	0.40
Met/FM7	0.030	0.017	0.053
FM7/Y	0.005	0.003	0.009
Met/Y	1.28	0.65	2.79
Met/Met	5.27	2.09	19.4
FM7/Y	0.02	0.01	0.05
Met/FM7	0.13	0.06	0.26

[a] LD_{50} values are given in nl ZR-2008/larva (upper grouping) or nl JH III/larva (lower grouping).

DISCUSSION

We have demonstrated that resistant mutants can be directly selected from mutagenized flies. It is assumed that Met is a point mutation. Since point mutations occur naturally, there is no reason to assume that mutants are being generated which do not also occur in nature. Therefore, we believe that our method of estimating the potential for resistance in an insect population to a particular insecticide is valid. One advantage of our method is that a variety of induced mutations can be tested; however, it suffers from an inability to evaluate polygenic resistance, which is known to be important in some resistant insect populations (3).

We have also demonstrated that mutants resistant to methoprene can be recovered. One mutant which is nearly 100X more resistant than susceptible strains was recovered and preliminarily characterized in this paper. Even though highly resistant to exogenous hormone, Met females have apparently normal oogenesis, suggesting that they still possess the ability to regulate endogenous JH. Raising Met larvae on methoprene food has a relatively small effect on fecundity and seemingly little effect on fertility of the adults. This suggests that a naturally occurring resistant population of Met could become established quickly under methoprene pressure in the field. Therefore, a more thorough understanding of this mutant is warranted from both basic and applied biological viewpoints.

ACKNOWLEDGMENTS

We are grateful to the Zoecon Corporation for its generous gift of methoprene.

REFERENCES

1. Georghiou GP, Mellon RB (1983). Pesticide resistance in time and space. In Georghiou GP, Sato T (eds): "Pest Resistance to Pesticides," New York: Plenum, p 1.
2. Harrison CM (1951). Inheritance of resistance to DDT in the housefly, Musca domestica L. Nature 167:855.
3. Crow JF (1954). Analysis of a DDT-resistant strain of Drosophila. J Econ Entomol 47:393.
4. Staal GB (1975). Insect growth regulators with juvenile hormone activity. Ann Rev Entomol 20:417.
5. Williams CM (1967). Third-generation pesticides. Sci Am 217:13.
6. Cerf DC, Georghiou GP (1972). Evidence of cross-resistance to a juvenile hormone analogue in some insecticide-resistant houseflies. Nature 239: 401.
7. Brown TM, Brown AWA (1974). Experimental induction of resistance to a juvenile hormone mimic. J Econ Entomol 67:799.
8. Wilson TG, Fabian J. A Drosophila melanogaster mutant resistant to a chemical analogue of juvenile hormone. (submitted for publication).
9. Wilson TG, Chaykin D (1985). Toxicity of methoprene to Drosophila melanogaster (Diptera: Drosophilidae): A function of larva culture density. J Econ Entomol 78:1208.
10. Lewis EB, Bacher F (1968). Method of feeding ethyl methane sulfonate to Drosophila males. Drosophila Inform Serv 43:193.

11. Vogel E, Natarajan AT (1979). The relation between reaction kinetics and mutagenic action of monofunctional alkylating agents in higher eukaryotic systems; I. Recessive lethal mutations and translocations in Drosophila. Mutation Res 62: 51.
12. Sankaranarayanan K, Sobels FH (1976). Radiation genetics. In Ashburner M, Novitski E (eds): "The Genetics and Biology of Drosophila," New York: Academic, vol 1c, p 1089.
13. Jowett T, Postlethwait JH (1980). The regulation of yolk polypeptide synthesis in _Drosophila_ ovaries and fat body by 20-hydroxyecdysome and a juvenile hormone analogue. Dev Biol 80:225.
14. Bownes M (1982). Hormonal and genetic regulation of vitellogenesis in Drosophila. Quart Rev Biol 57: 247.
15. Ashburner M (1970). Effects of juvenile hormone on adult differentiation of _Drosophila melanogaster_. Nature 227:187.
16. Baker FC, Lanzrein B, Miller CA, Tsai LW, Jamieson GC, Schooley DA (1984). Detection of _only_ JH III in several life-stages of _Nauphoeta cinerea_ and _Thermobia domestica_. Life Sci 35:1553.
17. Sehnal F, Zdarek J (1976). Action of juvenoids on the metamorphosis of cyclorrhaphous diptera. J Insect Physiol 22:673.

Molecular Entomology, pages 189–199

ESTROGENS AND ANDROGENS IN INSECTS[1]

D.L. Denlinger[2], R.W. Brueggemeier[3], R. Mechoulam[4], N. Katlic[3], L.B. Yocum[2], and G.D. Yocum[2]

Department of Entomology[2] and College of Pharmacy[3], Ohio State University, Columbus, Ohio 43210, and Faculty of Medicine[4], Hebrew University, Jerusalem 91 120, Israel

ABSTRACT Estrogens and androgens recently have been identified by gas chromatography - mass spectrometry in the flesh fly Sarcophaga bullata, and radioimmunoassays indicate the presence of estrogen- and androgen-like substances in a wide range of other insects. These steroids are present in both sexes at nearly the same concentrations. Highest concentrations in the flesh fly coincide with the onset of mating and the most rapid period of yolk deposition, but the steroids are detectable in all developmental stages. High levels of aromatase, the enzyme complex that converts androgens to estrogens, are observed in the reproductive system, nervous tissue and fat body. A hemolymph protein shows specific binding of estradiol, with an approximate K_D of 4.9×10^{-6} M and a concentration of 3.6 binding sites/mg protein as determined from Scatchard analysis. The role of estrogens and androgens in insects remains to be discovered.

INTRODUCTION

Estrogens and androgens appear to be very ancient and highly conserved molecules. Not only are these steroids well documented among vertebrates (1), they have been detected in several invertebrate groups including molluscs (2), starfish (3), platyhelminthes (4), and crustaceans (5), and more recently, estrogen and estrogen-binding protein have been found even in yeast (6).

[1]This work was supported by NIH grant AI21321.

Though only occassional references suggest the possibility of sex hormones in insects (7,8), estrogens and androgens have recently been reported in several insect species. High concentrations of estrogens, androgens, and pregnanes are used as defense secretions in several dytiscid and carrion beetles (9,10). Lower concentrations of estrogens and androgens, more compatible with a physiological function, are present in the flesh fly _Sarcophaga bullata_ (11,12,13) and a range of species including _Periplaneta americana_, _Manduca sexta_, _Tenebrio molitor_, _Oncopeltus fasciatus_ (13), and _Bombyx mori_ (14). Testosterone is a constituent of royal jelly (15).

In the flesh fly, estradiol, estriol (13), testosterone, and several related androgens (11,12) have been identified by gas chromatography - mass spectrometry. Absence of estrogens and androgens in the larval food (beef liver) suggests that these steroids are synthesized by the flies, presumably from steroid skeletons obtained from the food (13).

Several intermediates and enzymes in the biosynthetic pathway from cholesterol to androgen and estrogen have been reported. _Manduca sexta_ can cleave the side chain of cholesterol to produce a steroid conjugate, 5-pregnen-3B,20β-diol glucoside (16). Pregnenalone and progesterone have been identified from _Tenebrio molitor_ (17), although the possibility that these steroids came directly from the food can not be excluded (18). Several key steroids in androgen and estrogen biosynthesis are present in _S. bullata_: progesterone, 17α -hydroxyprogesterone, and androstenedione (11,12). In addition, _S. bullata_ contains several metabolites normally associated with degradation of estrogens and androgens.

Many of the "mammalian" steroids can be transformed readily by insect tissue. Cockroaches and crickets convert pregnenolone to progesterone (19), progesterone to 20 α- and 20 β-hydroxy-pregn-4-ene-3-one (20), and testosterone to androstenedione (21). The capacity to reduce and oxidize many of these steroids has also been demonstrated in _Schistocerca gregaria_ (22,23). Among these Orthoptera species highest activity is found in the gonads. Though substrate specificity has not been defined in most cases, the preliminary results suggest the potential of insect tissue to transform many of the steroids in the androgen and estrogen pathway.

We are aware of only one report suggesting a biological effect of mammalian sex hormones on insect development, and that is the rather bizarre case of European rabbit flea, Spilopsyllus cuniculi, a species that utilizes the presence of estrogen and corticosteroids in the blood of the pregnant host to trigger egg maturation (24). In turn, progestins and luteinizing hormone, present in high concentrations in postpartum does, trigger regression of the flea's ovary. These cues thus enable the flea to produce her progeny in synchrony with birth of the young rabbits.

In this report we summarize our current research on insect estrogens and androgens. Since estrogens (13) and androgens (11,12) have already been chemically identified from Sarcophaga, most of our work continues to focus on the flesh fly. We monitor estrogen and androgen titers throughout development, describe an estrogen-binding protein, and examine a variety of tissues for aromatase activity. In addition, we discuss some of our attempts to affect insect development, reproduction, and behavior with exogenous estrogens and androgens.

DEVELOPMENTAL PROFILE

The flesh fly S. crassipalpis was used to monitor developmental changes in concentrations of estrogen- and androgen-like radioimmunoactive substances. Samples of 40 individuals each were homogenized and extracted as described (13), with the exception that anhydrous ether was used instead of dichloromethane. The aqueous phase remaining after ether extraction was acidified to pH 2.0 with HCl overnight and re-extracted. Further details of sample preparation and the radioimmunoassay (RIA) procedures are given elsewhere (Yocum et al, in preparation).

The estradiol RIA antisera (University Hospitals, Ohio State University) cross-reacts 0.5% with estrone and 0.1% with estriol, while the testosterone RIA antisera (Steranti Research Ltd., England) cross-reacts 25.0% with dihydrotestosterone, 28.0% 19-nortestosterone, 31.0% 11α-hydroxytestosterone, and $<$ 0.01% 17β-estradiol. No cross-reactivity is observed with 20-hydroxyecdysone at

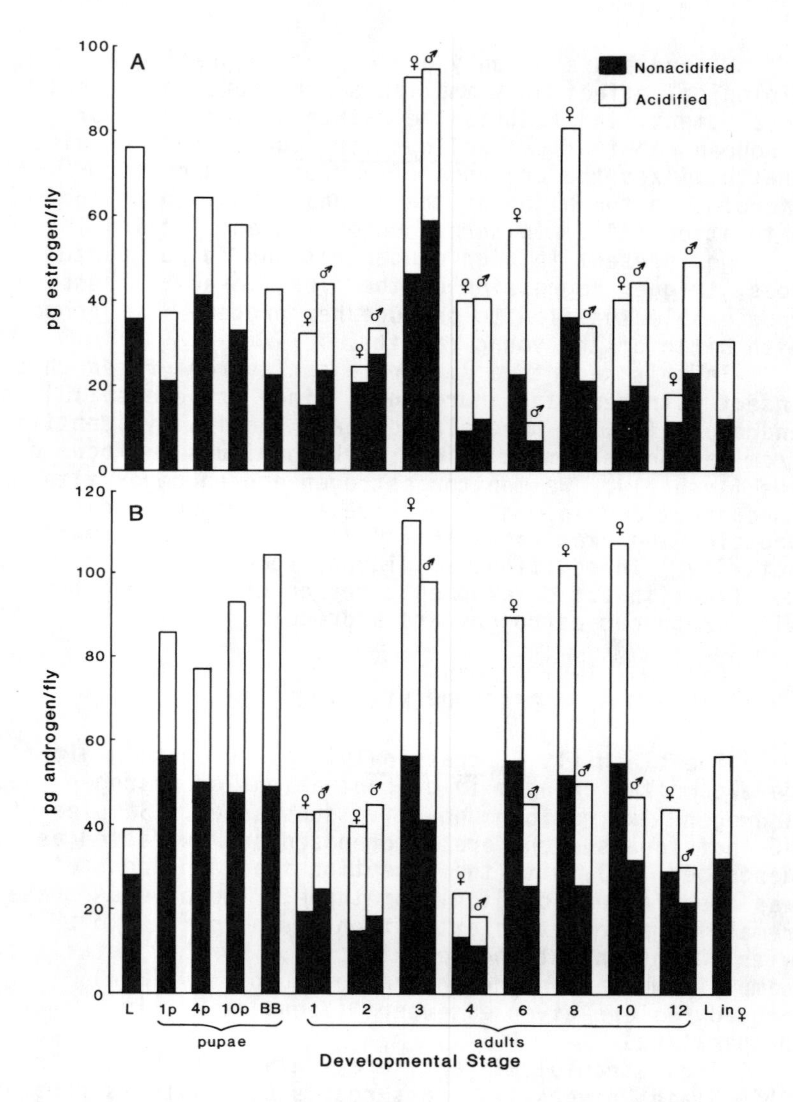

FIGURE 1. RIA determinations of estrogen (A) and androgen (B) in acidified and nonacidified extracts of the flesh fly Sarcophaga crassipalpis at different developmental stages. L = wandering phase of third larval instar, 1p = day of pupariation, 4p = 4d after pupariation, 10p = 10d after pupariation, BB = black bristle stage, L in ♀ = first instar larvae in uterus of female. N = 3-6 samples of 40 flies each.

concentrations ranging from 10 pg/ml to 10 ng/ml with either antisera.

By acidifying and re-extracting the samples, yields of estrogen- and androgen-like substances detected by RIA usually doubled (Fig. 1), suggesting that large amounts of these steroids are present in conjugated forms, possibly as glucosides, glucuronides or sulfates.

Estrogen (Fig. 1A) and androgen (Fig. 1B) profiles were similar, but concentrations of androgen-like substances detected were higher than levels of estrogens at comparable stages. In most age categories, males and females contained nearly the same concentrations. Physiological levels of estrogens (1.2 - 9.9 pg/mg protein) and androgens (1.5 - 16.5 pg/mg protein) were present throughout development and in both sexes during adult life. Correction of RIA results for extraction efficiency would elevate estrogen concentrations 6.3 times and androgen levels 2.5 times.

The conspicuous peak of both estrogens and androgens on day 3 of adult life coincides with two significant physiological events: the time of most rapid yolk deposition and the onset of mating. A role for estrogens in vitellogenesis is an attractive scenario based on vertebrates, but the presence of a peak in both males and females makes this suggestion less likely. A link with mating, however, remains a possibility.

In this ovoviviparous species, eggs are ovulated on the 5th day of adult life and the embryos remain in a sac-like uterus until larviposition on day 11. Very high concentrations of both steroids (9.2 pg estrogen/mg protein, 16.5 pg androgen/mg protein) were found in first instar larvae dissected from the uterus of the female, but whether these steroids are synthesized by the developing embryo or are maternal products packaged into the egg remains unresolved.

BIOCHEMISTRY OF INSECT ESTROGEN

Biochemical examination of endogenous estrogens in insects has focused on two aspects. The first is the examination of estrogen biosynthesis. In order to elucidate the physiological relevance of estrogens, identification of tissues which biosynthesize these steroids is critical. The second aspect is the determination of high affinity binding proteins specific for estrogens. The central dogma of estrogenic effects on cellular function involves the formation of steroid-protein complexes of high affinity, which influence genomic expression and result in altered cellular processes. Presence of estrogen receptors in certain tissues will provide leads on the role(s) of estrogens in insects. In addition, the presence of transport/storage proteins in circulating fluids supports the hypothesis that these steroids act as hormonal agents.

Estrogens are formed from cholesterol, with the final biosynthetic step being the conversion of androgens to estrogens by an enzymatic complex termed aromatase. Aromatase activity has been assayed in a variety of insect tissues. Organs of the male and female adult insects were surgically removed and placed in Grace's media containing 5% BSA. Tissues from 5 individuals were pooled in one dish and experiments were performed in triplicate. Two radiolabeled enzyme assays for aromatase were utilized. One assay involved the incubation of the cultures with [7-^{3}H]-testosterone, followed by extraction of the media with ether and reverse-phase HPLC isolation of [7-^{3}H]-estradiol. The second assay employed [1β-^{3}H]-androstendione and determined the amount of ^3H_2O released during the aromatization process. The media was extracted with chloroform and an aliquot of the aqueous solution was counted by LSC. Dishes of pooled organs were incubated with radiolabeled androgen (500 nM) at room temperature under a 95% O_2 atmosphere for 16 hours.

High levels of aromatization were observed in gonads, reproductive tract, accessory glands, brain, and fat body in 3-day and 5-day adult flies. The amount of estrogens formed ranged from 2.38 to 3.22 pmol per fly over 16 hours (0.145 to 0.202 pmol/fly/hr) in these tissues in the 3-day adult females and ranged from 2.10 to 2.82 pmol per fly over 16 hours (0.132 to 0.176 pmol/fly/hr) in the 5-day adult females, as shown in Fig. 2. Low levels were observed in the thoracico-abdomenal ganglia, corpora cardiaca/corpora allata complex, flight muscles, and digestive tract. Similar results were observed in 3-day and 5-day male insects. When these results from 5-day females are expressed per mg of protein, the reproductive, brain and ganglionic tissues produce significant quantities of estrogen. The data suggest that the estrogens may be produced locally in the

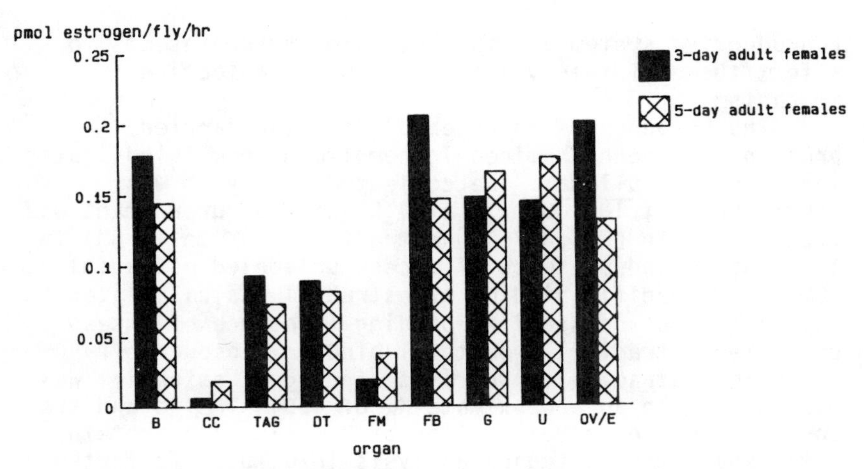

FIGURE 2. Aromatase activity present in select tissues of 3-day and 5-day old females of S. bullata. B = brain, CC = corpora cardiaca/corpora allata complex, TAG = thoracico-abdominal ganglia, DT = digestive tract, FM = flight muscles, FB = fat body, G = accessory glands, U = uterus, OV/E = ovaries containing maturing eggs.

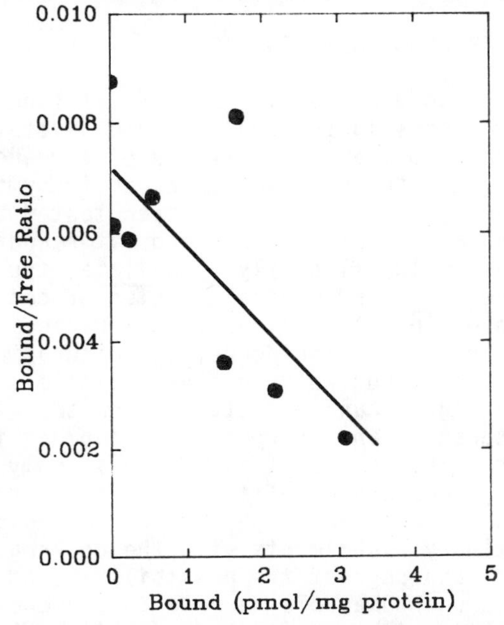

FIGURE 3. Scatchard plot for the binding of estradiol to hemolymph protein from S. bullata.

reproductive system and in the brain of the insects to affect these tissues via a paracrine or autocrine mechanism.

The presence of estrogen binding (or carrier) proteins has been examined in hemolymph from third instar larvae of S. bullata. The collected hemolymph was centrifuged at 100,000 x g at 4°C and the supernatant was incubated with [2,4,6,7-^{3}H]-estradiol (1 nM to 10 μM) in the absence and presence of excess unlabeled estradiol (60 μM). Specific binding of estradiol was calculated by subtraction of nonspecific binding (presence of excess unlabeled estradiol) from total binding (absence of unlabeled estradiol). Specific binding of estradiol was observed, with an approximate K_D of 4.9×10^{-6} M and the concentration of binding sites of 3.6 sites/mg protein determined from Scatchard analysis (Fig. 3). To further characterize this binding activity, ammonium sulfate was added to the 100,000 x g supernatant to form a 30% saturated solution and centrifuged at 10,000 x g. The precipitated protein fraction was redissolved and estradiol binding activity observed.

BIOLOGICAL SIGNIFICANCE

In an attempt to discover the biological function of estrogens and androgens in insects, we have devoted considerable effort to examining effects of exogenous estrogens, androgens, their analogs, and anti-hormones such as tamoxifen. These agents have been tested by injection, implantation, and as food and water additives on various stages of the flesh fly S. bullata, the mosquito Aedes aegypti, and several species of cockroaches. We looked for alterations in sex ratio, developmental rate, mating incidence, egg maturation, and dominance hierarchies, but we've had very little success in detecting any significant effects. Likewise, attempts to alter egg maturation and embryonic development in Bombyx mori with estradiol have failed (25): only a modest and erratic depression of the oviposition rate could be detected with a very high dose of estradiol.

Very preliminary experiments with the cockroach Periplaneta americana suggest the possibility that estrogens and androgens may play a role in the dominace hierarchy of this species. We've had some initial success in disrupting the hierarchy by implanting estradiol into alpha males and concurrently implanting testosterone into beta and gamma males, but our sample size remains small, and additional replicates are essential to validate these early results.

Thus, at this point, we have very little insight into the functional significance of estrogens and androgens in insects. Their presence does not necessarily imply that they function as sex hormones. The fact that both males and females contain rather high levels of both estrogens and androgens already suggests a significant departure from the usual mammalian pattern. Though most of our effort has focused on estradiol and testosterone, other closely related steroids may be of greater biological significance in insects, but our present assays are highly specific for estradiol and testosterone, and amounts of related C-18 and C-19 steroids are still unknown.

The possibility that estrogens and androgens serve primarily a defensive role in insects seems unlikely. Our observations are more compatible with a physiological function for several reasons: 1) The steroids are present in diverse taxa, 2) Concentrations are compatible with a physiological function, 3) Systematic changes in steroid titers correlate with behavioral and physiological events such as mating and larviposition, 4) The presence of an estrogen binding protein in the hemolymph suggests that estrogen serves a functional role, and 5) High aromatase activity is distributed in nervous and reproductive tissue, suggesting some tissue specificity for estrogen biosythesis.

In addition to their well known function as the sex hormones of vertebrates, estrogens and androgens are also known from vertebrate studies to affect cell proliferation, general metabolism, nerve growth, synaptic organization, gross morphology of the CNS, the immune response, and behavior. Now that these steroids are known to be present in many nonvertebrates, it may become possible to trace the evolutionary history of these important molecules. Indeed, it is quite possible that the original function of estrogens and androgens had little to do with the conspicuous sexual role now observed in mammals. Possibly these steroids were only secondarily captured by vertebrates as sex hormones. If so, insect studies may provide insightful information on the history and function of this important class of steroids.

REFERENCES

1. Callard IP, Ho SM, Gapp DA, Taylor S, Danko S, Wulczyn G (1980). Estrogens and estrogenic actions in fish, amphibians, and reptiles. In McLachlan JA (ed): "Estrogens in the Environment," New York: Elsevier/ North Holland, p 213.
2. Gottfried H, Dorfman RI, Wall PE (1967). Steroids of invertebrates: production of oestrogens by an accessory reproductive tissue of the slug Arion ater rufus (Linn.). Nature 215:409.
3. Hathaway RR (1965). Conversion of estradiol-17β by sperm preparations of sea urchins and oysters. Gen. Comp. Endocrin. 5:514.
4. Sander T, Sonea S, Mehdi AZ (1975). The possible role of steroids in evolution. Amer. Zool. 15(Suppl.1):227.
5. Burns BG, Sangalang GB, Freeman HC, McMenemy M (1984). Isolation of testosterone from the serum and testes of the American lobster (Homarus americanus). Gen. Comp. Endocrin. 54:429.
6. Feldman D, Tokes LG, Stathis PA, Miller SC, Kurz W, Harvey D (1984). Identification of 17B-estradiol as the estrogenic substance in Sarcharomyces cerevisiae. Proc. Natl. Acad. Sci. USA 81:4722.
7. Naisse J (1966). Controle endocrinien de la différenciation sexuelle chez l'insecte Lampyris noctiluca (Coleoptere Malacoderme Lampyride). I. Rôle androgène des testicules. Arch. Biol. (Liege). 77:139.
8. Engelmann F (1970). "The Physiology of Insect Reproduction," Oxford: Pergamon, 307 p.
9. Schildknecht H (1970). The defensive chemistry of land and water beetles. Angew. Chem. Intl. Ed. 9:1.
10. Meinwald J, Roach B, Hicks K, Alsop D, Eisner T (1985). Defensive steroids from a carrion beetle (Silpha americana). Experientia 41:516.
11. DeClerck D, Eechaute W, Leusen I, Diederik H, DeLoof A (1983). Identification of testosterone and progesterone in hemolymph of larvae of the fleshfly Sarcophaga bullata. Gen. Comp. Endocrin. 52:368.
12. DeClerck D, Diederik H, DeLoof A (1984). Identification by capillary gas chromatography-mass spectrometry of eleven non-ecdysteroid steroids in the hemolymph of larvae of Sarcophaga bullata. Insect Biochem. 14:199.

13. Mechoulam R, Brueggemeier RW, Denlinger DL (1984). Estrogens in insects. Experientia 40:942.
14. Ohnishi E, Ogiso M, Wakabayashi K, Fujimoto Y, Ikekawa N (1985). Identification of estradiol in the ovaries of the silkworm, Bombyx mori. Gen. Comp. Endocrin. 60:35.
15. Vittek J, Slomiany BL (1984). Testosterone in royal jelly. Experientia 40:104.
16. Thompson MJ, Svoboda JA, Lusby WR, Rees HH, Oliver JE, Weirich GF, Wilzer KR (1985). Biosynthesis of a C_{21} steroid conjugate in an insect. J. Biol. Chem. 260:15410.
17. Smissman EE, Jenny NA, Beck SD (1964). Sterol metabolism in larvae of the confused flour beetle, Tribolium confusum. J. Pharm. Sci. 53:1515.
18. Dubé J, Lemonde A (1970). The origin of progesterone in the confused flour beetle (Tribolium confusum). Experientia 26:543.
19. Lehoux J-G, Chapdelaine A, Sandor T (1970). L'oxydation de la prégnènolone en progestérone par des préparations de tissus des Orthoptères in vitro. Can. J. Biochem. 48:407.
20. Lehoux J-G, Sandor T. Lanthier A, Lusis O (1968). Metabolism of exogenous progesterone by insect tissue preparations in vitro. Gen. Comp. Endocrin. 11:481.
21. Lehoux J-G, Sandor T (1969). Conversion of testosterone to Δ^4-androstene-3,17-dione by house cricket (Gryllus domesticus) male gonad preparations in vitro. Endocrinology 84:652.
22. Dubé J, Villeneuve J-L, Lemonde A (1968). Metabolisme in vitro de la progesterone 7α-^{3}H dans l'ovaire de Schistocerca gregaria (Orthoptere). Arch. Intl. Physiol. Biochim. 76:64.
23. Dubé J, Lemonde A (1970). Transformations des steroides par la femelle adulte d'un insecte orthoptere, Schistocerca gregaria Forskal. Gen. Comp. Endocrin. 15:158.
24. Rothschild M, Ford B (1966). Hormones of the vertebrate host controlling ovarian regression and copulation of the rabbit flea. Nature 211:261.
25. Ogiso M, Ohnishi E (1986). Does estradiol play a role in ovarian maturation or embryonic development of the silkworm? Gen. Comp. Endocrin. 61:82.

Molecular Entomology, pages 201–210

GENOMIC CLONING OF DEVELOPMENTALLY REGULATED LARVAL CUTICLE GENES FROM THE TOBACCO HORNWORM, *MANDUCA SEXTA*[1]

John Rebers, Frank Horodyski, Robert Hice, and Lynn M. Riddiford

Department of Zoology, University of Washington
Seattle, WA 98195

ABSTRACT A genomic library was prepared from DNA of a fifth instar *Manduca sexta* larva. A number of clones were selected using a cDNA clone which encodes a 14.6 kD Manduca larval cuticle protein and a *Drosophila* genomic clone, I-11, which encodes a third instar *Drosophila* larval cuticle protein. All of the clones hybridized to a radioactive cDNA probe made using RNA isolated from day 2 *Manduca* fifth instar epidermis. 1C4-G11, selected using the *Manduca* cDNA clone, and I11-B311, selected using the *Drosophila* genomic clone, both hybridized to a single band on Northern blots of day 2 RNA and were expressed in the fourth and fifth instars at times appropriate for larval cuticle synthesis.

INTRODUCTION

The insect cuticle is a complex structure comprised of chitin and proteins secreted by the underlying epidermis. Changes in its form occur both at molts, particularly the metamorphic molts, and during the intermolt due primarily to changes in protein synthesis (1). Most of these changes are under hormonal control with 20-hydroxyecdysone regulating the level of expression of the cuticular genes, and juvenile hormone (JH) preventing the expression of new genes while allowing the re-expression of previously active genes (2). To elucidate the nature of this control by these two

[1]This work was supported by NIH grants to LMR and JR.

hormones, one needs to isolate specific cuticular genes.

In *Drosophila* several larval (3) and pupal (4) genes have been characterized, and several of the pupal genes have been shown to be regulated differentially by the changing ecdysteroid titer (4). However, the morphogenetic action of JH has not been clearly demonstrated in *Drosophila*. Therefore, we have begun the characterization of larval cuticle genes of the tobacco hornworm, *Manduca sexta*, where the developmental action of JH has been well studied (2; for review, see 6).

Initial studies on clones coding for three different larval cuticle proteins isolated from a cDNA library to day 2 fifth instar epidermal mRNA showed that all three genes were expressed during the feeding phase of both the 4th and 5th larval instars, but were not present during the molt (5). Also, expression declined rapidly on day 3, the final day of feeding before the onset of metamorphosis. Two of these were found to be larval-specific. The third (encoding a 14.6 kD protein) was expressed in both pharate pupal and adult epidermis, producing the flexible cuticle of the intersegmental membrane regions.

We have prepared a *Manduca* genomic library and used these clones as well as the *Drosophila* larval cuticle gene I-11 (3) to screen the library for clones containing cuticle genes. The use of this *Drosophila* clone was prompted by the discovery that several of the *Manduca* larval cuticle proteins cross-reacted with *Drosophila* larval cuticle antiserum in an immunoblot (Wolfgang and Riddiford, unpublished). This paper describes the preparation of the library and characterization of several clones selected by the *Manduca* and *Drosophila* probes.

METHODS

Construction and Screening of a *Manduca* Genomic Library.

DNA was isolated from a single fifth instar *Manduca* larva by dissecting out the gut, pulverizing the carcass in liquid nitrogen, and treating the resulting powder with proteinase K and phenol as described by Maniatis (7). This DNA was digested with Sau 3A I under conditions which gave the maximum number of molecules 15-20 kb long, treated with alkaline phosphatase to prevent ligation of non-adjacent fragments, and run on a sucrose gradient to select fragments 12-22 kb long. EMBL 3 DNA (8) was digested with Bam HI and Eco RI and isopropanol precipitated to eliminate the 9 bp Bam HI/ Eco RI fragment which joins the phage arms to the nonessential central portion of the phage. The phage arms were ligated to the genomic DNA, packaged using a freeze/sonic lysate, and used to infect the host CES 200. This host does not permit genetic selection of phage with inserts, but is rec BC^-. The library was plated for screening without amplification to further reduce the chance for loss of recombinant clones.

150,000 recombinant plaques (approximately 4 genome equivalents) were plated, absorbed to nitrocellulose filters (9) and hybridized to nick-translated insert from the clone pE2-1C4, which encodes a 14.6 kD larval cuticle protein (5) Hybridization was done at moderate stringency (65°,4X SSC, washed to 1X SSC at 65°). Filters were dried and exposed to X-ray film.

The filters were then washed and hybridized at low stringency (5X SSC, 43% formamide, 37°, washed to 0.1X SSC at 50°) to nick-translated insert from the *Drosophila* cuticle gene I-11 and exposed for autoradiography as above.

Genomic Southern Blots

Manduca genomic DNA from the same larva as used for library construction (see above) was digested with Hind III or Sal I and transferred to nitrocellulose (10). The filters were hybridized at 65° using the same hybridization and washing conditions used for screening the library.

Expression of the Genomic Clones

Cloned DNA was digested with restriction enzymes as described in the figures and transferred to nitrocellulose (10). The filters were hybridized at 65° (see above) to insert from the clones used for selection of the genomic clones or to radioactive cDNA made using reverse transcriptase (7). Ten ug of total RNA from day 2 (15:00 AZT) larvae was used for probe synthesis. RNA dot blots to quantitate expression during the fourth and fifth larval instars were made and hybridized as described previously (5), except that some of the RNAs were from pooled samples.

RESULTS

Construction and Screening of a Manduca Genomic Library.

The library was initially screened using nick-translated insert from the cDNA clone pE2-1C4 (5) under moderately stringent hybridization conditions. Four clones were selected, three of which are overlapping, while the fourth, 1C4-H11, comes from a different region of the genome (Table 1).

A 1.8 kb Hind III fragment subcloned from 1C4-G11 and the insert from pE2-1C4 were used to probe genomic Southern blots (Rebers and Ridiford, unpublished). Both probes hybridized to a 1.8 kb Hind III fragment and a 6.2 kb Sal I fragment. The 1C4-G11 probe also hybridized to a 3 kb band in both Hind III and Sal I digests.

The filters used for screening the genomic library were washed and hybridized to nick-translated insert from the Drosophila larval cuticle gene I-11 (3) using low hybridization stringency. Four hybridizing clones were found (Table 1). Restriction mapping is in progress to determine if these clones are from the same or different regions of the genome.

Identification of Transcribed Regions in the Genomic Clones

Southern blots of the four genomic clones selected with pE2-1C4 were probed with insert from pE2-1C4, washed, and reprobed with radioactive cDNA made using RNA from day 2 (15:00 AZT) fifth instar larvae. Fig. 1 shows that the same bands hybridize to pE2-1C4 and to radioactive cDNA. An additional hybridizing band is seen on the blots probed with radioactive cDNA, which may be from the same transcript as the pE2-1C4 hybridizing band or from a neighboring transcript which is less abundant. Transcript mapping is in progress to distinguish between these possibilities.

To compare the relative expression of the clones selected with the Manduca cDNA clone pE2-1C4 and the Drosophila clone I-11, all four of the pE2-1C4 homologous clones and two of the I-11 homologous clones were transferred to nitrocellulose and probed with radioactive cDNA. All clones except 1C4-H11 hybridize with roughly equal intensity (Fig. 2). A transcript which is not very abundant or poorly homologous hybridizes to clone IC4-H11

Table 1
Manduca Genomic Clones

Clone	Selected With	Insert Size
1C4-D11	pE2-1C4	14.1 kb
1C4-G11	pE2-1C4	14.2 kb
1C4-H11	pE2-1C4	12.5 kb
1C4-H31	pE2-1C4	12.9 kb
I11-B311	Dm I-11	12.6 kb
I11-D211	Dm I-11	13.9 kb
I11-F11	Dm I-11	13.0 kb
I11-H71	Dm I--11	12.2 kb

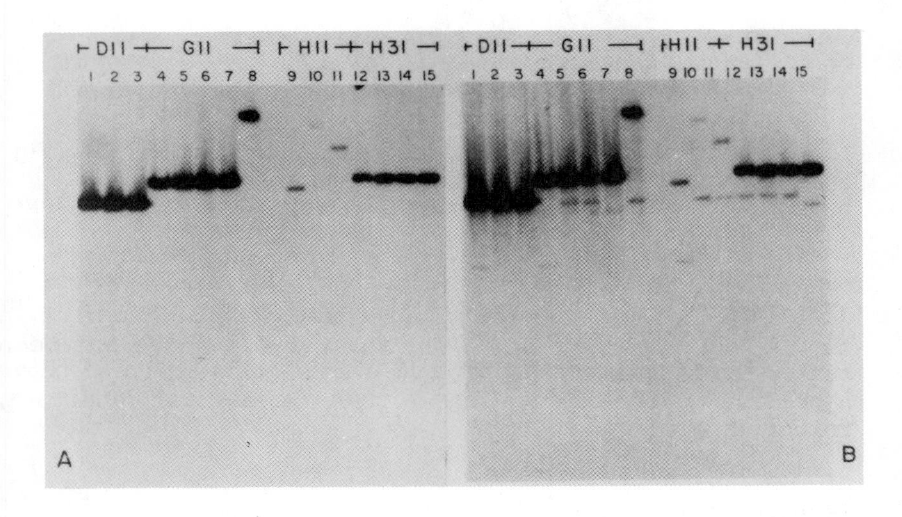

Figure 1. pE2-1C4 (1A) and cDNA (1B) hybridization to genomic clones. Clones 1C4-D11, 1C4-G11, 1C4-H11, and 1C4-H31 were digested with Eco RI + Sal I (lanes 1,4,9,12), Sal I (lanes 2,6,10,14), Xho I + Sal I (lanes 3,5,11,13), Bam HI + Sal I (lanes 7,15) or Bam HI (lane 8).

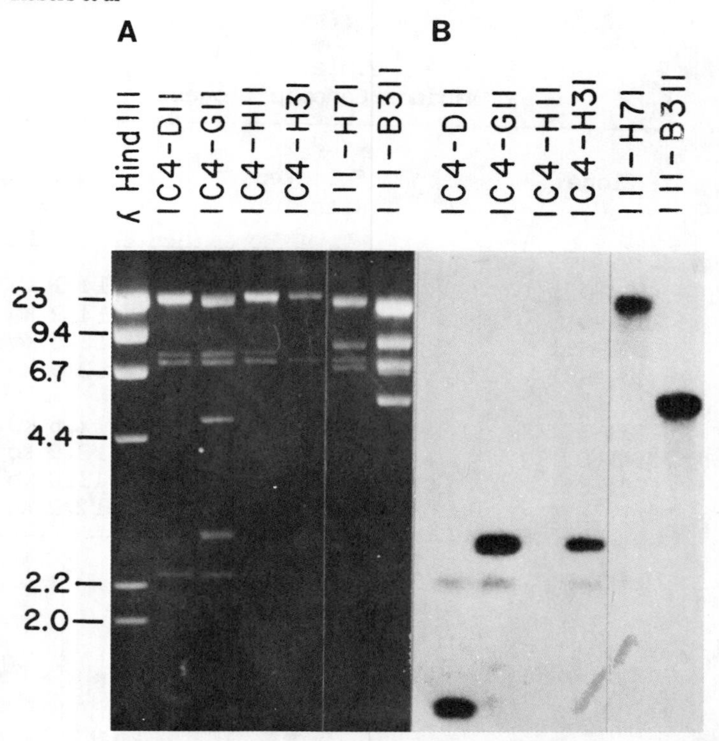

Figure 2. Radioactive cDNA hybridization to 1C4 and I11 genomic clones. Clones were digested with Bgl II (Fig 2A), transferred to nitrocellulose and hybridized to radioactive cDNA made using day 2 (15:00 AZT) epidermal RNA (Fig 2B).

[see Fig. 1, also seen on a longer exposure of the blot in Fig. 2 (not shown)]. I11-D211 and I11-F111 have not been tested using Southern blots, but plaque grids of these phage hybridize to radioactive day 2 cDNA (Rebers and Riddiford, unpublished).

Developmental Expression of the Genomic Clones

A 1.8 kb Hind III fragment from clone 1C4-G11 hybridizes strongly to pE2-1C4 and to radioactive cDNA. This fragment was subcloned and used to probe a Northern blot of RNA isolated from fifth instar epidermis. A single hybridizing band was seen (Rebers and Riddiford,

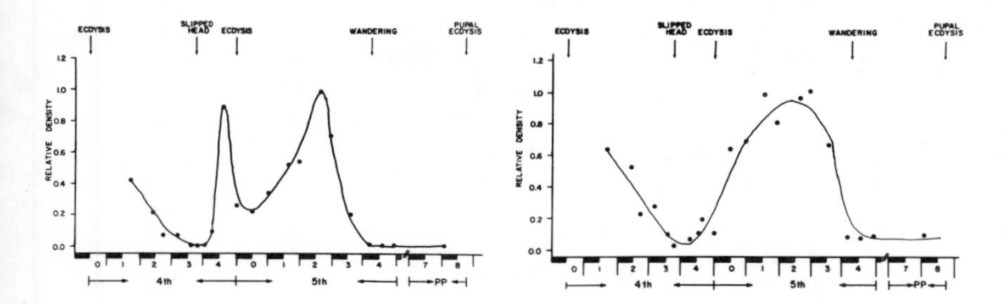

Figure 3. Developmental expression of 1C4-G11 and I11-B311. Five ug of total epidermal RNA from different stages was transferred to nitrocellulose and hybridized to 1C4-G11 (left) or I11-B311. The circles represent averages of replicate samples of each RNA prep. Days of the 4th and 5th larval instars, as well as the pharate pupal (PP) stage are indicated below.

unpublished), so the same probe was used for an RNA dot blot of epidermal RNA isolated from fourth and fifth instar Manduca larvae. Fig. 3 shows that the expression profile for 1C4-G11 during these stages, which is very similar to that found for pE2-1C4 (5,6).

To determine the developmental expression of the genomic region selected by the Drosophila cuticle gene, the 5.2 kb Bgl II fragment from I11-B311 which hybridizes to radioactive day 2 cDNA was used to probe a Northern blot of fifth instar epidermal RNA. A single hybridizing band was observed (Rebers and Riddiford, unpublished). This probe was then used to test expression in fourth and fifth instar larval epidermis (Fig 3). RNA is found early in the fourth instar, is turned off before ecdysis, and is found again early in the fifth instar. The expression pattern is similar, but not identical, to that of pE2-1C4 and 1C4-G11 (5,6).

Proteins Encoded by the Genomic Clones

Hybrid selection experiments (Rebers and Riddiford, unpublished) have shown that the genomic clone 1C4-G11 selects an RNA which encodes a 14.6 kD protein, identical in size to the protein obtained when the cDNA clone pE2-1C4 is used for hybrid selection. This protein is precipitated by *Manduca* larval cuticle antiserum. The genomic clone I11-B311 (homologous to the *Drosophila* genomic clone I-11) selects an RNA which encodes a 14 kD protein, which is precipitated by *Manduca* larval cuticle antiserum. Therefore, both of the genomic clones described above contain *Manduca* larval cuticle genes.

DISCUSSION

These studies have identified three genomic clones which encode the 14.6 kD larval cuticle protein of *Manduca*. All three of these clones hybridize strongly to pE2-1C4, a cDNA clone for the 14.6 kD protein, and both cDNA and genomic clones hybridize to the same bands on genomic Southerns. The genomic clone 1C4-G11 hybridizes to an RNA of the same size as detected by the cDNA clone. Moreover,the expression pattern of 1C4-G11 is very similar to that described for pE2-1C4, although more data are necessary to show that the patterns are identical. Hybrid-selection experiments have shown that the RNA detected by 1C4-G11 in the expression experiments described in Fig. 3 encodes the expected 14.6 kD larval cuticle protein.

Although both larval (3) and pupal (11) cuticle genes from *Drosophila* have been cloned, these clones are the first non-*Drosophila* genomic cuticle genes to be isolated. It will be interesting to compare the organization of the *Manduca* cuticle genes to that of *Drosophila* to see if any controlling elements have been conserved. One way to explore this question will be to transform the *Manduca* cuticle genes into *Drosophila*, as was done by Mitsialis and Kafatos (12) for silkmoth chorion genes.

Expression of the clone I11-B311, selected by the *Drosophila* cuticle gene I-11, occurs at times and amounts appropriate for a larval cuticle gene (Fig. 3). Hybrid selection experiments confirm that this clone encodes a cuticle gene. Restriction mapping is in progress to determine if the three other *Manduca* genomic clones selected with the *Drosophila* clone I-11 are overlapping fragments from the same genomic region as I11-B311.

It will be interesting to map the regions of homology between I11-B311 and the *Drosophila* gene I-11. Comparison of the *Drosophila* cuticle genes reveals a conserved 5' non-coding element, as well as homology within the genes (3). Since a large *Drosophila* genomic fragment was used in screening the library, either of these elements, or even some flanking non-transcribed region, could have give rise to the cross-hybridization detected.

The activity of the *Manduca* larval cuticle genes is controlled by 20-hydroxyecdysone and juvenile hormone. Some of the cuticle genes are permanently turned off after the final larval instar, while the gene for the 14.6 kD protein is active in the flexible intersegmental cuticle of pupal and adult epidermis (5). One way to explore these changes in gene activity will be to investigate the changes in nuclease sensitivity of the cuticle genes as the epidermis progresses through development. In some cases (13,14) nuclease sensitivity is retained in tissues which are not expressing previously active genes, while in others (15,16), the nuclease sensitivity is lost. With the cloned *Manduca* cuticle genes described here, we will be able to investigate the hormonal control of nuclease sensitivity of these genes in *Manduca* epidermis.

REFERENCES

1. Silvert DJ (1985) Cuticular proteins during postembryonic development. In: Kerkut GA and Gilbert LI (eds) "Comprehensive Insect Physiology, Biochemistry and Pharmacology", Oxford: Pergamon, p 239.
2. Riddiford LM (1985) Hormone action at the cellular level. In: Kerkut GA and Gilbert LI (eds) "Comprehensive Insect Physiology, Biochemistry and Pharmacology", Oxford: Pergamon, p 37.
3. Snyder M, Hunkapiller M, Yuen D, Silvert D, Fristrom J, Davidson N (1982). Cuticle protein genes of Drosophila melanogaster: structure, organization and evolution of four clustered genes. Cell 29:1027.
4. Fristrom JW, Alexander S, Brown E, Doctor J, Fechtel K, Fristrom D, Kimbrell D, King DS, Wolfgang WJ (1986). Ecdysone regulation of cuticle gene expression in *Drosophila*. Arch Insect Biochem Physiol, in press.
5. Riddiford LM, Baeckmann A, Hice RH, Rebers J (1986). Developmental expression of three larval cuticle genes of the tobacco hornworm, Manduca sexta. Submitted for publication.

6. Riddiford LM (1986). Hormonal control of sequential gene expression in insect epidermis. In: Law J (ed): "Molecular Entomology. UCLA Symposia on Molecular and Cellular Biology, New Series, vol 49," New York: Alan R. Liss, in press.
7. Maniatis T, Fritsch EF, Sambrook J (1982). "Molecular Cloning: A Laboratory Manual." Cold Spring Harbor, New York: Cold Spring Harbor Laboratory.
8. Frischauf A-M, Lehrach H, Poustka A, Murray N (1983). Lambda repacement vectors carrying polylinker sequences. J Mol Biol 170:827.
9. Benton WD, Davis RW (1977). Screening of λgt recombinant clones by hybridization to single plaques in situ. Science 196:180.
10. Southern E (1975). Detection of specific sequences among DNA fragments separated by gel electrophoresis. J Mol Biol 98:503.
11. Henikoff S, Keene M, Fechtel K, Fristrom J (1986). Gene within a gene: nested Drosophila genes encode unrelated proteins on opposite DNA strands. Cell 44:33.
12. Mitsialis SA, Kafatos FC (1985). Regulatory elements controlling chorion gene expression are conserved between flies and moths. Nature 317:453.
13. Wu C (1980). The 5' ends of Drosophila heat shock genes in chromatin are hypersenstive to DNase I. Nature 286:854.
14. Groudine M, Weintraub H (1982). Propagation of globin DNAase I-hypersensitive sites in absence of factors required for induction: a possible mechanism for determination. Cell 30:131.
15. Groudine M, Kohwi-Shigematsu T, Gelinas G, Stamatoyannopoulos G, Papayannopoulou T (1983). Human fetal to adult hemoglobin switching: changes in chromatin structure of the β-globin gene locus. Proc Nat Acad Sci USA 80:7551.
16. Kok K, Snippe L, Geert AB, Gruber M (1985). Nuclease-hypersensitive sityes in chromatin of the estrogen-inducible apoVLDL gene of chicken. Nucleic Acids Res 13:5189.

Molecular Entomology, pages 211–222

HORMONAL CONTROL OF SEQUENTIAL GENE EXPRESSION IN INSECT EPIDERMIS[1]

Lynn M. Riddiford

Department of Zoology, University of Washington
Seattle, Washington 98195

ABSTRACT The hormonal regulation of 4 larval cuticle genes of the tobacco hornworm, *Manduca sexta* has been studied both *in vivo* and *in vitro* using cDNA clones. The 3 Class I cuticle genes were expressed both in the 4th and 5th larval instars but not during the molting period. Two were larval-specific, whereas the third was expressed in regions of flexible cuticle in all stages. The Class II cuticle genes were first expressed just before the onset of metamorphosis causing the formation of a stiffer cuticle. The expression of these genes was initiated by low concentrations of 20HE in the absence of JH and suppressed by high ecdysteroid. Thus, ecdysteroid serves both to turn on and to turn off larval cuticular gene expression, depending on its concentration. JH modulates both actions.

INTRODUCTION

A major question in developmental biology is how a cell becomes committed to a particular differentiative pathway and how expression of that commitment is controlled. Most differentiated cells, such as muscle or neurons, make only a certain set of specific products. Other cells, such as the insect ovarian follicular cell which produces the vitelline membrane, then the chorion (1), make a sequential series of products. How such sequential expression is regulated on a genomic level is not yet understood (1). The insect epidermis proves to be

[1]This work was supported by grants from NIH and NSF.

an excellent system in which to study this regulation since both stage-specific commitment and the sequential expression of genes is regulated by hormones (2, 3).

Molting and metamorphosis in insects are governed primarily by two hormones: 20-hydroxecdysone (20HE) which initiates and orchestrates the molt and juvenile hormone (JH) which determines whether the molt will be progressive or not (2). If JH is present during a rise in ecdysteroid, then the stage is repeated with the same genes being expressed. If JH is absent at the time of ecdysteroid action, new genes can be expressed. During larval life JH is continuously present, although fluctuations in its titer may occur. During the final larval instar its titer falls; the subsequent rise(s) in 20HE then initiate metamorphosis causing first a commitment of the cells to pupal or adult differentiation, then expression of that commitment.

The insect epidermis produces the overlying cuticle, a complex structure composed of chitin and many different proteins (4). In holometabolous insects larval, pupal, and adult cuticles are usually quite different--a flexible growing larval cuticle, a hardened pupal cuticle, a tanned adult cuticle. A recent study of the proteins of the 3 types of cuticle in the moth, *Hyalophora cecropia*, has shown that although some proteins are stage-specific, others are specific to particular types of cuticle, i.e., flexible versus hard, and thus are found in more than one stage (5). Also, within a particular stage, changes in cuticular protein synthesis may occur leading to changes in cuticular structure (6, 7). The same epidermis in most insects produces these different types of cuticle, and the major switches in gene expression appear to be controlled by 20HE and JH (2, 3, 6, 7).

To study this hormonal control, we have utilized the tobacco hornworm, *Manduca sexta*, where the endocrine control of molting and metamorphosis has been well studied (8, 9) (Fig. 1). During growth of *Manduca* larvae, the epidermis continuously deposits endocuticle to provide mechanical support to the expanding body (10). The same major endocuticular proteins (Class I) are made during each larval instar but are not made during the molting period until shortly before ecdysis (7, 11, 12). These Class I cuticular proteins contribute to the formation of 1μ thick lamellae during the first few days of the 5th (final) larval instar, but on the final day of feeding (day 3) synthesis of most of these proteins ceases (7). Instead a new set of endocuticular proteins (Class II) appears, and thin (0.1-0.2μ) cuticular lamellae are formed (7, 10).

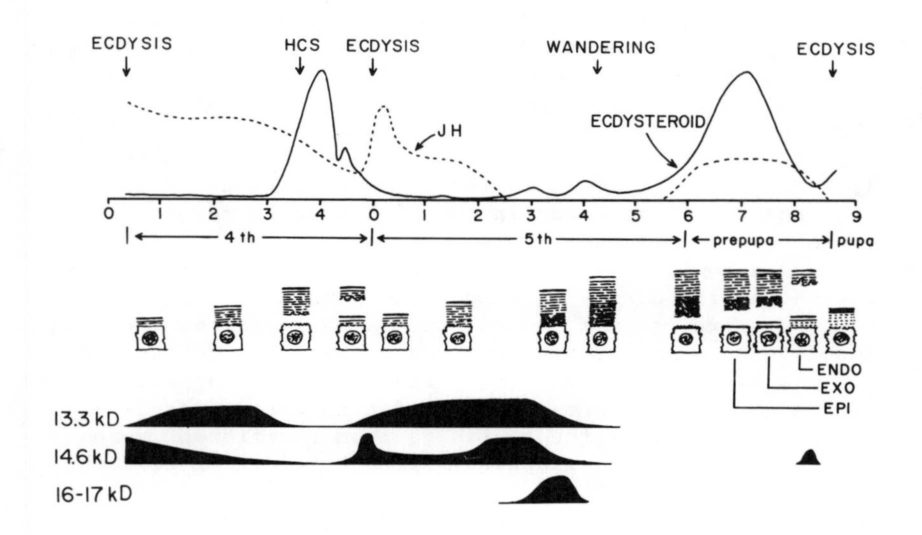

FIGURE 1. Schematic summary of the ecdysteroid and JH titers of Manduca, the cuticular events that they control, and the expression of Class I (13.3 and 14.6 kD) and Class II (16-17 kD) cuticle genes.

On day 3 the cells are exposed to a small rise in ecdysteroid in the absence of JH and become pupally committed by the following day (wandering stage) (13) (Fig. 1). Neither Class I nor Class II cuticular proteins are synthesized by the pupally committed cell (7, 11), although this cell synthesizes several new high molecular weight cuticular proteins (7), a 34 kD "pupal commitment" protein (11), and several unidentified proteins. Pupal cuticular proteins are synthesized beginning on the 3rd day after wandering in response to the prepupal ecdysteroid rise. First are the epicuticular proteins followed by the exocuticular proteins; finally endocuticular proteins are synthesized during the final 19 hrs before ecdysis (11; Boring & Riddiford, unpublished) (Fig. 1).

CLASS I CUTICULAR GENES

To study the hormonal control of the Class I

endocuticular genes, we isolated 3 cDNA clones coding for different cuticular proteins (12.8, 13.3, and 14.6 kD) from a cDNA library prepared from day 2 5th instar epidermal mRNA (14). Analysis of the partial sequences of these cDNA clones using a 65% homology of at least 15 linear base pairs shows no cross-homology among the 3 clones or within the sequenced protein coding regions of the 5 *Drosophila* larval cuticle genes (15). The cDNA coding region for the 12.8 kD protein however shares a 75% homologous 17 bp sequence with a sequence in the 5' noncoding regions of *Drosophila* larval cuticle genes I and II (14). The meaning of such a homology is unclear, and further analysis awaits the sequencing of the corresponding genomic clones.

Putative genomic clones containing genes for the 12.8 and 14.6 kD proteins have been selected from the *Manduca* genomic library (16) and are currently being mapped. Three of the 4 genomic clones for the 14.6 kD protein are overlapping whereas the fourth is from a different genomic region (16). The overlapping region hybridizing with the cDNA is flanked by 9 and 11 kb on either side, indicating that we likely have both the gene and its 5' and 3' flanking regions.

Northern and dot blot hybridization analysis of the developmental expression of these 3 genes showed that all are expressed during the feeding stage of the 4th instar. The RNAs then disappear during the early part of the molt as the ecdysteroid titer increases (14) (Fig. 1). Although they all reappear beginning about 14 hrs before ecdysis as endocuticle begins to be deposited (7), each shows a somewhat different time course of increase with the RNA for the 14.6 kd protein having a pronounced peak shortly before ecdysis. All decline rapidly on day 3 of the 5th instar when the cells become pupally committed, again each with its own time course. The mRNAs for the 12.8 and 13.3 kD proteins do not reappear in epidermal cells during the prepupal period or during synthesis of the pre- or post-ecdysial pupal or adult cuticle. Consequently, they are larval-specific. By contrast, the mRNA for the 14.6 kD protein reappears in the pharate pupa and the pharate adult, but only in the *inter*segmental epidermis (14) (Fig. 1). Therefore, it seems to be expressed by cells producing flexible cuticle as will be discussed in more detail below.

Culture of 4th instar epidermis *in vitro* showed that exposure to 2.5 μg/ml 20HE {equivalent to the molting surge of ecdysteroid (17)} in the presence of JH for 18 hrs followed by culture in hormone-free media for 30 hrs resulted in the formation of a new larval cuticle. These mRNAs disappeared by 18 hrs (14) and subsequently reappeared during the formation of the new cuticle in the absence of 20HE (unpublished). When day 2 5th instar epidermis was cultured 24 hr with 100 ng/ml 20HE {about the concentration of the commitment peak on day 3 (7)}, 75% of the cells became pupally committed and all 3 Class I cuticular RNAs declined (14). The RNA for the 14.6 kD protein showed the most pronounced decline as expected from its more rapid decrease on day 3. The addition of JH prevented the change to pupal commitment and slowed the 20HE-induced decline in the RNAs for the 2 lower molecular weight proteins (14). JH alone however suppressed expression of the gene for the 14.6 kD protein so that its effects on the 20HE-induced suppression could not be assessed.

These studies together show that Class I cuticular genes are expressed as the larva grows during the intermolt period but then are turned off in response to 20HE. When they are exposed to ecdysteroid in the presence of JH (as in a larval molt), the turning-off is transient. By contrast, ecdysteroid acting in the absence of JH may result in the permanent inactivation of larval-specific genes.

CLASS II CUTICLE GENES

During the deposition of the thin lamellae on the final day of feeding in the last larval instar, at least 3 major new cuticular proteins (16, 17 and 27 kD) (designated Class II) are produced (7). At this time the cuticle becomes much stiffer (18), apparently in preparation for burrowing. Through a combination of *in vivo* and *in vitro* experiments using both the appearance of thin lamellae and the appearance of the 27 kD protein, we found that this switch was hormonally controlled by a small rise in ecdysteroid to 30 ng/ml 20HE equivalents after the JH titer fell on day 2 (7). It could be mimicked *in vitro* only by exposure of day 1 epidermis to 10-100 ng/ml 20HE in the absence of JH for 18-24 hrs (3, 7). The levels of RNAs for the Class I 12.8 and 13.3 kD proteins were relatively unaffected by these concentrations of 20HE, whereas that for the 14.6 kD protein showed a dose-dependent decrease (3).

To study this ecdysteroid-induced change further, we isolated day 3-specific clones from a λgt10 cDNA library to day 3 epidermal RNA, 2 of which each hybridized with RNA that coded for both the 16 and 17 kD cuticular proteins (Horodyski and Riddiford, unpublished). In Northern hybridization analysis, the 2 RNAs were not separable. Whether there are 2 highly homologous genes involved or whether alternative splicing may account for this phenomenon is presently under study. Preliminary sequence analysis of the nearly full length cDNA from one clone indicates the presence of a 126 codon open reading frame that would code for a 15 kD protein. A screening of the genomic library (16) with the cDNA clone selected 10 clones, 6 of which have been characterized by restriction mapping and Southern hybridization and represent at least two different genomic regions. Both regions contain 3 hybridizing sub-regions.

Northern and RNA dot blot hybridization analysis showed that these two RNAs first appear on day 2 of the 5th instar coincident with the small rise in ecdysteroid, then attain their maximal expression on day 3 and are absent by the end of the wandering stage (Horodyski and Riddiford, unpublished). These 2 RNAs also appeared in day 1 epidermis cultured with 25 ng/ml 20HE but not in the absence of hormone, indicating that they, just as the gene for the 28 kD protein, require not only the decline of JH but also the subsequent small rise in ecdysteroid for their expression. These RNAs were expressed in day 2 epidermis cultured in hormone-free or JH-containing medium, but not after 24 hrs exposure to 1 μg/ml 20HE in either the presence or absence of JH (19). Thus, once they are turned on by low ecdysteroid in the absence of JH, JH alone can no longer prevent their expression.

FLEXIBLE CUTICLE GENES

Cox and Willis (1985) have shown that some cuticular proteins seem to be a component of a certain type of cuticle such as the flexible cuticle of the larva and that of the intersegmental membrane of the adult. In *Manduca* the Class I larval cuticle gene that codes for the 14.6 kD protein appears to be one specific for flexible cuticle. This gene is equally expressed in the intra- and intersegmental regions of the larval abdominal epidermis. At metamorphosis it is shut off in the intrasegmental region and henceforth expressed only in the intersegmental region (14; Fig. 2). Moreover, in the latter it is expressed only for a very short period early during the pre-ecdysial deposition of both pupal and adult endocuticle. The timing of this expression coincides with the first peak of its expression during a larval molt. *In situ* hybridization studies are now necessary to localize the particular cells that might be producing this RNA in abundance at these times to provide a clue as to the role of this protein.

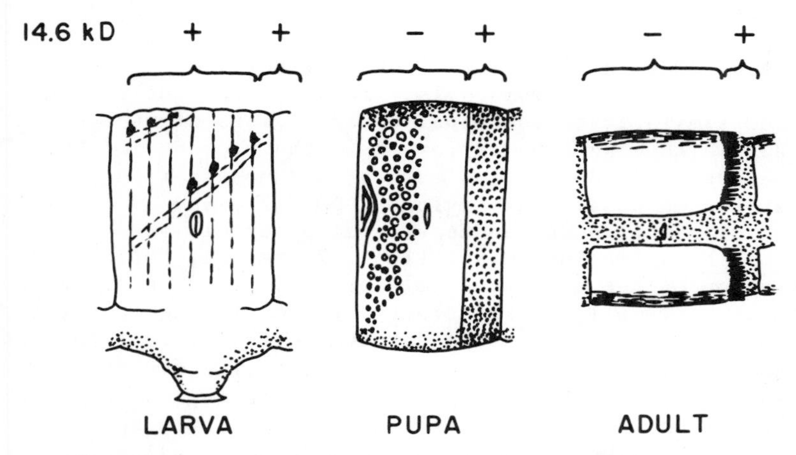

FIGURE 2. Schematic summary of the regional expression of the "flexible cuticle" gene in the larva, pupa and adult.

This regional specification at metamorphosis indicates that the regulation of this gene must be more complex than that of the other Class I genes. At a larval molt ecdysteroid in the presence of JH causes a transient cessation of transcription all over the segment. At metamorphosis ecdysteroid acting in the absence of JH must permanently repress the gene in the intrasegmental region, but only transiently repress it in the intersegmental region. Then subsequently when the molting surges of ecdysteroid fall, it can be turned on at a specified time.

This gene is also aberrant in its regulation by JH in that exposure to JH alone causes a loss of this RNA. Whether this repression is a direct action of JH on gene expression or whether JH may affect the stability of the mRNA is not known. Whatever its action, this appears to be an action of JH that does not require the presence of ecdysteroid and thus is different from the other known morphogenetic actions of JH which are only observed when ecdysteroid initiates a developmental event (2).

CONCLUSIONS

The Class I larval cuticle genes are expressed during the feeding, growing phase of each instar, but are inactive during the molt when growth ceases and the epicuticle is deposited (7). The cessation of mRNA synthesis appears to be dependent on the rising titer of ecdysteroid since high levels of 20HE *in vitro* for 18 hrs severely depressed these levels in 4th instar epidermis. These genes thus respond to 20HE similarly to the pupal cuticle genes of *Drosophila* (6, 20) which cannot be expressed in the presence of 20HE. High ecdysteroid during the molt also suppresses the expression of dopa decarboxylase (DDC) (21) and insecticyanin (12) in *Manduca* larval epidermis and of DDC in *Drosophila* imaginal discs (22). Therefore, during the initiation of the molt, the role of ecdysteroid appears to be to shut off all nonessential syntheses in the epidermis so that all resources may be devoted to the production of a new epicuticle. This suppressive effect of ecdysteroid is likely similar to this hormone's action in repressing transcription of the salivary gland glue protein genes (23). Whether these effects are all at the level of transcription or are also at the level of mRNA stability is not yet known.

As the ecdysteroid titer falls at the end of the larval molt, the suppressed genes again are expressed. All three Class I larval cuticle RNAs reappear about the same time, but thereafter the mRNA for the 14.6 kD protein shows a completely different pattern of expression. Likely the fall of ecdysteroid to some threshold level is critical for the onset of expression of these genes as has been found in *Drosophila* imaginal discs (6, 20, 22). The subsequent modulation of expression then may either be due to differing sensitivities to fluctuating hormone titers or to an autonomous program of expression as seems to be involved in chorion deposition (1).

The role of JH in the maintenance of the larval condition is two-fold. Firstly, it prevents the permanent cessation of the expression of larval-specific genes until the time of metamorphosis as nicely illustrated by its effects on the Class I cuticle genes. Secondly, JH prevents the ecdysteroid-induced expression of certain larval cuticle (Class II) genes that apparently have their primary function at the initiation of metamorphosis. To elucidate how JH modulates the action of ecdysteroid on this genes, we must first determine how ecdysteroid acts on these cuticle genes.

The action of 20HE on the glue protein genes seems to be a direct action of the ecdysteroid-receptor complex on these genes (23). By contrast, its action on these cuticle genes is apparently not a primary action since it does not occur in the presence of cycloheximide (24) although a change in the stability of the mRNA has not been ruled out in these experiments. The isolation of genomic clones for two of the Class I and two of the Class II genes will allow us to compare the gene structure and putative 5' (and other) regulatory regions which could provide clues as to how these genes could be regulated by both hormones.

DNAse I hypersensitive sites are characteristic of active genes (25) and can be lost, thereby signalling a change in the chromatin conformation when a gene is permanently repressed as in the case of the fetal hemoglobin gene in the adult (26). Thus, one attractive possibility is that JH prevents the permanent repression of larval cuticle genes during a larval molt by preventing the loss of certain hypersensitive sites. Similarly, its presence in the cell could prevent the ecdysteroid-induced change in chromatin conformation that is necessary for the expression of the Class II genes. Other mechanisms are of course possible for this modulatory morphogenetic action of JH. The isolation of specific genes that JH modulates is an exciting beginning to the search for these molecular mechanisms.

ACKNOWLEDGEMENTS

I thank Robert Hice and Janell Green for technical assistance and Professor James Truman for critical comments on this manuscript.

REFERENCES

1. Regier JC and Kafatos FC (1985) Molecular aspects of chorion formation. In: Kerkut GA and Gilbert LI (eds)

"Comprehensive Insect Physiology, Biochemistry and Pharmacology", Oxford: Pergamon, p. 113.

2. Riddiford LM (1985) Hormone action at the cellular level. In: Kerkut GA and Gilbert LI (eds) "Comprehensive Insect Physiology, Biochemistry and Pharmacology", Oxford: Pergamon, p. 37.
3. Riddiford LM (1986) Hormonal regulation of sequential larval cuticular gene expression. Arch Insect Biochem Physiol: in press.
4. Silvert DJ (1985) Cuticular proteins during postembryonic development. In: Kerkut GA and Gilbert LI (eds) "Comprehensive Insect Physiology, Biochemistry and Pharmacology", Oxford: Pergamon, p. 239.
5. Cox DL and Willis JH (1985) The cuticular proteins of *Hyalophora cecropia* from different anatomical regions and metamorphic stages. Insect Biochem 15: 349.
6. Doctor J, Fristrom D and Fristrom JW (1985) The pupal cuticle of Drosophila: biphasic synthesis of pupal cuticle proteins *in vivo* and *in vitro* in response to 20-hydroxyecdysone. J Cell Biol 101: 189.
7. Wolfgang WJ and Riddiford LM (1986) Larval cuticular morphogenesis in the tobacco hornworm, *Manduca sexta* and its hormonal regulation. Devel Biol 113: 305.
8. Bollenbacher WE, Smith SL, Goodman W and Gilbert LI (1981) Ecdysteroid titer during larval-pupal-adult development of the tobacco hornworm, *Manduca sexta*. Gen Comp Endocrin 44: 302.
9. Sedlak BJ, Marchione L, Devorkin B and Davino R (1983) Correlations between endocrine gland ultrastructure and hormone titers in the fifth larval instar of *Manduca sexta*. Gen Comp Endocrin 52: 291.
10. Wolfgang WJ and Riddiford LM (1981) Cuticular morphogenesis during continuous growth of the final instar larva of a moth. Tiss Cell 13: 757.
11. Kiely ML and Riddiford LM (1985) Temporal programming of epidermal cell protein synthesis during the larval-pupal transformation of *Manduca sexta*. Wilh Roux's Arch Devel Biol 194: 325.
12. Riddiford LM, Kiely ML and Wolfgang WJ (1986) Hormonal regulation of larval-specific protein synthesis in *Manduca* epidermis. In: Kurstak E (ed) "Techniques in Cell Biology, Vol. 2: In vitro invertebrate hormones and genes", Amsterdam: Elsevier/North Holland, in press.

13. Riddiford LM (1978) Ecdysone-induced change in cellular commitment of the epidermis of the tobacco hornworm, Manduca sexta, at the initiation of metamorphosis. Gen Comp Endocrin 34: 438.
14. Riddiford LM, Baeckmann A, Hice RH and Rebers J (1986) Developmental expression of 3 larval cuticle genes of the tobacco hornworm, Manduca sexta. Submitted for publication.
15. Snyder M, Hunkapillar M, Yuen D, Silvert D, Fristrom J and Davidson N (1982) Cuticle protein genes of Drosophila: Structure, organization, and evolution of four clustered genes. Cell 29: 1027.
16. Rebers J, Horodyski FL, Hice RH and Riddiford LM (1986) Genomic cloning of developmentally regulated larval cuticle genes from the tobacco hornworm, Manduca sexta. In: Law J (ed) "Molecular Entomology. UCLA Symp Molec Cell Biol, New Series, vol. 49", New York: Alan R. Liss, in press.
17. Curtis AT, Hori M, Green JM, Wolfgang WJ, Hiruma K and Riddiford LM (1984) Ecdysteroid regulation of the onset of cuticular melanization in allactectomized and black mutant Manduca sexta larvae. J Insect Physiol 30: 597.
18. Wolfgang WJ (1984) "The Basis of Structural Morphogenesis in the Larval Cuticle of the Tobacco Hornworm, Manduca sexta". Ph.D. Thesis, Univ. of Washington, Seattle.
19. Riddiford LM (1982) Changes in translatable mRNAs during the larval-pupal transformation of the epidermis of the tobacco hornworm. Develop Biol 92: 330.
20. Fristrom JW, Alexander S, Brown E, Doctor J, Fechtel K, Fristrom D, Kimbrell D, King DS and Wolfgang WJ (1986) Ecdysone regulation of cuticle protein gene expression in Drosophila. Arch Insect Biochem Physiol, in press.
21. Hiruma K and Riddiford LM (1985) Hormonal regulation of dopa decarboxylase during a larval molt. Devel Biol 110: 509.
22. Clark WC, Doctor J, Fristrom JW and Hodgetts RM (1986) Differential responses of the dopa decarboxylase gene to 20-OH-ecdysone in Drosophila melanogaster. Devel Biol 114: in press.
23. Crowley TE and Meyerowitz EM (1984) Steroid control of RNAs transcribed from the Drosophila 68C polytene chromosome puff. Devel Biol 102: 110.

24. Riddiford LM (1981) Hormonal control of epidermal cell development. Amer Zool 21: 751.
25. Weisbrod S (1982) Active chromatin. Nature 297: 289.
26. Groudine M, Kohwi-Shigematsu T, Gelinas R, Stamatoyannopoulos G, and Papayannopoulous T (1983) Human fetal to adult hemoglobin switching: changes in chromatin structure of the β-globin gene locus. Proc Nat Acad Sci US 80: 7551.

Molecular Entomology, pages 223–231

BLASTODERM-SPECIFIC GENES IN DROSOPHILA[1]

Judith A. Lengyel

Department of Biology and
Molecular Biology Institute
UCLA
Los Angeles, Calif. 90024

Transitions in nuclear replication and RNA transcription, as well as determination of cells with regard to ectodermal segment identity, occur by the end of the cellular blastoderm stage of *Drosophila* embryogenesis. Genes active specifically at this stage are expected to be involved in these processes. I review here what is currently known regarding blastoderm-specific genes. Of those which have been identified on the basis of their mutant phenotype, two are required for segment pattern determination, and two for normal gastrulation. Blastoderm-specific genes have also been identified by molecular screening (competition hybridization) of a genomic DNA library. The homologies revealed by sequence analysis suggest that some blastoderm-specific genes may be regulators of the activity of other genes.

INTRODUCTION

More than any other insect, *Drosophila* is the subject of detailed molecular and genetic scrutiny of its early embryogenesis. Analysis of the processes occurring during *Drosophila* early

[1]This work was supported by NIH Grant HD09948 to J.A.L., and NSF Grant Grant PCM21830 to J.A.L. and John Merriam.

embryogenesis should provide insight into the cellular and determinative events which set the stage for all of later development, both in Drosophila and in other organisms.

The unfolding of events that occurs during early embryogenesis is the result of a complex interaction between the maternal program and zygotic gene activity. Gene activity during oogenesis leads to the formation of the large and highly organized egg. Many thousands of gene products, in the form of maternal mRNA and protein, are present in the mature egg (reviewed in ref. 1). After fertilization, transcription from the zygotic genome plays an important role in replacing decaying maternal transcripts. In addition to this homeostatic role, transcription from the zygotic genome also introduces new genetic information. The interaction of the new informational molecules with the maternally encoded program results in greater organizational complexity of the embryo.

The events which occur during early embryogenesis suggest that novel zygotic gene activity occurs by the end of the cellular blastoderm stage of Drosophila embryogenesis. After fertilization the nuclei undergo 9 syncytial cleavages in the center of the embryo. They then migrate to the surface of the embryo where they undergo 4 more syncytial cleavages; this constitutes the syncytial blastoderm stage (2). During the subsequent cellular blastoderm stage (2.5-3.5 hrs post-fertilization) dramatic transitions occur (reviewed in ref 3). These include: 1) a transition in nuclear multiplication, from rapid and synchronous to slow and asynchronous, 2) activation of RNA transcription, to the highest rate during embryogenesis, 3) cellularization and the beginning of the first true cell cycle, and 4) the determination of cells to give rise to different ectodermal segment derivatives of the larva and adult. The movements of gastrulation ensue at the end of the cellular blastoderm stage.

IDENTIFICATION AND CHARACTERIZATION OF BLASTODERM-SPECIFIC GENES

The occurrence of unique events at the cellular blastoderm stage suggests that there should be genes specifically active at this time which control these processes. I define blastoderm-specific genes as those with the great preponderance of their activity at the blastoderm stage; there may be a small amount of activity in subsequent hours of embryogenesis.

Blastoderm-specific Genes Identified by Mutant Phenotype.

Genes that affect the segmental pattern of the larval cuticle might be expected to act at the cellular blastoderm stage, when segment determination is occurring. Both molecular and genetic analysis of the segment pattern genes _fushi tarazu_ (_ftz_) and _Krüppel_ (_Kr_) reveal that they encode transcripts whose expression is largely confined to the blastoderm stage (4-8, Table 1). Genes required for the initiation of gastrulation, which occurs immediately after cellularization of the blastoderm, might also be expected to have a blastoderm-specific expression pattern. The genes _zerknüllt_ (_zen_) and _twisted gastrulation_ (_tsg_), which are required for normal gastrulation, also appear to be essentially blastoderm-specific in their expression (4,9-11, Table 1). Thus in some cases the phenotype of a gene suggests that it may be specifically active at the blastoderm stage.

Blastoderm-specific Genes Identified by Molecular Screening.

The involvement of a particular gene in a process specific to the blastoderm stage does not mean that a defect in such a gene will necessarily result in a phenotype that can be predicted _a priori_. Thus it may be difficult to identify all blastoderm-specific genes by genetic screening of mutant phenotypes. An alternative approach is to identify blastoderm-specific genes on the basis of a molecular phenotype, i.e. the production of a blastoderm-specific transcript. ^{32}P-kinased blastoderm stage poly(A)$^{+}$ RNA was used to screen a *Drosophila* genomic library in the presence of an excess of unlabeled poly(A)$^{+}$ RNA from newly laid eggs. Putative positive clones were rescreened differentially with either cDNA or RNA probes. Finally, each putative clone was used to probe a developmental RNA gel blot. As a result of these screens, four blastoderm-specific genes (*bsg*'s) were identified (12,13). These have been given designations according to their chromosomal location (e.g. *bsg25D* maps to 25D). One locus (*bsg99D*), identified in an earlier screen due to its proximity to a ribosomal protein gene, has been named *serendipity* (*sry*) (14,15). The characteristics of the *bsg*'s are summarized in Table 1.

Each *bsg* locus encodes an RNA which is at least 50-100 fold more abundant at the blastoderm than any other stage examined (12-15). The proteins encoded by these transcripts are thus most likely to carry out their primary function at the blastoderm stage. The chromosome locations of the *bsg*'s suggest that they do not correspond to genes which have been identified by mutational analysis as affecting the embryonic pattern (16-18). The *bsg*'s therefore appear to define new genes involved in events at the blastoderm stage.

Sequence Homologies to DNA-binding Proteins.

For a number of the blastoderm-specific genes, sequence analysis has revealed homologies which suggest possible functions. The *ftz* gene and, most likely, the *zen* gene each encodes a homeodomain homology unit (10,19,20). The 60 amino acid homeodomain was first detected in certain *Drosophila* genes, but has now been found in a number of genes in other organisms, including vertebrates (reviewed in refs. 21,22). Both its homology to bacterial and yeast gene regulatory proteins (23), and direct experimental evidence (24) suggest that it is a DNA-binding domain.

TABLE 1
BLASTODERM-SPECIFIC GENES

Gene		Chrom. Locat.	Sequence Homology	Ref.
ftz		84AB	homeodomain	(4,6,19,20,23)
Kr		60F	DNA-binding finger TFIIIA	(5,7,8)
zen		84AB	homeodomain[a]	(4,10)
tsg		11A	n.d.[b]	(9,11)
bsg25A		25A	n.d.[b]	(13)
bsg25D		25D	n.h.[c]	(12,27)
bsg75C	λ[d]	75C	DNA binding finger TFIIIA	(12,26)
	σ[d]		" " " "	
sry	δ[d,e]	99D	DNA binding finger TFIIIA	(12,14, 15)
	β[d,e]		" " " "	
	α[d]		n.h.[c]	

[a]Two homeodomains have been mapped to the zen region (10).
[b]Sequence not determined.
[c]Significant homology to other proteins not detected.
[d]Greek letters refer to different transcripts from a particular bsg locus.
[e]The δ and β transcription units are part of a blastoderm-specific locus, in that they closely flank the blastoderm-specific α transcription unit (14,15) and are expressed maximally at the blastoderm stage (12); they are not, however, blastoderm-specific genes in the strict sense.

Homology to a different DNA binding domain is found in the predicted proteins encoded by several other blastoderm-specific loci. The Kr gene, and two transcripts each from the sry and bsg75C loci encode proteins which contain one to seven repeats of a motif with homology to the proposed "DNA-binding finger" of Xenopus transcription factor IIIA (8,15,25,26). This motif is 21-25 amino acids long and contains two cysteine and two histidine residues, which are postulated to act as ligands for zinc (26). The high proportion of positively charged residues in the "finger" are believed to mediate binding to DNA (26).

Genetic Analysis of Genes Identified in Molecular Screens.

The phenotypes of the blastoderm-specific genes ftz and Kr, and zen and tsg, suggest that these genes are involved in segment determination and gastrulation, respectively. If mutations could be obtained in the bsg's identified by molecular cloning, their phenotypes might provide insight into the function of these genes. Because of the richness of Drosophila genetics, it is possible (if the gene in question is mutable to a lethal or visible phenotype) to obtain mutations in a gene isolated on the basis of a molecular, rather than a genetic phenotype.

Saturation mutagenesis screens have been initiated in the chromosomal regions to which bsg75C and sry map. By the use of P-factor transformation rescue, 4 lethal mutations, one of which has an embryonic lethal phase, have been mapped to the δ transcription unit of the sry locus (28). Thus at least one of the transcripts from a blastoderm-specific locus identified by molecular screening has been shown to be required for normal embryogenesis.

CONCLUSIONS AND FUTURE PROSPECTS

The blastoderm stage is a time of crucial transitions in early embryogenesis. Genes specific to, and thus likely to be controlling processes which occur at this stage, have been identified by both mutational and molecular approaches.

It is striking that all the homologies found in the proteins encoded by the blastoderm-specific loci are to DNA-binding domains. These homologies suggest that the products of some of the blastoderm-specific loci are likely to regulate the expression of other genes. The appearance of new DNA regulatory proteins at the cellular blastoderm stage could be involved in changes in the cell cycle and in transcription rate, as well as in the cell determination which occur at this time.

For those genes which have been identified only by molecular phenotype, the localization of the gene products in the embryo by means of _in situ_ hybridization and antibody staining should provide additional important information about their role in embryogenesis. The isolation of embryonic lethal mutations in the _sry_ δ gene suggest that at least some of the _bsg_ loci are required for embryogenesis. Characterization of the phenotypes of these and other mutants in blastoderm-specific loci should provide insight into the function of these genes in early embryogenesis.

ACKNOWLEDGEMENTS

I thank Richard Baldarelli, Paul Mahoney, Teresa Strecker, Kritaya Kongsuwan and John Merriam for helpful comments on the manuscript, and the many individuals, cited in the References, who allowed me to refer to their results prior to publication.

REFERENCES

1. Hough-Evans BR Anderson DM (1981). in Buckingham ME (ed): "Biochemistry of Cellular Regulation, vol. III Development and Differentiation", CRC Press, p 83.
2. Foe VE, Alberts BM (1983). J Cell Sci 61:31.
3. Lengyel JA, Roark M, Kongsuwan K, Mahoney PA, Boyer PD, Merriam JR (1985). in Sawyer RH, Showman RM (eds) "Cellular and Molecular Biology of Invertebrate Development", South Carolina: University of South Carolina Press, p 239.
4. Wakimoto BT, Turner FR, Kaufman, TC (1984). Dev Biol 102:147.
5. Wieschaus E, Nüsslein-Volhard C, Kluding H (1984). Dev Biol 104:172.
6. Weiner AJ, Scott MP, Kaufman TC (1984). Cell 37:843.
7. Preiss A, Rosenberg UB, Kienlin A, Seifert B, Jäckle H (1985). Nature 313:27.
8. Rosenberg UB, Schröder C, Preiss A, Kienlin A, Côté, Riede I, Jäckle H (1986). Nature 319:336.
9. Zusman SB, Wieschaus EF (1985). Dev. Biol 111:359.
10. Doyle H, Hoey T, Harding K, Levine M. manuscript submitted.
11. Goralski T, Konrad K, Mahowald A, personal communication.
12. Roark M, Mahoney PA, Graham ML, Lengyel JA 85). Dev. Biol 109:476.
13. Waite P, Snyder M, Lengyel JA, manuscript in preparation.
14. Vaslet CA, O'Connell P, Izquierdo M, Rosbash M (1980). Nature 285:674.
15. Vincent A, Colot HV, Rosbash M (1985). J Mol Biol 186:149.
16. Nüsslein-Volhard C, Wieschaus E, Kluding H (1984). Roux's Arch Dev Biol 193:267.
17. Jürgens G, Wieschaus E, Nüsslein C, Kluding H (1984). Roux's Arch Dev Biol 193:283
18. Wieschaus E, Nüsslein-Volhard C, Jürgens G (1984). Roux's Arch Dev Biol 193:296.
19. Scott MP, Weiner AJ (1984). Proc Natl Acad Sci USA 81:4115.

20. McGinnis W, Levine MS, Hafen E, Kuroiwa A, Gehring WJ (1984). Nature 308:428.
21. Gehring W (1985). Cell 40:3.
22. Manley JL, Levine MS (1985). Cell 43:1.
23. Laughon A, Scott MP (1984). Nature 310:25.
24. Desplan C, Theis J, O'Farrell PH (1985). Nature 318:630.
25. Baldarelli R, Boyer P, Chang MF, Gustavson E, Salas F, Roark M, Lengyel JA, manuscript in preparation.
26. Miller J, McLachlan AD, Klug A (1985) EMBO J 4:1609.
27. Boyer PD, Lengyel JA, manuscript in preparation.
28. Kongsuwan K, Vincent A, Lengyel JA, Merriam JR, manuscript in preparation.

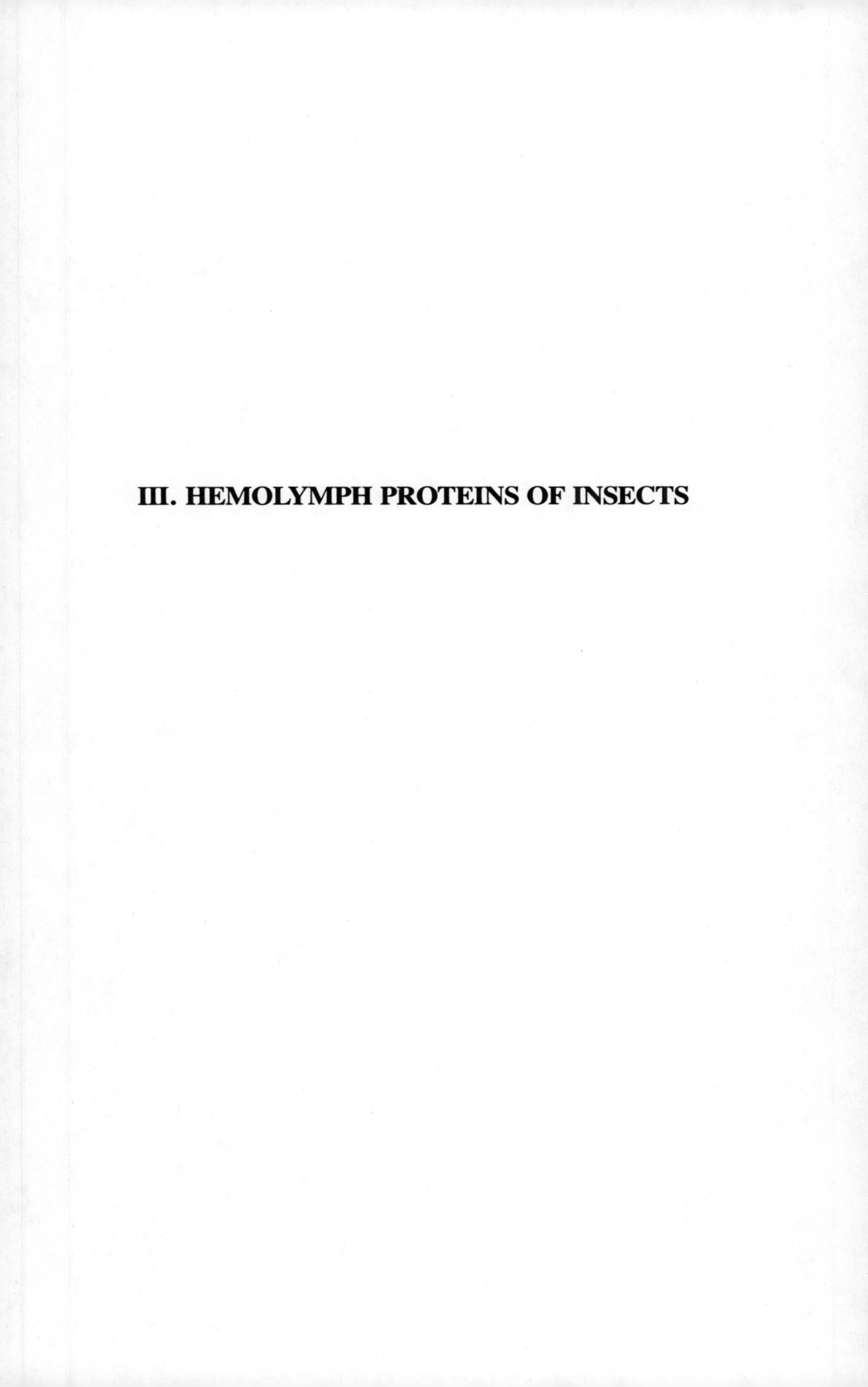

III. HEMOLYMPH PROTEINS OF INSECTS

Molecular Entomology, pages 235–245

APOLIPOPHORIN III IN LOCUSTS, IN RELATION TO THE ACTION OF ADIPOKINETIC HORMONE

HARUO CHINO

Biochemical Laboratory, Institute of Low Temperature Science, Hokkaido University, Sapporo, Japan.

ABSTRACT Three molecular species of apolipophorin III are present in adult locust hemolymph and named apo-III-a, apo-III-b, and apo-III-c, respectively. They are indistinguishable by SDS-polyacrylamide gel electrophoresis (PAGE), immunodiffusion, and in amino acid composition; however, they have different isoelectric points (5.43 for a, 5.11 for b, and 4.98 for c) and, therefore, can be separated by native- or urea-PAGE. All three apo-IIIs are glycoproteins and contain fucose, mannose, and glucosamine. The total sugar content amounts to about 11% for each of the three apo-IIIs. The molecular weight of apo-III determined by SDS-PAGE is approximately 20,000 which is almost equivalent to the native molecular weight (approximately 19,000) estimated by the sedimentation-equilibrium method, thereby indicating that the locust apo-III exists in hemolymph as a monomeric form. It was also demonstrated that a total 9 moles apo-III (2 moles apo-III-a, 6 moles apo-III-b, and 1 mole apo-III-c) associate with each mole of lipophorin in response to the action of locust adipokinetic hormone.

INTRODUCTION

Lipophorin is the major lipoprotein in the hemolymph of most insects (1), and serves as a reusable shuttle to transport various lipids including diacylglycerol, hydrocarbons, cholesterol and carotenoids between tissues (2-7). Insect adipokinetic hormone (AKH), first discovered in locusts by Mayer and Candy (8), is released from the corpora cardiaca during flight and stimulates the loading by lipophorin of diacylglycerol from the fat body, resulting in the formation of larger lipophorin particles with increased diacylglycerol content (9-11). Injection of AKH into adult locusts causes the association of a low molecular weight, non-lipid-containing protein with lipophorin (12-15). Similar observations have been extended to the adult tobacco hornworm, *Manduca sexta*, by Shapiro and Law (16) who proposed the term apolipophorin III (apo-III) for the low molecular weight protein that becomes associated with lipophorin in response to AKH injection. The purification of apo-III from the hemolymph of *M. sexta* has been recently achieved in their laboratory (17).

We (18) extended these observations to the locust; the injection of AKH promotes the association of free apo-III in hemolymph with lipophorin and the loading of diacylglycerol from the fat body by lipophorin to produce larger, lower density lipophorin particles. These structural changes of lipophorin in response to AKH are completely reversible, so that within 24 hr of AKH injection, the apo-III dissociates from lipophorin and the size and density of lipophorin return to the original values observed in resting locusts.

The mechanism by which the above structural changes of lipophorin are induced under the action of AKH still remains unresolved. The resolution of this mechanism requires the availability of purified apo-III. We have recently achieved the purification of locust apo-III. This paper reports the physico-chemical properties of locust apo-III in relation to the action of adipokinetic hormone. More detailed informations are given in other communication (19).

MATERIALS AND METHODS

Animals and collection of hemolymph

Adult locusts, *Locusta migratoria*, (3-5 weeks after the final molt) were taken from colonies maintained in this laboratory.

Preparation of activated lipophorin

The activated lipophorin associated with apo-III was prepared by potassium bromide density-gradient ultracentrifugation from the hemolymph of male locusts following injection of synthetic AKH (10 pmoles/insect) (16).

Purification of apolipophorin III from resting locust hemolymph

The lipophorin and vitellogenin-free supernatant was first prepared from hemolymph of resting male and female locusts (2), which was then used as starting material for the purification of apo-III. The purification procedures include the precipitation by ammonium sulfate, gel filtration chromatography on Ultrogel ACA 44, and ion exchange chromatography on DEAE Toyopearl M650. The detailed procedures are described in other communication (19). Three molecular species of apo-IIIs were separated at the final step of purification on the DEAE column, and named apo-III-a, apo-III-b, and apo-III-c, respectively.

RESULTS AND DISCUSSION

SDS-polyacrylamide gel electrophoresis of apolipophorin III

SDS-PAGE of locust apo-IIIs is illustrated in Fig. 1 and demonstrates that both apo-III-b and apo-III-c are homogeneous and contain no other proteins (Fig. 1, D, E). However, the apo-III-a fraction is not homogeneous and contaminated with a lower molecular weight protein (Fig. 1, B). The crude apo-III-a fraction was rechromatographed on

Ultrogel ACA 44 to eliminate the contaminating protein. As shown in Fig.1 C, the rechromatographed apo-III-a is sufficiently pure and the minor contaminant is completely eliminated. Fig. 1 also demonstrates that the three apo-IIIs are not distinguishable on SDS-PAGE. The molecular weight determined on SDS-PAGE was estimated to be 20,000.

The homogeneity of apo-III-a (rechromatographed), apo-III-b, and apo-III-c was also demonstrated by immunodiffusion in which IgG prepared against apo-III-b was used as an antibody (data not shown). Both apo-III-b and apo-III-c cross-react with anti-apo-III-b, indicating that the three apo-IIIs are not immunologically distinguishable.

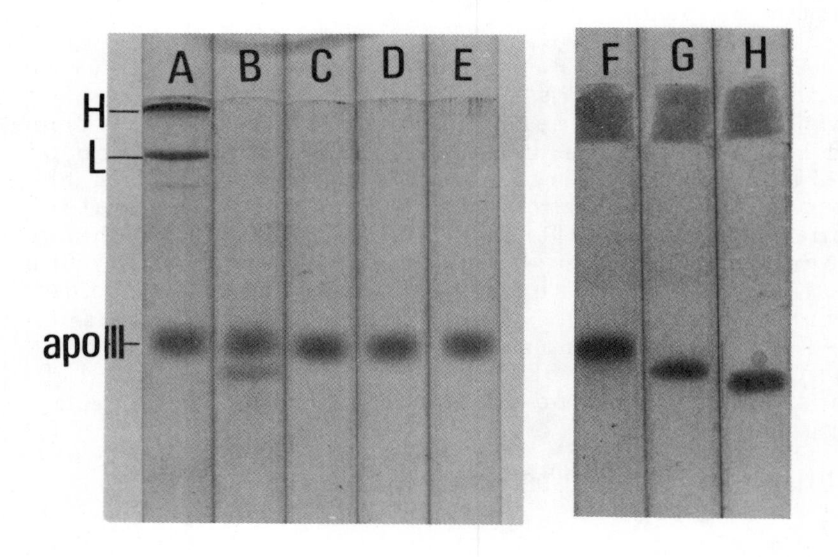

FIGURE 1. Left, SDS-PAGE (10% gel) of three apo-IIIs. A, activated lipophorin as reference; B, crude apo-III-a; C, rechromatographed apo-III-a; D, apo-III-b; E, apo-III-c. H, heavy chain; L, light chain. Resting lipophorin comprises heavy chain and light chain only (2). Right, native-PAGE (10% gel) of apo-IIIs. F, apo-III-a; G, apo-III-b; H, apo-III-c.

Electrophoretic behavior of three apo-IIIs on native- or urea-PAGE and their isoelectric points

The data presented in the previous section indicate that the three apo-IIIs are indistinguishable on the basis of their mobility on SDS-PAGE. However ,as illustrated in Fig. 1 , the three apo-IIIs are separable on native-PAGE; they migrate rapidly in the order of apo-III-c, apo-III-b, and apo-III-a.

Given that there are three molecular species of apo-III, it is important to know which apo-III becomes associated with lipophorin in response to AKH injection and, indeed, if all the apo-IIIs can associate. In order to resolve this problem, activated lipophorin was run on urea-PAGE together with the three apo-IIIs for reference. The results are illustrated in Fig. 2 and clearly demonstrate

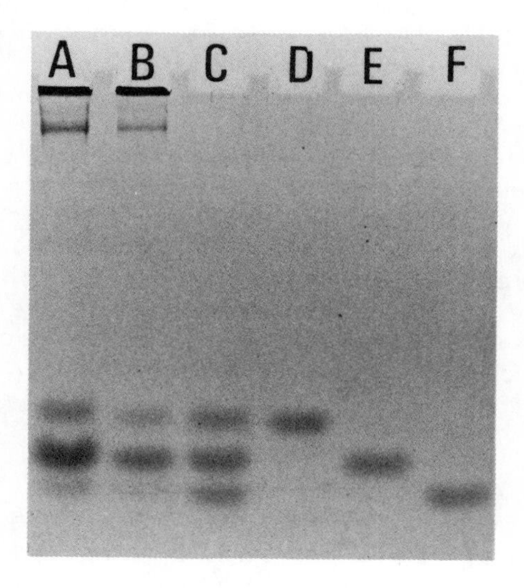

FIGURE 2. Urea-PAGE (5% gel). A, activated lipophorin (10 ug); B, activated lipophorin (5ug); C, mixture of three apo-IIIs; D, apo-III-a; E, apo-III-b; F, apo-III-c. The heavy and light chains of activated lipophorin hardly migrate in this system.

the presence of the three apo-IIIs in activated lipophorin, thereby indicating that all three apo-IIIs can associate with lipophorin in response to AKH injection.

The above observations suggest that the three apo-IIIs have different electric charges. This possibility was tested by the use of isoelectric focusing to determine the isoelectric points of the three apo-IIIs. On the basis of data from three determination, the isoelectric points were estimated to be pH 5,43 ± 0.04, 5.11 ± 0.03, and 4.98 ± 0.03 for apo-III-a, apo-III-b, and apo-III-c, respectively. These values are consistent with the elution profile of three apo-IIIs on ion-exchange chromatography and the mobilities on native- and urea-PAGE.

How many molecules of apo-III associate with lipophorin ?

Various amounts of lyophilized apo-III-b were applied to SDS-PAGE and, after staining, the gels were scanned in a chromatographic scanner to provide a standard curve. A known amount (determined by Biuret or Lowry's method) of activated lipophorin was also run on SDS-PAGE, stained, and scanned. The amount of apo-III associated with lipophorin was then determined by reference to the standard curve. Since the molecular weights of heavy chain, light chain, and apo-III are 250K, 85K, and 20K, respectively, the number (x) of apo-III molecules associated with lipophorin can be calculated:

$$\frac{B}{A - B} = \frac{20\,x}{250 + 85}$$

Where, A represents the amount of activated lipophorin originally applied to the gel, and B represents the amount of apo-III associated with lipophorin as determined by the above procedure. Thus, the number of apo-III molecules associated with lipophorin was calculated to be 8 or 10/each lipophorin molecule; the value varied in accordance with the method by which the amount of activated lipophorin applied to gel was determined (8 for Biuret and 10 for Lowry's method).

Urea-polyacrylamide gel of activated lipophorin (Fig. 2) was further scanned to determine the molecular ratio of respective apo-IIIs associated with lipophorin. The ratio was estimated to be 2 : 6 : 1 for apo-III-a : apo-III-b : apo-III-c. This ratio is consistent with the above values and, therefore, indicates that 2 moles apo-III-a, 6 moles apo-III-b, and 1 mole apo-III-c associate with each mole of lipophorin in response to AKH injection.

TABLE 1

AMINO ACID COMPOSITIONS OF LOCUST APO-IIIS

Amino acids	Apo-III-a	Apo-III-b	Apo-III-c	[a]Apo-III of M. sexta
		(mol/1000 mol)		
Asp/Asn	123	125	126	120
Thr	70	71	72	48
Ser	53	53	54	78
Glu/Glm	200	201	200	184
Pro	28	30	27	18
Gly	28	28	27	30
Ala	179	181	179	138
Val	49	49	50	60
Cys/2	1	1	1	0
Met	0	0	0	12
Ileu	36	37	38	12
Leu	106	108	110	72
Tyr	0	0	0	6
Phe	15	15	14	48
Lys	49	50	51	138
His	37	37	37	24
Arg	9	8	8	12
Trp	7	6	6	0

[a] Modified from the data of Kawooya *et al* (17).

Compositions of amino acids, amino sugar and neutral sugar of three apo-IIIs

The amino acid compositions of three apo-IIIs are presented in Table 1 and demonstrate that the three apo-IIIs are not distinguishable in terms of amino acid composition. The amino acid composition of apo-III from *M. sexta* is also given in this table for comparison. It is evident that the amino acid composition of locust apo-III differs greatly from that of *M. sexta* apo-III (17).

A preliminary test for the presence of sugars in the three apo-IIIs using the anthrone-method was positive and, therefore, the locust apo-III is a glycoprotein. Further analyses of sugar and amino sugar were carried out by gas-liquid chromatography and a colorimetric method. The result demonstrates the presence of fucose, mannose and glucosamine. The quantitative data indicate that the sum of neutral sugar and glucosamine amounts to approximately 11% of the total weight of apo-III, and that there is little difference among the three apo-IIIs.

Native molecular weights of three apo-IIIs

The molecular weights of three apo-IIIs were determined by a sedimentation-equilibrium method. The molecular weights of the three apo-IIIs were similar and estimated to be 19,100 ± 400, which is almost equivalent to the molecular weight (20,000) determined by SDS-PAGE, thereby indicating that locust apo-IIIs exist in hemolymph as monomer.

The present study provides several lines of evidence for the existence of three molecular species of apo-III in locust hemolymph. The three apo-IIIs are indistinguishable by SDS-PAGE, immunodiffusion, and in amino acid composition, but their isoelectric points differ and, therefore, they are separable by native- or urea-PAGE.

The current study also reveals that locust apo-III and *M. sexta* apo-III differ markedly in physico-chemical properties. The locust apo-III is evidently a glycoprotein and exists as a monomeric form, whereas the *M. sexta* apo-III is reported to be a non-glycosylated protein which exists as a dimeric form (17). The apo-IIIs from the two insect species also exhibit considerable differences in amino acid composition (Table 1) and in molecular weights as determined by SDS-PAGE; 20,000 for locust apo-III and 17,000 for M. sexta (17). Apo-III isolated recently from hemolymph of the mesquite bug, Thasus acutagulus, (20) resembles the M. sexta apo-III in physico-chemical natures thus, interspecific difference may exist in the nature of apo-III.

ACKNOWLEDGMENTS

This study was partly supported by a research grant (60440004) from the Japanese Ministry of Education.

REFERENCES

1. Chino, H., R. G. H. Downer, G. R. Wyatt, and L. I. Gilbert.(1981) Lipophorins, a major class of lipoprotein of insect hemolymph. Insect Biochem. 11: 491.
2. Chino, H., and K. Kitasawa. (1981) Diacylglycerol-carrying lipoprotein of hemolymph of the locust and some insects. J. Lipid Res. 22: 1042-1052.
3. Katase, H., and H. Chino.(1982) Transport of hydrocarbons by the lipophorin of insect hemolymph. Biochem. Biophys. Acta. 710: 341-348.
4. Katase, H., and H. Chino.(1984) Transport of hydrocarbons by the lipophorin in Locusta migratoria. Insect Biochem. 14: 1-6.
5. Downer, R. G. H., and H. Chino.(1985) Turnover of protein and diacylglycerol components of lipophorin in locust hemolymph. Insect Biochem. 15: 627-630.
6. Pattnaik, K. M., E. C. Mundall, B. G. Trambsti, J. H. Low, and F. J. Kezdy.(1979) Isolation and charaterization of larval lipoprotein from the hemolymph of Manduca sexta. Comp. Biochem. Physiol. 63B: 469-476.

7. Shapiro, J. P. , P. S. Keim, and J. H. Law.(1984) Structural studies on lipophorin, an insect lipoprotein. J. Biol. Chem. 259: 3680-3685.
8. Mayer, R. J., and D. J. Candy.(1969) Control of hemolymph lipid concentration during locust flight: An adipokinetic hormone from the corpora cardiaca. J. Insect Physiol. 15: 611-620.
9. Mwangi. R. W., and G. J. Goldsworthy. (1977) Diglyceride-trasporting lipophorins in _Locusta_. J. Comp. Physiol. 114: 177-190.
10. Van Der Horst D. J., J. M. Van Doorn, and A. M. Beenakkers.(1979) Effects of the adipokinetic hormone on the release and turnover of hemolymph diglycerides andon the formation of the diglyceride-transporting lipoproteins and flight in _Locusta migratoria_. Insect Biochem. 11: 717-723.
11. Mwangi, R. W., and D. J. Goldsworthy. (1981) Diacylglycerol-transporting lipoprotein system during locust flight. Insect Biochem. 27: 47.
12. Van Der horst, D. J., J. M. Van Doorn, A. N. De keijzer, and A. M. Beenakkers.(1981) Interconversions of diacylglycerol-trsnsportorting lipoproteins in the hemolymph of _Locusta migratoria_. Insect Biochem. 11: 717-723.
13. Van Der Horst, D. J., J. M. Van Doorn, and A. M. Beenakkers.(1984) Hormone-induced rearrangement of locust hemolymph lipoproteins. The involvement of glycoprotein C_2. Insect Biochem. 14: 495-504.
14. Wheeler, C. H., and D. J. Goldsworthy.(1983) Qualitative and quantitative changes in _Locusta_ hemolymph proteins and lipoproteins during ageing and adipokinetic hormone action. J. Insect Physiol. 29: 339-347.
15. Wheller, C. H., and D. J. Goldsworthy. (1983) Protein and lipoprotein interaction in the hemolymph of _Locusta_ during the action of adipokinetic hormone: The role of CL proteins. J. Insect Physiol. 29: 349-354.
16. Shapiro, J. P., and J. H. Law.(1983) Locust adipokinetic hormone stimulates lipid mobilization in _Manduca sexta_. Biochem. Biophys. Res. Comm. 115: 924-931.

17. Kawooya, J. K., P. S. Keim, R. O. Ryan, J. P. Shapiro, P.Samaraweera, and J. H. Law.(1983) Insect apolipophorin III. J. Biol. Chem. 259: 10733-10737.
18. Chino, H., R. G. Downer, and K. Takahashi. (1986) Effect of adipokinetic hormone on the structure and properties of lipophorin in locusts. J. Lipid Res. 27: 21-29.
19. Chino, H., and M. Yazawa.(1986) Apolipophorin III in locusts :Purification and characterization. J. lipid Res. inpress.
20. Wells, M. A., R. O. Ryan, S. V. Prasad, and J. H. Law. (1985) A novel procedure for the purification of apolipophorin III. Insect Biochem. 15: 565-571.

Molecular Entomology, pages 247–256

ADIPOKINETIC HORMONE-INDUCED LIPOPHORIN TRANSFORMATIONS DURING LOCUST FLIGHT

Dick J. Van der Horst, Miranda C. Van Heusden, Thomas K.F. Schulz, and Ad M.Th. Beenakkers

Department of Experimental Zoology, University of Utrecht
8 Padualaan, 3508 TB Utrecht, The Netherlands

ABSTRACT Adipokinetic hormone induces association of resting lipophorin (density ∿ 1.11 g/ml) with large amounts of a 20,000 dalton protein (C_2 or apo-III) in locust hemolymph. The resulting lipoprotein (A^+) has a higher molecular weight, while density is lower (∿ 1.04 g/ml) due to its high diacylglycerol loading at the fat body site. Using monoclonal antibodies specific for the various lipoprotein apoproteins, it was shown that apo-III is exposed in A^+, which is compatible with an involvement in flight muscle cell membrane recognition and/or lipoprotein lipase activation. However, in addition, despite the considerable increase in size and lipid loading, epitopes of apo-I and apo-II remain exposed like in lipophorin. From *in vitro* experiments it is inferred that at the flight muscle site, diacylglycerol of A^+ is hydrolyzed by a specific lipoprotein lipase; its protein constituents (lipophorin, C_2), however, are recovered and may re-associate and reload lipid at the fat body site. Thus, the reversible hormone-induced lipophorin conversions act as an efficient lipid shuttle during locust flight.

INTRODUCTION

In the migratory locust, *Locusta migratoria*, which provides an attractive model system for metabolic regulation, lipid stored in the fat body is mobilized and constitutes the principal fuel for the flight muscles during long-term flight. Flight-induced stimulation of lipid mobilization is predominantly controlled by the decapeptide adipokinetic hormone

(AKH) released from the corpus cardiacum, and results in a considerable increase in both the level and the turnover of the hemolymph diacylglycerol. In addition, significant changes are evoked in the hemolymph lipoproteins which function in loading the lipid at the fat body site and deliver it to the flight muscles, apparently involving a mechanism quite different from vertebrates, as has been reviewed recently (1-3). These changes in lipoproteins specifically include the AKH-induced association of the main lipoprotein present in the resting condition, lipophorin or A_{yellow} (A_y), with large amounts of a non-lipid containing protein (C_2 or apolipophorin-III) (4). This results in the formation of a new carrier lipoprotein (A^+) with a larger particle size, apparently capable of reversibly loading the increased amount of diacylglycerol for transport to the flight muscles (5-7) (Fig. 1). Similar changes in lipoproteins have also been demonstrated in the sphinx moth, *Manduca sexta*, upon injection of AKH (8-10).

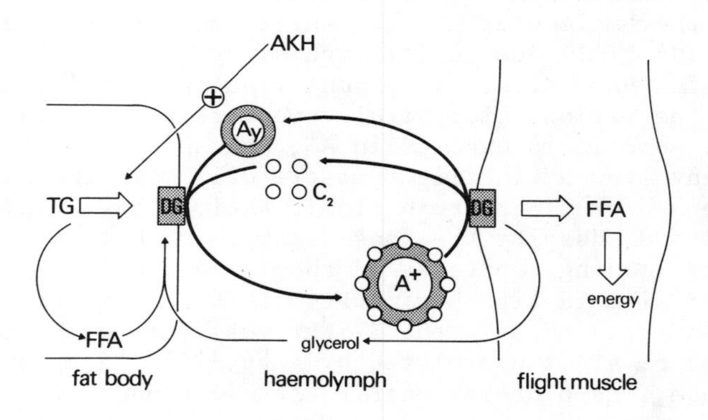

FIGURE 1. AKH-induced lipophorin transformation in the locust during flight.

In the present paper, evidence for the recycling of protein components as suggested in Fig. 1 is discussed. Previously, however, some novel aspects on the structure of locust lipophorins are presented.

RESULTS AND DISCUSSION

Isolation and Physical Properties of Locust Lipophorins

Lipophorin (A_y) and lipoprotein A^+ were routinely isolated by gel filtration chromatography (4,6). Since lipophorin (MW ∿ 450,000) is eluted at the leading edge of a protein fraction (A), the trailing edge of which consists of a blue fraction (A_{blue} or A_b), an additional ion exchange procedure is applied to obtain a complete separation between A_y and A_b (4). However, alternatively, the KBr density-gradient ultracentrifugation as described by Shapiro *et al.* (9) provided excellent results (Fig. 2). The density of locust lipophorin is about 1.11 g/ml, indicating that it is a high-density lipoprotein. In the electron microscope, the mean diameter of the globular lipophorin particles is 17 nm (3, 11).

From density-gradient ultracentrifugation of whole hemolymph it is clear that injection of locusts with 10 pmol AKH results in a shift of the lipophorin protein to a lower density (1.03-1.05 g/ml) (Fig. 3). Thus, the lipid loaded lipoprotein A^+, MW of which was estimated as approximately 3,500,000 by gel filtration chromatography (6) while size of the globular particle is increased to about 30 nm (3,11), is a low-density lipoprotein which is very similar in density to the low-density lipophorin described in adult *Manduca sexta* (8).

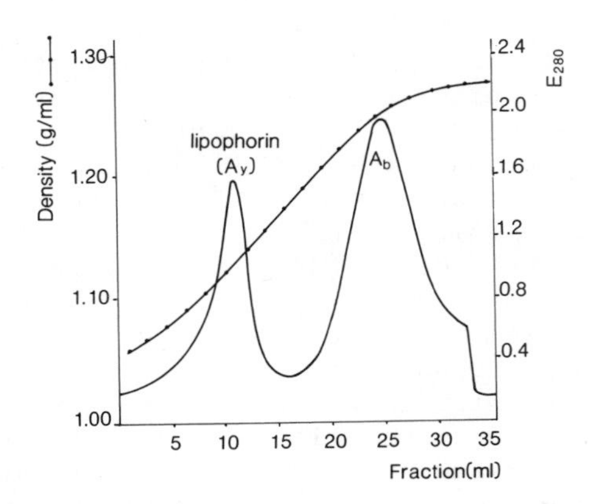

FIGURE 2. Ultracentrifuge density-gradient profile of lipophorin (A_y) and the blue protein fraction (A_b) from protein fraction A obtained by gel filtration chromatography of locust hemolymph on Ultrogel AcA 22.

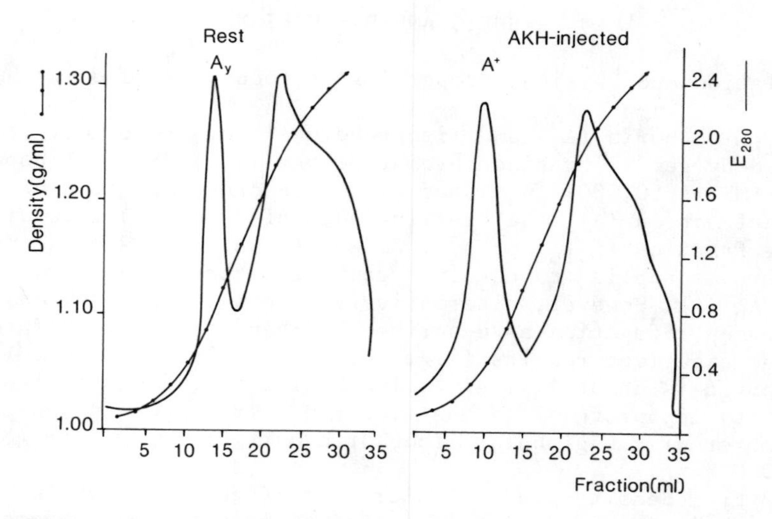

FIGURE 3. Ultracentrifuge density-gradient profiles of hemolymph from locusts at rest, and 90 min after injection of 10 pmol AKH.

Structural Organization of the Lipophorins

Subunit structure of lipophorin and lipoprotein A^+ is shown in Fig. 4. The protein from a sodium dodecyl sulfate (SDS)-polyacrylamide gradient gel has been transferred electrophoretically to nitrocellulose paper and stained subsequently (12); similar blots were also used for the immunological procedures described below. Protein C_2 consists of a single polypeptide chain (MW ∿ 20,000) which is also recovered in the subunit structure of A^+, however hardly in the resting lipophorin (4). Several authors have reported that lipophorin contains a heavy (MW ∿ 250,000) and a moderate (MW ∿ 80,000) polypeptide (apo-I and apo-II, respectively), as has been reviewed recently (3). In gradient gels, however, we observe a double band of apo-I, which is relatively weakly transferred to the nitrocellulose; staining in the gel is more intensely. Both apo-I and apo-II are also recovered in A^+; besides, lesser amounts of an additional polypeptide (MW ∿ 50,000) are present in the lipophorins irrespective of the isolation method.

Fluorescent isothiocyanate (FITC)-conjugated lectins revealed the presence of covalently bound carbohydrate in

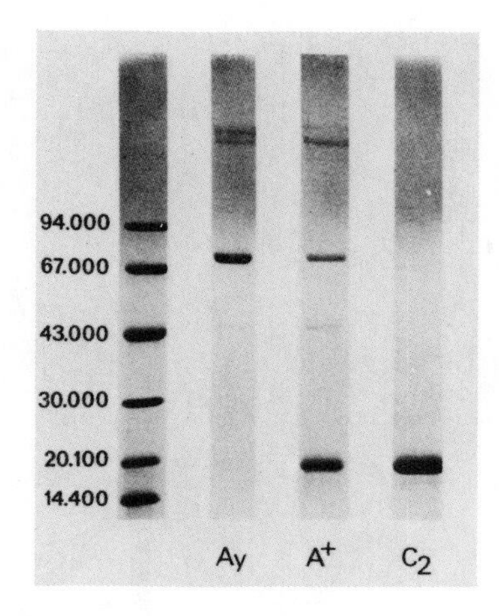

FIGURE 4. Electrophoretic blot of apoproteins separated by SDS-polyacrylamide gradient gels (7-30%). Hemolymph protein fractions were isolated from resting locusts (lipophorin, C_2) or AKH-injected insects (A^+) by gel permeation chromatography.

both apo-I and apo-II (Fig. 5). The binding to concanavalin A indicates terminal mannosyl or glucosyl residues in the apoproteins, the wheat germ agglutinin is specific for N-acetylglucosaminyl residues. Although apo-III is not reacting with both these lectins, the protein contains some 12.5% carbohydrate (4), the biantennary chain structure of which terminates with N-acetyl-neuraminic acid residues (13).

Immunological Characteristics of Locust Lipophorins

Immunoblotting with monoclonal antibodies specific for the different locust apoproteins has demonstrated that the apoproteins are not homologous (12). In addition, however, none of these antibodies showed cross-reactivity with the lipophorins of some other insects tested (cockroach, hawk moth, Colorado potato beetle), which seems at variance with the reported cross-reactivity between the polyclonal antiserum obtained against *M. sexta* apo-II and lipophorin apo-II of several other insect species (14).

Enzyme-linked immunosorbent assays (ELISA) using native lipophorins and the monoclonal antibodies specific for the

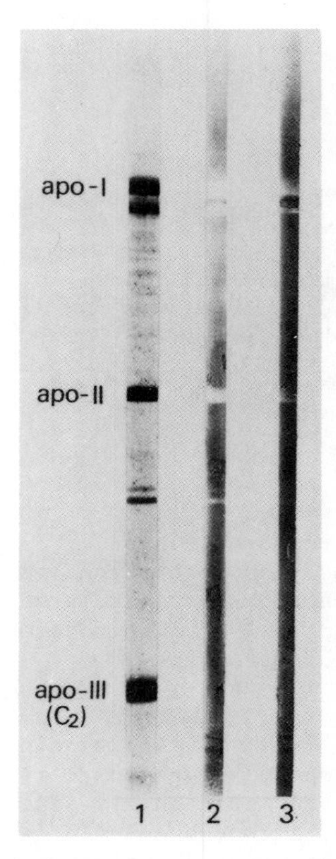

FIGURE 5. Staining of the apoprotein bands of lipoprotein A^+ on nitrocellulose paper for protein (lane 1) and carbohydrate (lanes 2 and 3). Binding to fluorescein isothiocyanate-conjugated concanavalin A (lane 2) and wheat germ agglutinin (lane 3) was visualized under UV light.

apoproteins indicate that both apoproteins in locust lipophorin are exposed, although the significantly lower reactivity of apo-II points to major localization inside of the particle, which is in agreement with the limited proteolytic degradation studies and immunological data on insect lipophorins reported by other investigators (9,15-17) (Fig. 6).

In the A^+ particle, apo-III is exposed, which is compatible with a possible involvement in flight muscle cell membrane recognition and/or lipoprotein lipase activation (3,12,

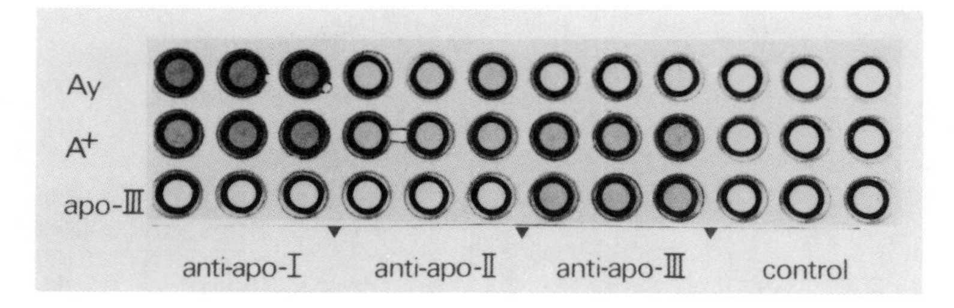

FIGURE 6. Enzyme-linked immunosorbent assay (ELISA) of native locust lipoproteins coated in multiwell trays, using the monoclonal antibodies specific for the apoproteins. Procedures used were described earlier (12).

18). However, in addition, apo-I and apo-II remain exposed as in lipophorin despite the dramatic increase in both lipid and apo-III loading, and thus may reflect recognition sites as well (Fig. 6).

Recycling of Lipophorin Protein Components

Formation of lipoprotein A^+ takes place at the fat body site, as demonstrated by *in vitro* incubation of fat body tissue in a medium containing both lipophorin and the C-proteins (19). Although, to some extent, association of the lipophorin with apo-III is apparently possible without the addition of AKH, the stimulatory action of the hormone was clearly shown in these *in vitro* assays.

At the flight muscle site, a lipoprotein lipase hydrolyzing the lipophorin-bound diacylglycerol has been identified (20-22). General features of this insect lipoprotein lipase are apparently completely different from vertebrate lipoprotein lipase, as reviewed recently (3), while both isolation and characterization proved rather difficult. Flight muscle homogenates incubated with either lipophorin or lipoprotein A^+ containing equal amounts of lipid demonstrated specificity for A^+ (7). Very recently, we have obtained indications for the recovery of the protein components constituting A^+ at the flight muscle site. Protein C_2 was labeled with ^{35}S and injected into locusts together with AKH. The resulting labeled A^+ was isolated and incubated *in vitro* with flight muscles (Fig. 7). The recovery of both radiolabeled protein C_2 and a substantial amount of lipophorin demonstrates release of the protein constituents of A^+, which in the *in vivo* situation

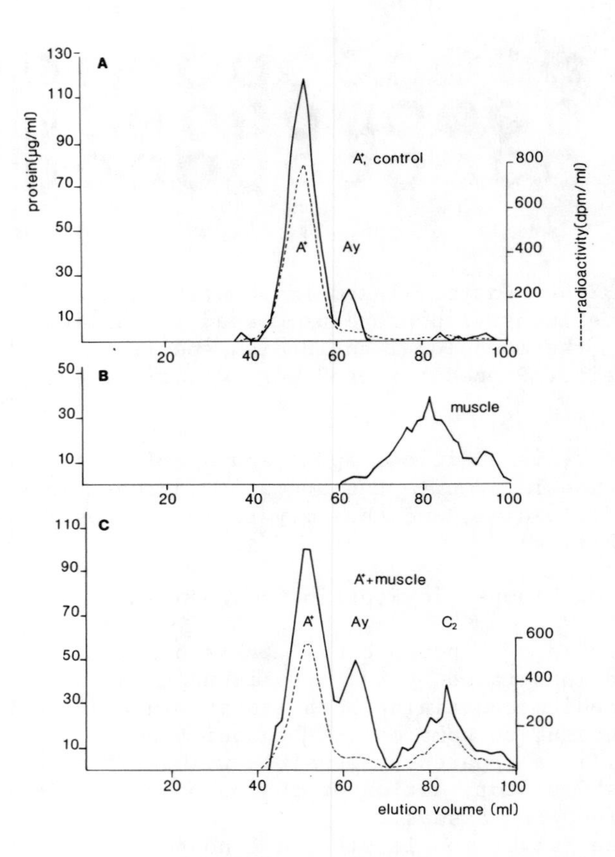

FIGURE 7. *In vitro* incubation of locust flight muscles with lipoprotein A^+, the apo-III of which was labeled with ^{35}S. The incubation media of both the control experiments (A: labeled A^+ without the addition of muscle; B: muscle without the addition of labeled A^+) and the final incubation (C) were analyzed by gel filtration chromatography after 1 h (A,C) or 2 h of incubation.

may return to the fat body and re-associate and reload lipid.

CONCLUSION

The proposed model for the reversible AKH-induced lipophorin conversions and lipid loading (Fig. 1) has gained convincing support, and for the locust and probably a wider spectrum of insect species using lipid for flight muscle energy generation it may be propound that this is a unique system acting as an efficient lipid shuttle.

ACKNOWLEDGMENT

The authors wish to thank Mr. Jan M. Van Doorn for his invaluable help in both the planning and performance of several experiments.

REFERENCES

1. Beenakkers AMTh, Van der Horst DJ, Van Marrewijk WJA (1984). Insect flight muscle metabolism. Insect Biochem 14:243.
2. Beenakkers AMTh, Van der Horst DJ, Van Marrewijk WJA (1985). Biochemical processes directed to flight muscle metabolism. In Kerkut GA, Gilbert LI (eds):"Comprehensive Insect Physiology, Biochemistry and Pharmacology," Oxford: Pergamon Press, p 451.
3. Beenakkers AMTh, Van der Horst DJ, Van Marrewijk WJA (1986). Insect lipids and lipoproteins, and their role in physiological processes. Prog Lipid Res, in press.
4. Van der Horst DJ, Van Doorn JM, Beenakkers AMTh (1984). Hormone-induced rearrangement of locust haemolymph lipoproteins: the involvement of glycoprotein C_2. Insect Biochem 14:495.
5. Mwangi RW, Goldsworthy GJ (1977). Diglyceride-transporting lipoproteins in *Locusta*. J comp Physiol 114:117.
6. Van der Horst DJ, Van Doorn JM, Beenakkers AMTh (1979). Effects of the adipokinetic hormone on the release and turnover of haemolymph diglycerides and on the formation of the diglyceride-transporting lipoprotein system during locust flight. Insect Biochem 9:627.
7. Van Heusden MC, Van der Horst DJ, Van Doorn JM, Wes J, Beenakkers AMTh (1986). Lipoprotein lipase activity in the flight muscle of *Locusta migratoria* and its specificity for haemolymph lipoproteins. Insect Biochem, in press.
8. Shapiro JP, Law JH (1983). Locust adipokinetic hormone stimulates lipid mobilization in *Manduca sexta*. Biochem Biophys Res Commun 115:924.

9. Shapiro JP, Keim PS, Law JH (1984). Structural studies on lipophorin, an insect lipoprotein. J biol Chem 259:3680.
10. Wells MA, Ryan RO, Prasad SV, Law JH (1985). A novel procedure for the purification of apolipophorin III. Insect Biochem 15:565.
11. Van Antwerpen HGAM, Linnemans WAM, Schulz TKF, Van der Horst DJ, Beenakkers AMTh (1986). The ultrastructure of locust lipoproteins and their immuno-goldlabelling *in vitro* and *in vivo*. Submitted.
12. Schulz TKF, Van der Horst DJ, Amesz WJC, Beenakkers AMTh (1986). Monoclonal antibodies specific for apolipoproteins from the migratory locust. Submitted.
13. Abbink JHM, Van der Horst DJ, Van Doorn JM, Beenakkers AMTh (1986). The carbohydrate moieties of locust haemolymph proteins involved in lipoprotein rearrangements. Submitted.
14. Ryan RO, Schmidt JO, Law JH (1984). Chemical and immunological properties of lipophorins from seven insect orders. Archs Insect Biochem Physiol 1:375.
15. Pattnaik NM, Mundall EC, Trambusti BG, Law JH, Kézdy FJ (1979). Isolation and characterization of a larval lipoprotein from the hemolymph of *Manduca sexta*. Comp Biochem Physiol 63B:469.
16. Robbs SL, Ryan RO, Schmidt JO, Keim PS, Law JH (1985). Lipophorin of the larval honeybee, *Apis mellifera* L. J Lipid Res 26:241.
17. Kashiwazaki Y, Ikai A (1985). Structure of apoproteins in insect lipophorin. Archs Biochem Biophys 237:160.
18. Van der Horst DJ, Schulz TKF, Beenakkers AMTh (1985). Lipoprotein rearrangements during locust flight. Proc 26th Int Conference on the Biochem of Lipids: 109.
19. Van Heusden MC, Van der Horst DJ, Beenakkers AMTh (1984). *In vitro* studies on hormone-stimulated lipid mobilization from fat body and interconversion of haemolymph lipoproteins of *Locusta migratoria*. J Insect Physiol 8:685.
20. Wheeler CH, Van der Horst DJ, Beenakkers AMTh (1984). Lipolytic activity in the flight muscles of *Locusta migratoria* measured with haemolymph lipoproteins as substrates. Insect Biochem 14:261.
21. Van Heusden MC, Van der Horst DJ, Van Doorn JM, Schulz TKF, Beenakkers AMTh, in preparation.
22. Wheeler CH, Goldsworthy GJ (1985). Specificity and localisation of lipoprotein lipase in the flight muscles of *Locusta migratoria*. Biol Chem Hoppe-Seyler 366:1071.

Molecular Entomology, pages 257–266

THE DYNAMICS OF LIPOPHORIN INTERCONVERSIONS IN *MANDUCA SEXTA*

Robert O. Ryan, Michael A. Wells and John H. Law

Department of Biochemistry, University of Arizona
Tucson, Arizona 85721

ABSTRACT Lipophorin of the tobacco hornworm, *Manduca sexta*, is a multifunctional lipid transport vehicle present in the hemolymph of all life stages. While the mechanism of lipid transport is not fully understood it appears that lipids, mainly diacylglycerol, may be added or removed from the particle without destruction of a basic lipoprotein matrix. Thus the several distinct forms of lipophorin which appear in hemolymph on a developmental schedule arise from an alteration of the lipid content and composition of preexisting particles. This phenomenon, as well as the dramatic change in adult lipophorin induced by the adipokinetic hormone, suggest the occurrence of facilitated lipid transfer and exchange processes in hemolymph. An *in vitro* lipid transfer assay was established and used in the purification of *M. sexta* lipid transfer protein. Lipid transfer protein exists as a very high density lipoprotein and catalyzes lipid transfer between lipophorins of different densities and between phosphatidylcholine liposomes.

INTRODUCTION

Lipids provide an important energy source for *Manduca sexta*. During any given life stage the bulk of lipid

[1]This work was supported by grants from the National Institutes of Health HL 34786, DMD 8416620 and GM 29238.

transport is mediated by the hemolymph lipoprotein, lipophorin (1). During larval and pupal life lipophorin is comprised to two apoproteins, apolipophorin-I (ApoLp-I) and apolipophorin-II (ApoLp-II) and lipid (2,3). ApoLp-I is a large glycosylated apoprotein (M_r = 250,000) with solubility characteristics similar to mammalian apolipoprotein B. ApoLp-II is a somewhat smaller (M_r = 80,000) glycosylated apoprotein that is antigenically distinct from apoLp-I (3). The lipid component of lipophorin includes mainly diacylglycerol (DG) and phospholipid (PL) and comprises between 30 and 50 percent of the particle weight. The large proportion of polar lipid in this particle suggests basic structural differences between insectan and mammalian (4) high density lipoproteins. Structural studies on lipophorin (2,5) using limited proteolysis and radioiodination lead to the conclusion that ApoLp-I is sensitive to these probes, while ApoLp-II is resistant. Furthermore the relatively polar nature of the predominant PL and DG imposes certain structural constraints on lipophorin. Thus a model of lipophorin structure in which ApoLp-II is shielded from the aqueous environment with DG and PL at or near the surface has been suggested.

An important feature of this model is the location of DG at the surface of the particle which would allow for DG exchange or transfer without destruction of the particle core. Indeed, unlike most mammalian lipoproteins, evidence has accumulated which suggests that lipophorin functions as a reusable lipid shuttle (6-8). Other studies (9) have shown that lipophorin can be loaded with lipid independent of *de novo* protein synthesis. The work described in this report is intended to further our understanding of the developmental changes in lipophorin and the mechanism whereby these changes occur.

Lipophorins of the Larval and Pupal Stages.

During larval development, ingested dietary lipids provide an important nutrient source. Lipophorin functions in the transport of dietary lipid from midgut tissue to sites of storage (fat body) or utilization (developing tissues). During the feeding stage of the fifth instar, lipophorin exists as high density lipophorin-larval (HDLp-L). HDLp-L has been extensively characterized with respect to its structural and immunological properties (2,3). HDLp-L contains 37 percent lipid with a predominance of DG and PL with lesser amounts of hydrocarbon, cholesterol and triacylglycerol (10). The particle has a density of 1.15 g/ml. HDLp-L persists in hemolymph during the feeding stage of the fifth instar. Upon cessation of feeding and initiation of wandering behavior, however, a new higher density lipophorin is found. This high density lipophorin-wanderer 2 (HDLp-W2)(10) possesses apoprotein components identical to HDLp-L but differs significantly in lipid content and composition. In comparison to HDLp-L, there is a reduction in the diacylglycerol content and a density increase (Table 1). Within 12 h of the appearance of HDLp-W2 a second wandering stage lipophorin form, high density lipophorin-wanderer 1 (HDLp-W1) replaces HDLp-W2 (10). Comparison of this particle with HDLp-W2 reveals identical apoprotein components but significantly higher DG and PL levels and a lower density (Table 1). Within three days of the appearance of HDLp-W1, just following the larval-pupal molt, yet another lipophorin, high density lipophorin-pupal (HDLp-P) is found in hemolymph. HDLp-P has a higher density and a different lipid composition compared to HDLp-W1 but the same apoprotein components.

TABLE 1
LIPOPHORINS OF *MANDUCA SEXTA*

Lipophorin form	% lipid	Density
HDLp-L	37	1.151
HDLp-W2	35	1.177
HDLp-W1	47	1.128
HDLp-P	46	1.135
HDLp-A	51	1.076
LDLp	62	1.030

Thus, during the period of larval-pupal metamorphosis, four distinct lipophorin forms appear in hemolymph. It is important to note that 1) the different lipophorin forms appear in hemolymph on a precisely timed developmental schedule; 2) only one lipophorin form is present in hemolymph at a given time; and 3) all lipophorin forms differ in lipid content and composition but retain identical apoprotein components.

Lipophorins of the Adult Stage.

The adult sphinx moth requires an efficient lipid transport system to supply developing eggs with sufficient nutrients and flight muscle with enough DG to support flight activity. Lipids mobilized from the fat body and transported via the adult form of lipophorin help meet these demands. High density lipophorin-adult (HDLp-A) has a lower density and a relatively higher DG content compared to larval lipophorins (Table 1). While containing apoprotein components similar to the larval forms (ApoLp-I and ApoLp-II), HDLp-A has two copies of a third low molecular weight (M_r = 17,000) apoprotein component, apolipophorin III, (ApoLp-III)(11). Large amounts of ApoLp-III exist in a non-lipophorin associated form in hemolymph and this free ApoLp-III plays a role in remarkable transformation of HDLp-A from high to low density lipophorin (LDLp). This change in HDLp-A can be induced by prolonged flight activity, injection of corpora cardiaca extract or the nonapeptide adipokinetic hormone (12,13). The observed decrease in lipophorin density is the result of specific uptake of DG by HDLp-A (14), which results in a doubling of the mass of the lipoprotein, largely by addition of DG, thus greatly increasing the capacity of lipophorin to transport DG from fat body to flight muscle (16). Concomitant with DG loading is an association of several molecules of ApoLp-III with the DG rich particle. The physical and surface properties of ApoLp-III have been extensively studied (15) and suggest that this apoprotein is well suited to stabilize the DG loaded LDLp-particle. It has also been suggested that ApoLp-III may serve as a recognition signal/activator of a flight muscle lipoprotein lipase but confirmation of this proposed role awaits experimental verification. An important feature of the HDLp-A to LDLp transformation is its apparent reversibility. LDLp is not catabolized during the process of delivering DG to the flight muscle. Instead, excess ApoLp-III dissociates, thereby regenerating HDLp-A and free ApoLp-III. HDLp-A can then return to the fat body and bind more DG and free ApoLp-III once again forming LDLp. In support of this cyclic interconversion between HDLp-A and LDLp is evidence provided by *in vivo* experiments with radioiodinated HDLp-A. Following administration of labeled HDLp-A to adult animals the radioactive lipophorin can be recovered as LDLp 1 h following treatment with synthetic AKH. On the other hand, if the animals are bled 16 h after AKH treatment the radiolabeled lipophorin has a density corresponding to the original HDLp-A (Ryan and Law, unpublished).

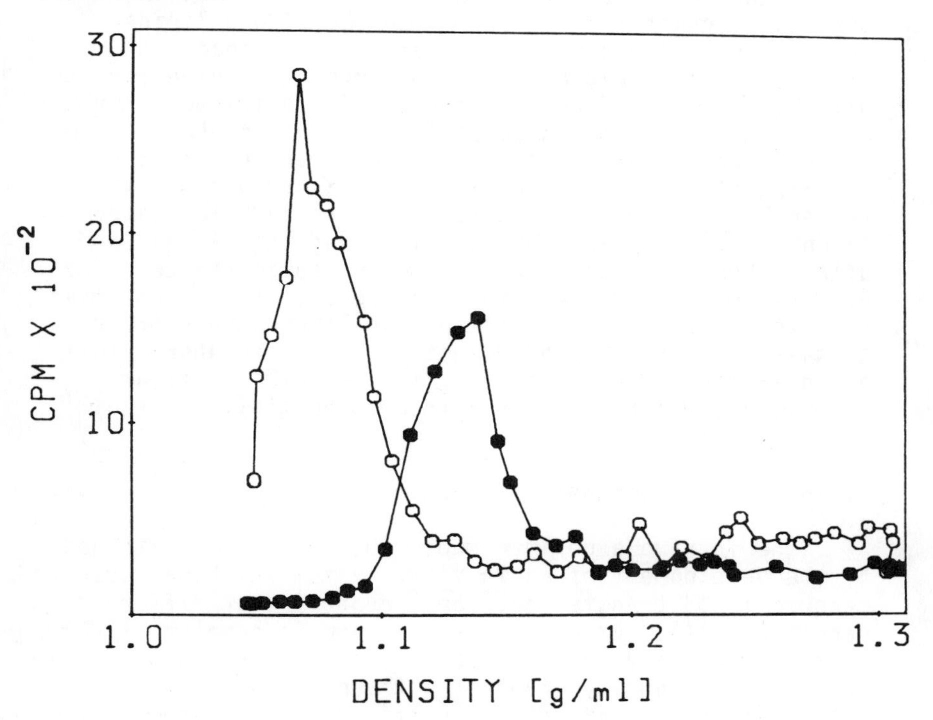

FIGURE 1. The fate of ^{125}I-LDLp following injection into *M. sexta* fifth instar larvae. Animals were injected with 6 μg radioiodinated LDLp (specific activity = 2.72×10^5 cpm/μg) and after 1 h the animals were bled and the hemolymph subjected to density gradient ultracentrifugation as previously described (3). The contents of the tube were then fractionated and the radioactivity and density in each fraction determined. The closed circles refer to LDLp injected into animals and open circles refer to control LDLp not injected into animals.

A Basic Lipophorin Matrix Structure.

It appears, therefore, that throughout the life cycle of _M. sexta_, lipophorin changes in response to different physiological lipid transport demands. Throughout all these alterations, however, ApoLp-I and ApoLp-II remain integral components of the lipoprotein particle. Thus it may be that there exists a basic lipoprotein matrix structure (comprised of ApoLp-I, ApoLp-II and lipids) whose structure has yet to be defined, and that alterations in lipid content and composition give rise to all the observed lipophorin forms. The existence of such a matrix particle would suggest that any lipophorin form found in hemolymph can be converted to another form by altering its exchangeable lipids. This hypothesis was tested by _in vivo_ experiments employing radioiodinated lipophorins. Labeled LDLp was injected into larvae and after 1 hour the larvae were bled and subjected to density gradient ultracentrifugation (Figure 1). It was observed that the density of the radioactive lipophorin shifted so that it now corresponded directly to the lipophorin form found in the recipient animal, i.e. HDLp-L. Similar experiments demonstrated the interconvertibility of all lipophorin forms.

Lipid Transfer Between Lipophorins _In Vitro_.

The _in vivo_ experiment described above was extended by the development of an _in vitro_ system which was used to measure lipid transfer between lipophorins of different densities (14). Briefly, when lipophorin samples of different density were incubated together and then subjected to density gradient ultracentrifugation, no change in the density of the particles was observed. Addition of lipophorin-deficient hemolymph (the infranatant fraction [density > 1.21 g/ml] obtained after density gradient ultracentrifugation of hemolymph) results in a change in the density and amount of the original lipophorins. _In vitro_ incubation of LDLp (donor particle) and HDLp-W2 (acceptor particle) with lipophorin-deficient hemolymph results in a decrease in the amount of LDLp present and the appearance of a new lipophorin form, intermediate in density between LDLp and HDLp-W2 (Lipophorin intermediate, Lp-I). Radiolabeling studies have demonstrated that Lp-I is comprised of lipid depleted LDLp donor particles and lipid enriched HDLp-W2 acceptor particles and results from net lipid transfer from donor to acceptor lipophorin (14). It has also been shown that Lp-I does not arise from lipoprotein fusion and that apoprotein exchange does not occur during the process of Lp-I formation. It can be concluded, therefore, that a lipid transfer factor exists in lipophorin-deficient hemolymph and is capable of catalyzing net lipid transfer between lipophorin particles. Furthermore, this lipid transfer factor may play a role in lipophorin interconversions *in vivo* and may function in the "shuttle" mechanism of lipophorin lipid transport whereby lipid can be transported without destruction of the lipoprotein particle.

M. sexta Lipid Transfer Protein (LTP).

The assay of the lipid transfer activity described above was used to monitor progress during purification of the factor responsible for the observed lipid transfer between lipophorins (17). Using this assay it could be shown that the lipid transfer factor was proteinaceous in nature and that the reaction progress was dependent upon the amount of LTP in the reaction medium. Gel permeation chromatography of a 0-45 percent ammonium sulfate cut of lipophorin-deficient hemolymph indicated LTP was a high molecular weight protein and substantial purification was achieved by chromatography on a column of Bio-Gel A 1.5. The LTP pool from the gel filtration column was subjected to the density gradient ultracentrifugation procedure developed by Haunerland and Bowers (18), which facilitates flotation of very high density lipoproteins (density = 1.22 - 1.28 g/ml). This step yielded a purified LTP preparation. Thus, LTP exists as a very high density lipoprotein (density = 1.23 g/ml) and can be readily purified by taking advantage of two of its unique physical properties; size and density. Examination of the apoprotein composition of LTP by sodium dodecyl sulfate polyacrylamide gel electrophoresis revealed the presence of two apoproteins, ApoLTP-I ($M_r \cong 320,000$) and ApoLTP-II ($M_r = 85,000$). The similarity in size between LTP apoproteins and the corresponding apoproteins of lipophorin prompted an investigation of their immunological relationship. Immunodiffusion of purified LTP versus antibodies directed against larval lipophorin or vitellogenin showed no evidence of immunological reactivity.

Role of LTP in Lipophorin Lipid Transport.

Lipophorin functions mainly in the transport of lipid between tissues (i.e., fat body and flight muscle, midgut and fat body) and thus apparently interacts with membranes in the process of accepting or delivering lipid. Since the LTP assay described above measures lipid transfer between lipoproteins, it cannot be used to assess LTP's ability to interact with membranes and perhaps facilitate lipid transfer from biomembrane to lipophorin. To test this possibility, we have examined the ability of LTP to catalyze the transfer of a fluorescent phospholipid between phosphatidylcholine (PC) liposomes. Parinaroyl-PC was prepared, incorporated into egg PC liposomes and employed in a lipid transfer assay according to Somerharju *et al*. (19). Addition of LTP to an incubation mixture containing parinaroyl-PC containing liposomes and unlabeled liposomes resulted in an increase in fluorescence intensity. The fluorescence intensity increase is a measure of LTP activity and demonstrates the ability of LTP to transfer PC between liposomes. Furthermore, it supports the possible role of LTP in the lipoprotein interconversions observed *in vivo*.

M. sexta LTP is the third of a family of structurally analogous lipoproteins that occur in hemolymph. The others are lipophorin itself, and vitellogenin, the egg yolk protein precursor, which also has large (M_r = 180,000) and smaller (M_r = 45,000) apoproteins and a similar lipid composition, but lower lipid content (20). The identification and purification of *M. sexta* LTP constitutes the first report of a lipid transfer protein from insect tissue. The availability of LTP should allow elucidation of its role in lipid transport and lipophorin transformations. It may also allow experiments that will give additional information on the structural arrangement of lipophorins and the mechanism of their interaction with tissues.

REFERENCES

1. Chino H (1985). Lipid transport: Biochemistry of hemolymph lipophorin. In Kerkut GA, Gilbert LI (eds): "Comprehensive Insect Physiology Biochemistry and Pharmacology," Oxford: Pergamon Press, Vol 10, p 115.

2. Pattnaik NM, Mundall EC, Tranbusti BG, Law JH, Kézdy FJ (1979). Isolation and characterization of a larval lipoprotein from the hemolymph of Manduca sexta. Comp Biochem Physiol 63B:469.
3. Shapiro JP, Keim PS, Law JH (1984). Structural studies on lipophorin: An insect lipoprotein. J Biol Chem 259:3680.
4. Smith LC, Pownall HJ, Gotto AM (1978). The plasma lipoproteins: Structure and metabolism. Ann Rev Biochem 47:751.
5. Mundall EC, Pattnaik NM, Trambusti BG, Hromnak G, Kézdy FJ, Law JH (1980). Structural studies on an insect high density lipoprotein. Ann New York Acad Sci 348:431.
6. Downer RGH, Chino H (1985). Turnover of protein and diacylglycerol components of lipophorin in locust hemolymph. Insect Biochem 15:627.
7. Chino H, Kitazawa K (1981). Diacylglycerol-carrying lipoprotein of hemolymph of the locust and some insects. J Lipid Res 22: 1042.
8. Van der Horst DJ, Van Doorn JM, De Keijzer AN, Beenakkers AMTh (1981). Interconversions of diacylglycerol transporting lipoproteins in the hemolymph of Locusta migratoria. Insect Biochem 11:717.
9. Peled Y, Tietz A (1973). Fat transport in the locust, Locusta migratoria: The role to protein synthesis. Biochim Biophys Acta 296:499.
10. Prasad SV, Ryan RO, Wells MA, Law JH (1986). Changes in lipoprotein composition during larval-pupal metamorphosis of an insect, Manduca sexta. J Biol Chem 261:558.
11. Kawooya JK, Keim PS, Ryan RO, Shapiro JP, Samaraweera P, Law JH (1984). Insect apolipophorin III. Purification and properties. J Biol Chem 259:10733.
12. Shapiro JP, Law JH (1983). Locust adipokinetic hormone stimulates lipid mobilization in Manduca sexta. Biochem Biophys Res Comm 115:924.
13. Ziegler R, Eckart K, Schwarz H, Keller R (1985). Amino acid sequence of Manduca sexta adipokinetic hormone elucidated by combined fast atom bombardment (FAB)/tandem mass spectrometry. Biochem Biophys Res Comm 113:337.
14. Ryan RO, Prasad SV, Henriksen EJ, Wells MA, Law JH (1986). Lipoprotein interconversions in an insect, Manduca sexta. J Biol Chem 261:563.

15. Kawooya JK, Meredith SC, Wells MA, Kézdy FJ, Law JH (1986). Physical and surface properties of insect apolipophorin III. J Biol Chem, submitted.
16. Haunerland NH, Bowers WS (1986). A larval specific lipoprotein: Purification and characterization of a blue chromoprotein from Heliothis zea. Biochem Biophys Res Comm 134:580.
17. Ryan RO, Wells MA, Law JH (1986). Lipid transfer protein from Manduca sexta hemolymph. Biochem Biophys Res Comm, in press.
18. Beenakkers AMTh, Van der Horst DJ, Van Marrewijk WJA (1985). Biochemical processes directed to flight muscle metabolism. In Kerkut GA, Gilbert LI (eds): "Comprehensive Insect Physiology Biochemistry and Pharmacology," Oxford: Pergamon Press, Vol 10, p 451.
19. Somerharju P, Brockerhoff H, Wirtz KWA (1981). A new fluorometric method to measure protein-catalyzed phospholipid transfer using 1-acyl-2 parinaroylphosphatidylcholine. Biochim Biophys Acta, 649:521.
20. Osir EO, Wells MA, Law JH (1986). Studies on vitellogenin from the tobacco hornworm, Manduca sexta. Arch Insect Biochem Physiol, in press.

Molecular Entomology, pages 267–273

LIPOPHORIN BIOSYNTHESIS DURING THE LIFE CYCLE OF THE TOBACCO HORNWORM, *MANDUCA SEXTA*[1]

Sarvamangala V. Prasad, Kozo Tsuchida,
Kenneth D. Cole, and Michael A. Wells

Biochemistry Department, Biological Sciences West
University of Arizona, Tucson, AZ 85721

ABSTRACT Lipophorin, the major lipoprotein in the hemolymph, is biosynthesized in the fat body. *In vitro* experiments showed that isolated fat body secretes lipophorin most actively during the first 3 days of the fifth instar. Levels of lipophorin mRNA in the fat body follows the same trend. Removal of the corpus allata from animals on the first day of either the fourth or fifth instar did not cause a significant alteration in the hemolymph concentration of lipophorin, suggesting that juvenile hormone levels do not control lipophorin biosynthesis.

INTRODUCTION

Lipophorin is the major lipoprotein in insect hemolymph (1,2). During the larval and pupal stages of the tobacco hornworm, *Manduca sexta*, lipophorin is comprised of approximately 60% protein and 40% lipid (3-5). The protein portion consists of two glycosylated apolipoproteins, ApoLp-I ($M_r \cong 250,000$) and ApoLp-II ($M_r \cong 80,000$), while the lipids are primarily phospholipid and diacylglycerol with small amounts of hydrocarbon, cholesterol, and triacylglycerol. During the transition from the last feeding larval stage (fifth instar) to the pupal stage, significant alterations occur in the properties of lipophorin (5). These changes are characterized by changes in density, which are the result of

[1]This work was supported by grants from NSF (DMB-8416620) and NIH (HD 10954).

alteration in lipid content and composition. All of these changes are accomplished without detectable synthesis of new apoproteins. In order to characterize these transitions further we have investigated lipophorin secretion and mRNA levels during the fifth instar and wandering stages.

METHODS

Animals were raised and bled as previously described (5). *In vitro* lipophorin synthesis and secretion was measured by incubating 200-300 mg of fat body in 0.5 ml of Grace's insect media containing 25 μCi of ^{35}S-met for 1 hr at 30°C. At the end of the incubation, fat body was removed for nucleic acid extraction and DNA estimation (6). The media was filtered and immunoprecipitated with anti-lipophorin antibody attached to immunobeads (Bio-Rad Laboratories, Richmond, CA) after an initial precipitation with normal rabbit serum and *S. aureus* cells to remove nonspecific precipitating material (7). The data are expressed as rates of lipophorin secretion per mg of DNA relative to the value found on day 2 of the fifth instar. RNA was isolated from freeze-clamped fat body by extraction with guanidine isothiocyanate (8). Poly A(+) RNA was isolated by oligo (dT)-cellulose chromatography (9). *In vitro* translation was measured using 1 μg of poly A(+) RNA and a rabbit reticulocyte lysate (Bethesda Research Laboratories, Gaithersburg, MD). The translated product was immunoprecipitated as described above.

The concentration of lipophorin in hemolymph was measured by single radial immunodiffusion in 0.75% agar containing 2 or 4% antiserum (10). The content of lipophorin per animal was calculated from the hemolymph concentration and the body weight, assuming the hemolymph volume is 33% of the body weight (11). The role of juvenile hormone in lipophorin biosynthesis was assessed by determining the effect of corpus allatectomy, performed within 1 hr after ecdysis to either the fourth or fifth instar, on the hemolymph concentration of lipophorin.

RESULTS

In order to determine the period of most active synthesis of lipophorin, the lipophorin content per animal was determined during the fourth and fifth instar and during the wandering stage. The results of these experiments are presented in Table 1.

TABLE 1
LIPOPHORIN CONTENT PER ANIMAL

Stage	Days after molt	Lipophorin (mg/animal)
4th Instar		
	1	0.11 ± 0.020
	2	0.26 ± 0.035
	3	1.08 ± 0.093
	4	1.56 ± 0.16
5th Instar		
	1	1.67 ± 0.14
	2	2.84 ± 0.21
	3	8.03 ± 0.83
	4	12.66 ± 0.55
Wanderer		
	1	13.29 ± 0.90
	2	12.22 ± 1.45
	3	12.24 ± 1.15

As can be seen in Table 1, the most active period of accumulation of lipophorin occurs at the beginning of the fifth instar. We therefore concentrated further efforts on the fifth instar and wandering stages. At all stages investigated, lipophorin mRNA represented 1.3% of the total translation product. For comparative purposes lipophorin mRNA levels were calculated by multiplying the amount of fat body total RNA/mg DNA by the per cent of poly A(+) and then by the per cent of the translation product corresponding to lipophorin. Lipophorin mRNA levels were then expressed relative to the value found on day 2 of the fifth instar. These data and the relative rates of lipophorin secretion _in vitro_ by fat body are presented in Table 2. These data show active secretion of lipophorin during the first 3 days of the fifth instar with little activity observed throughout the remainder of the fifth instar or during the wandering stage. These data are entirely consistent with the conclusions reached previously from _in vivo_ labeling studies (5). Although the levels of mRNA generally parallel these secretion rates, their level appears to remain elevated at times when the secretion rate is quite low. If these values can be confirmed using cDNA probes, it would suggest that lipophorin biosynthesis and secretion depends on unknown factors in addition to mRNA levels.

TABLE 2
RELATIVE RATES OF LIPOPHORIN SECRETION AND mRNA LEVELS

Stage	Secretion rate	mRNA level
5th Instar		
Day 1	0.33	0.76
Day 2	1.00	1.00
Day 3	0.29	0.49
Day 4	0.06	0.45
Wanderer		
Day 1	0.02	0.17
Day 2	0.02	0.05
Day 3	0.006	0.01

In order to assess the potential role of juvenile hormone in controlling lipophorin biosynthesis, we investigated the effect of corpus allatectomy on hemolymph lipophorin concentration. The corpus allata were removed within 1 hr after ecdysis to either the fourth or fifth instar. In the case of the fourth instar, experimental animals wandered during the third night after the corpus allatectomy, whereas control animals molt to the fifth instar at this time. In the fifth instar, experimental animals wander during the third night and pupate during the sixth night, while control animals wander during the fourth night and pupate during the eighth night. The results of these experiments are presented in Tables 3 and 4.

TABLE 3
EFFECT OF CORPUS ALLATECTOMY IN THE FOURTH INSTAR

Day[a]	Lipophorin Concentration (mg/ml)[b] Experimental	Control
1	2.31 ± 0.45	2.94 ± 0.58
2	3.74 ± 0.33	3.33 ± 0.41
3	4.37 ± 0.19	3.59 ± 0.41
4	5.24 ± 0.58	4.77 ± 0.41
5	5.89 ± 0.37	4.66 ± 0.16
6	4.95 ± 0.25	4.09 ± 0.43

[a]Days after allatectomy
[b]n=4-5 for each group

TABLE 4
EFFECT OF CORPUS ALLATECTOMY IN THE FIFTH INSTAR

Day[a]	Lipophorin Concentration (mg/ml)[b] Experimental	Control
1	2.85 ± 0.21	3.58 ± 0.27
2	4.62 ± 0.68	4.26 ± 0.44
3	5.82 ± 0.91	4.64 ± 0.20
4	8.79 ± 0.86	5.61 ± 0.38
5	9.83 ± 1.35	8.08 ± 0.96
6	12.35 ± 1.06	9.33 ± 0.88
7	12.29 ± 1.59	12.07 ± 0.91

[a]Days after allatectomy
[b]n=4-5 for each group

These data show that removal of the corpus allata at the beginning of either the fourth or fifth instar does not have a significant effect on the hemolymph concentration of lipophorin.

DISCUSSION

The results presented in this paper represent the first report of lipophorin secretion _in vitro_ by the fat body and the first measurements of lipophorin mRNA levels in the fat body. It should be noted that all attempts to demonstrate lipophorin apoprotein synthesis in midgut tissue have been unsuccessful. If these negative results can be confirmed using cDNA probes to lipophorin apoprotein mRNA (experiments in progress), they will provide the final proof that lipophorin biosynthesis occurs only in the fat body. These data also confirm and extend our earlier observations on the cessation of lipophorin biosynthesis at the end of the fifth instar (5). It is clear from the data in Table 2 that both lipophorin mRNA levels and fat body secretion of lipophorin reach a maximum on day 2 of the fifth instar and then fall to very low levels before initiation of wandering. Data not presented shows that pupal fat body has undetectable levels of lipophorin mRNA and does not secrete measurable amounts of lipophorin.

The data presented in Table 1 show that the lipophorin content per animal, which must represent newly synthesized material, increases 8-10 fold during both the fourth and fifth instar. This observation would suggest activation of lipophorin biosynthesis occurs at the beginning of each feeding instar. The hormonal factor(s) responsible for this activation are unknown at present. Corpus allatectomy immediately after ecdysis to either the fourth or fifth instar has no effect on the activation of lipophorin biosynthesis, suggesting that the continual presence of juvenile hormone is not required to achieve activation of lipophorin biosynthesis. We cannot eliminate the possibility that a pulse of juvenile hormone, present during the molting process the night before allatectomy, may be responsible for activation of lipophorin biosynthesis. It is also possible that ecdysone may be the responsible hormone. Experiments are underway to further characterize the factor(s) responsible for the cyclic nature of lipophorin biosynthesis during the feeding instars and to determine the mechanism whereby lipophorin biosynthesis is shut off at the end of the fifth instar.

ACKNOWLEDGMENTS

The authors thank Mary Gonzales for animal care and Dr. John H. Law for many helpful discussions.

REFERENCES

1. Gilbert LI, Chino H (1974). Transport of lipids in insects. J Lipid Res 15:439.
2. Chino H, Downer RGH, Wyatt GR, Gilbert LI (1981). Lipophorins, a major class of lipoproteins of insect hemolymph. Insect Biochem 11:491.
3. Pattnaik NM, Mundall EC, Trambusti BG, Law JH, Kézdy FJ (1979). Isolation and characterization of a larval lipoprotein from the hemolymph of Manduca sexta. Comp Biochem Physiol 63B:469.
4. Shapiro JP, Keim PS, Law JH (1984). Structural studies on lipophorin, an insect lipoprotein. J Biol Chem 259:3680.
5. Prasad SV, Ryan RO, Law JH, Wells MA (1986). Changes in lipoprotein composition during larval-pupal metamorphosis of an insect, Manduca sexta. J Biol Chem 261:558.
6. Linzen B, Wyatt GR (1964). The nucleic acid content of tissues of cecropia silkworm pupae. Relations to body size and development. Biochem Biophys Acta 87:185.
7. Harnish DG, Wyatt GR, White BN (1982). Insect vitellins: Identification of primary products of translation. J Exp Zool 220:11.
8. Chirgwin JM, Przybyla AE, MacDonald RJ, Rutter WJ (1979). Isolation of biologically active ribonuleic acid from sources enriched in ribonuclease. Biochem 18:5294.
9. Maniatis T, Fritsch EF, Sambrook J (1982). "Molecular cloning, a laboratory manual." Cold Spring Harbor: Cold Spring Harbor Laboratory, p 197.
10. Ouchterlony D, Nilsson LA (1973). Immunodiffusion and immunoelectrophoresis. In Weir DM (ed): "Handbook of experimental immunology," Oxford: Blackwell Scientific Publications, p 19.1.
11. Beckage NE, Riddiford LM (1982). Effects of parasitism by Apanteles congregatus on the endocrine physiology of the tobacco hornworm Manduca sexta. Gen Comp Endocrinol 47:308.

Molecular Entomology, pages 275–283

CLONING AND EXPRESSION OF THE GENE FOR APOLIPOPHORIN III FROM LOCUSTA MIGRATORIA[1]

M.R. Kanost, H.L. McDonald, J.Y. Bradfield[2], J. Locke[3], and G.R. Wyatt

Department of Biology, Queen's University, Kingston, Canada K7L 3N6

ABSTRACT Apolipophorin III (apoLp-III), a 20,000 dalton protein that becomes associated with lipophorin upon stimulation by adipokinetic hormone, is present at much greater concentration in mature adults than in larvae of *Locusta migratoria*. We have purified apoLp-III from locust hemolymph, and have cloned a gene coding for this protein, as steps in studying the regulation of its synthesis. A genomic clone that contains a gene for ApoLp-III hybridizes with a 1200 nucleotide RNA that is abundant in adult male and female locust fat body and present at low levels in fifth instars. This mRNA accumulates in adult fat body after treatment *in vivo* with the juvenile hormone analog, methoprene.

INTRODUCTION

Lipophorin, the major lipoprotein of insect hemolymph, functions as a shuttle for the transport of lipids between tissues (for reviews, see 1,2,3,4). All lipophorins that have been studied contain a large apoprotein, apoLp-I (M_r ~250,000) and a smaller apoprotein, apoLp-II (M_r ~78,000) (5). In species such as *Locusta migratoria* and *Manduca sexta*, which use lipid as the major fuel for flight muscles, the peptide adipokinetic hormone (AKH) causes lipid stored

[1]This work was supported by NSERC Canada and U.S. NIH.
[2]Present address: Department of Entomology, Texas A&M University, College Station, TX 77843.
[3]Present address: Fox Chase Cancer Center, 7701 Burholme Avenue, Philadelphia, PA 19111.

in fat body to be loaded onto lipophorin, resulting in a larger, less dense lipoprotein particle (1,3,6). Associated with this low density lipophorin of adults is a third apoprotein, apoLp-III (M_r ~20,000). ApoLp-III has been purified from Manduca (7), a hemipteran, Thasus acutangulatus (8), and Locusta (9, 10) and has similar properties in each of these species. In both Manduca and Locusta, apoLp-III is present at very low concentrations in larval hemolymph, whereas it is one of the most abundant hemolymph proteins in mature adults (7, 11).

This change in apoprotein composition of lipophorin has been described as the metamorphosis of a protein (2), a remodeling of an existing larval protein to take on altered functions in the adult. ApoLp-III differs in its pattern of expression from vitellogenin, another hemolymph protein that has been proposed as a model for metamorphosis (12). In locusts, vitellogenin is induced by juvenile hormone in adult females, but is not normally expressed in males or larvae (13) whereas apoLp-III is present at low levels in larvae and at high concentrations in both male and female adults. As first steps in studying the regulation of apoLp-III synthesis in Locusta, we have purified this protein and cloned a gene which codes for it.

RESULTS

Purification and characterization of apoLp-III.

We took advantage of the exceptional heat stability of apoLp-III to develop a simple method for its purification (Fig. 1). Heating cell free hemolymph at 100°C precipitated nearly all of the proteins except apoLp-III. A gel filtration step removed a few minor contaminants and very small peptides. Reversed phase high performance liquid chromatography (HPLC) yielded a very pure preparation of apoLp-III.

Characterization of this protein (M. Kanost and H. McDonald, unpublished data) showed that it has properties similar to protein C_2 of Van der Horst et al.(9), protein C-I of Goldsworthy et al.(10), and to Manduca apoLp-III (7). It has a M_r=19,000 determined by SDS polyacrylamide gel electrophoresis, M_r=37,000 determined by gel filtration on Sephadex G-100, and pI=5.0. The amino acid composition shows that our preparation of apoLp-III is the equivalent of protein C_2 (9).

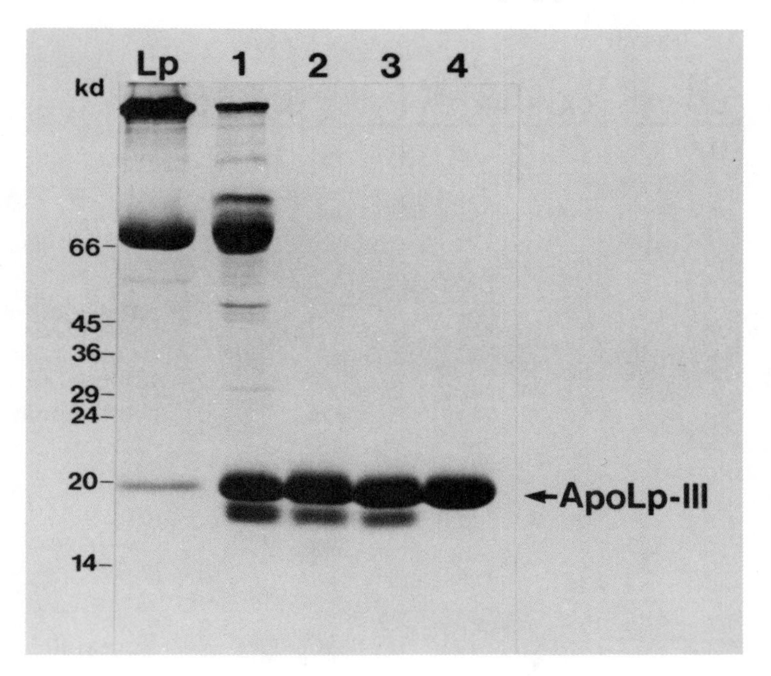

Figure 1. Purification of apoLp-III. Steps in the purification are shown by SDS polyacrylamide gel electrophoresis. Lane Lp-lipophorin isolated from hemolymph by KBr density gradient centrifugation (14); lane 1-adult male hemolymph; lane 2-supernatant after heat precipitation (boiling water bath, 10 min); lane 3-after Sephacryl S-200 chromatography; lane 4-after reversed phase HPLC (μ-Bondapak C-18 column, Waters, eluted with a linear gradient of 0 to 64% acetonitrile in 0.1% trifluoroacetic acid).

ApoLp-III is present at low levels in fourth and fifth instar and first day adult female locusts, but increases greatly during adult maturation (Fig. 2). Immunodiffusion assay (15) showed apoLp-III concentration of 7 mg/ml in newly ecdysed adult females and 35 mg/ml by day 10. In the same time period, the concentration of arylphorin decreases dramatically, vitellogenin appears on day 8, and some proteins such as apoLp-I maintain a relatively constant level.

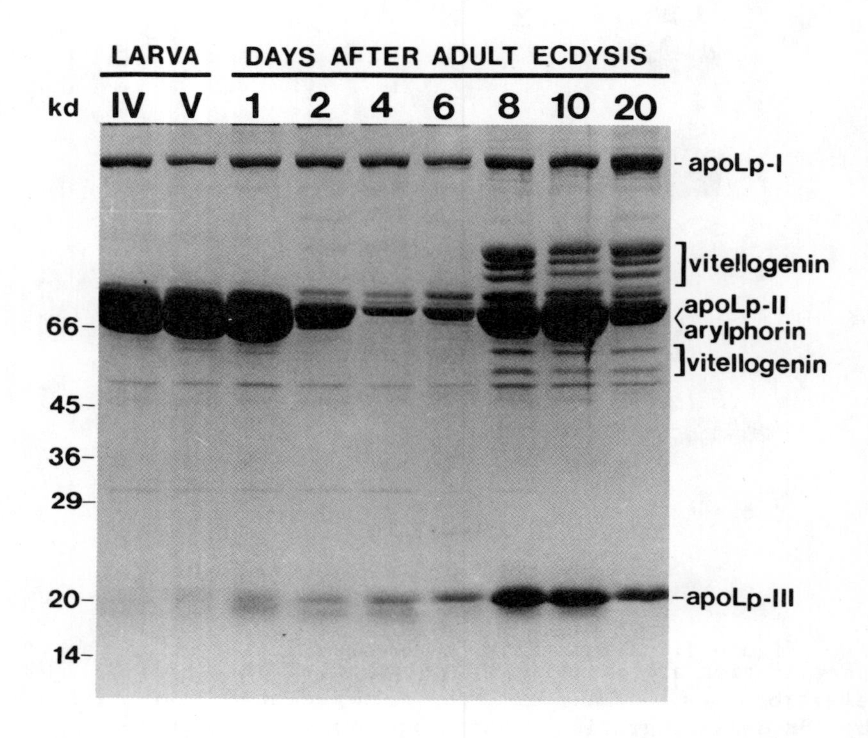

Figure 2. Accumulation of apoLp-III in hemolymph during maturation of adult female locusts. Hemolymph was collected from female locusts after one hour of feeding on wheat seedlings and diluted 1:10 in PBS. Samples (10 µl) were electrophoresed on an SDS polyacrylamide (8-15%) gradient gel.

Cloning of the apoLp-III gene

To gain an understanding of the mechanisms that regulate apoLp-III concentration during its "metamorphosis" we have isolated a gene which codes for the protein. A Locusta genomic DNA library (partially Sau 3A digested DNA in lambda Charon 30) screened with cDNA to adult female fat body RNA (16), yielded a clone, lambda Sau C, which hybridizes to a 1200 nucleotide fat body RNA. This RNA is abundant in adult male and female locusts and present at low concentration in fifth instars (Fig. 3), a pattern of expression consistent with that of ApoLp-III. Lambda Sau C was shown by hybrid selection of mRNA (17) to code for the apoLp-III protein (M.R. Kanost, K.E. Cook, G.R. Wyatt, in preparation). Hybridization of a subclone (pLpIII 1.7) from Sau C with adult fat body RNA selected a mRNA with a translation product of M_r=19,000 that was precipitated by antibody to apoLp-III.

We can use the cloned apoLp-III gene as a hybridization probe to measure levels of apoLp-III mRNA at different stages of development and after experimental treatments. One factor that might be involved in stimulating apoLp-III synthesis during early adult development is juvenile hormone (JH), which rises in titer during the first 10 days after adult ecdysis in both male (18) and female (19) locusts. An experiment to test the effects of JH on apoLp-III expression was performed. Corpora allata of newly ecdysed adult females were inactivated with ethoxyprecocene (20) and after two weeks the insects were injected with 300 µg of the JH analog, methoprene, in mineral oil. After a lag period of 18 hr, apoLp-III mRNA began to accumulate in fat body and by 42 hr was double the level observed in control locusts injected with mineral oil.

DISCUSSION

Lipophorin from adult Locusta contains three apoproteins, with a variable amount of apoLp-III depending on stimulation by AKH (10). It was previously reported from this laboratory that the 250,000 dalton protein (apoLp-I) was observed only after precipitation of lipophorin (21), but we now believe that this was a misinterpretation of data. Figure 1 shows a sample of lipophorin isolated by KBr density gradient centrifugation in which the 250,000 dalton apoLp-I is clearly present. Furthermore, a recent report (22) has shown that apoLp-I and apoLp-II have different peptide maps, establishing that the 250,000 dalton protein is not an aggregate of apoLp-II.

The heat stability of apoLpIII provides a simple method of purifying the protein from hemolymph for structural analysis and antibody production. We do not know whether apoLp-III prepared by this procedure is physiologically active.

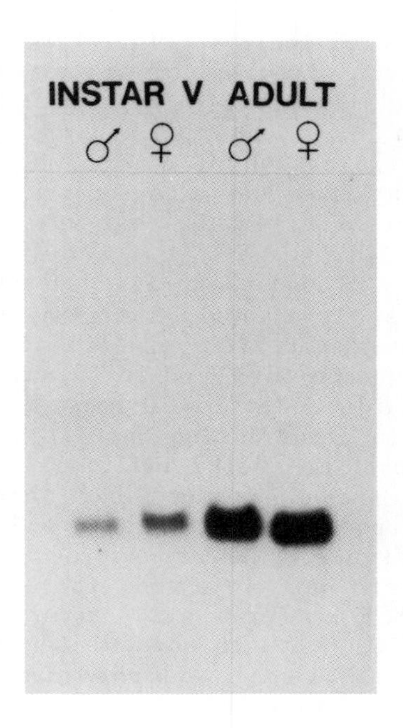

Figure 3. Detection of apoLp-III mRNA in locust fat body. Samples of fat body RNA (5 μg) from fifth instar or mature adult locusts were electrophoresed in the presence of formaldehyde, transferred to GeneScreen Plus (New England Nuclear) and probed with pLpIII 1.7.

We have cloned a genomic DNA sequence that codes for the apoLp-III protein, shown by hybrid selection of mRNA and immunoprecipitation of translation products. Mapping the structure of the apoLp-III gene within this cloned DNA is in progress. In addition, we have recently isolated cDNA clones for apoLp-III (M.R. Kanost, unpublished data). Analysis of the sequence of these cDNA clones may provide some clue to the function of the apoLp-III protein.

ApoLp-III levels in locust hemolymph increase during the first days of adult maturation when JH titer is increasing (Fig. 2). ApoLp-III is present in the hemolymph and its mRNA is present in fat body of precocene-treated locusts, but a JH analog can cause elevation of apoLp-III mRNA in such individuals. These data suggest that the ApoLp-III gene is constitutively expressed at low levels and may be positively regulated by the high JH titer in adult locusts. The increased concentration of JH in the adult induces a variety of cellular and molecular responses in fat body. These include DNA replication leading to higher ploidy levels (23), proliferation of rough endoplasmic reticulum and Golgi complexes (24), stimulation of rRNA synthesis (25), total protein synthesis (21) and the induction of vitellogenin mRNA (25). Whether the elevation of apoLp-III mRNA reflects a specific effect of JH on the apoLp-III gene or is a consequence of other responses in the fat body cells is not yet known. Experiments are continuing to determine whether JH has a significant role in regulating the apoLp-III gene during adult maturation.

ACKNOWLEDGEMENTS

We thank K.E. Cook for excellent technical assistance and V.K. Walker and R. Ziegler for critical reading of the manuscript.

REFERENCES

1. Goldsworthy GJ (1983). The endocrine control of flight metabolism in locusts. Adv Insect Physiol 17:149-204.
2. Ryan RO, Law JH (1984). Metamorphosis of a protein. Bioessays 1:250-252.
3. Beenakkers AMTh, Van der Horst DJ, Van Marrewijk WJA (1985). Biochemical processes directed to flight muscle metabolism. In Kerkut GA, Gilbert LI (eds): "Comprehensive Insect Physiology Biochemistry and Pharmacology, Vol. 10" New York: Pergamon Press, pp 451-486.
4. Chino H (1985). Lipid transport: Biochemistry of hemolymph lipophorin. In Kerkut GA, Gilbert LI (eds): "Comprehensive Insect Physiology Biochemistry and Pharmacology, Vol. 10" New York: Pergamon Press, pp 115-135.
5. Ryan RO, Schmidt JO, Law JH (1984). Chemical and immunological properties of lipophorins from seven insect orders. Arch Insect Biochem Physiol 1:375-383.
6. Shapiro JP, Law JH (1983). Locust adipokinetic hormone stimulates lipid mobilization in _Manduca sexta_. Biochem Biophys Res Comm 115:924-931.

7. Kawooya JK, Keim PS, Ryan RO, Shapiro JP, Samaraweera P, Law JH (1984). Insect apolipophorin III. Purification and properties. J Biol Chem 259:10733-10737.
8. Wells MA, Ryan RO, Prasad SV, Law JH (1985). A novel procedure for the purification of apolipophorin III. Insect Biochem 15:565-571.
9. Van der Horst DJ, Van Doorn JM, Beenakkers AMTh (1984). Hormone-induced rearrangements of locust haemolymph lipoproteins. The involvement of glycoprotein C_2. Insect Biochem 14:495-504.
10. Goldsworthy GJ, Miles CM, Wheeler CH (1985). Lipoprotein transformations during adipokinetic hormone action in *Locusta migratoria*. Physiol Entomol 10:151-164.
11. Wheeler CH, Goldsworthy GJ (1983). Qualitative and quantitative changes in *Locusta* haemolymph proteins and lipoproteins during ageing and adipokinetic hormone action. J Insect Physiol 29:339-347.
12. Kunkel JG (1981). A minimal model of metamorphosis: Fat body competence to respond to juvenile hormone. In Bhaskaran G, Friedman S, Rodrigues JG (eds): "Current Topics in Insect Endocrinology and Nutrition," New York: Plenum, pp 107-129.
13. Dhadialla TS, Wyatt GR (1983). Juvenile hormone-dependent vitellogenin synthesis in *Locusta migratoria* fat body: Inducibility related to sex and stage. Dev Biol 96:436-444.
14. Shapiro JP, Keim PS, Law JH (1984). Structural studies on lipophorin, an insect lipoprotein. J Biol Chem 259:3680-3685.
15. Vaerman J (1981). Single radial immunodiffusion. Meth Enzymol 73:291-305.
16. Wyatt GR, Locke J, Bradfield JY, White BN, Deeley RG (1981). Molecular cloning of vitellogenin gene sequences from *Locusta migratoria*. In Pratt GE, Brooks GT (eds): "Juvenile Hormone Biochemistry," Amsterdam: Elsevier/North Holland, pp 299-307.
17. Miller JS, Patterson BM, Ricciardi RP, Cohen L, Roberts BE (1983). Methods utilizing cell-free protein-synthesizing systems for the identification of recombinant DNA molecules. Meth Enzymol 101:650-674.
18. Johnson RA, Hill L (1973). Quantitative studies on the activity of the corpora allata in adult male *Locusta* and *Schistocerca*. J Insect Physiol 19:2459-2467.

19. Reimbold H (1985). Mikroanalytische Bestimmung von Juvenilhormonen. Chimia 39:348-354.
20. Chinzei Y, Wyatt GR (1985). Vitellogenin titre in haemolymph of Locusta migratoria in normal adults, after ovariectomy, and in response to methoprene. J Insect Physiol 31:441-445.
21. Gellissen G, Wyatt GR (1981). Production of lipophorin in the fat body of adult Locusta migratoria: comparison with vitellogenin. Can J Biochem 59:648-654.
22. Kashiwazaki Y, Ikai A (1985). Structure of apoproteins in insect lipophorin. Arch Biochem Biophys 237:160-169.
23. Nair KK, Chen TT, Wyatt GR (1981). Juvenile hormone-stimulated polyploidy in adult locust fat body. Dev Biol 81:356-360.
24. Couble P, Chen TT, Wyatt GR (1979). Juvenile hormone-controlled vitellogenin synthesis in Locusta migratoria fat body: cytological development. J Insect Physiol 25:327-337.
25. Chinzei Y, White BN, Wyatt GR (1982). Vitellogenin mRNA in locust fat body: identification, isolation and quantitative changes induced by juvenile hormone. Can J Biochem 60:243-251.

Molecular Entomology, pages 285–293

THE FUNCTION OF THE MAJOR LARVAL SERUM PROTEINS OF DROSOPHILA MELANOGASTER: A REVIEW[1]

David B. Roberts

Genetics Laboratory, Department of Biochemistry
South Parks Road
Oxford OX1 3QU

ABSTRACT LSP-1 is the major serum protein of Drosophila larvae. It is very abundant and it is evolutionarily conserved. In spite of this it is a dispensible protein and stocks lacking LSP-1 have been maintained for over two years. It does not appear that the presence of LSP-1 affects the survival of the organism but its absence does have a considerable adverse effect on the fecundity of flies.

INTRODUCTION

The larval serum proteins are the major proteins found in the serum of the final larval instar of dipterans and similar proteins have been found in lepidopterans and orthopterans (1).

The synthesis of Larval Serum Protein 1 (LSP-1) of *Drosophila melanogaster* is both temporally and spatially regulated. Synthesis starts at the beginning of the 3rd larval instar and the initiation of synthesis is transcriptionally controlled (2). Synthesis ceases as the wandering larval stage begins and cessation of synthesis is translationally controlled (3). The protein is synthesised only by the larval fat body (4).

It is an extremely abundant protein; at pupariation 0.7% of the wet weight is LSP-1 and in *Calliphora vicina* 5% of the wet weight of the wandering larva is the homologous protein calliphorin.

[1]This work was supported by the SERC

LSP-1 and its homologues are coded for by a multi-gene family and their polypeptides are thought to associate at random to form a family of hexamers (5). In *Calliphora*, *Lucilia* and *Sarcophaga* and in *Drosophila* of the subgenera Drosophila and Dorsilopha the genes are in a tandem array (6, 7, 8 and 9). There are about fifteen genes in the array in *Lucilia* (7). In the Sophophorans this array breaks down. *D. willistoni* and *D. saltans* have a tandem array; *D. pseudoobscura* has at least two genes in tandem and at least one gene at a second site (9); *D. melanogaster* and its sibling species have three LSP-1 genes α, β and γ which are dispersed on the 1st, 2nd and 3rd chromosomes respectively (9). The X-linked gene is not dosage compensated (10).

Immunological studies show the protein to be more conserved than most dipteran proteins and at the DNA level not only is there cross hybridization between the three LSP-1 genes but their DNAs hybridize with the homologous sequences from other *Drosophila* species and from *Calliphora*. This degree of cross species hybridization is greater than that for most other cloned *melanogaster* sequences I have tested.

The three *Drosophila* genes have been cloned (11); conserved sequences have been found at the 5' ends of the three genes (12) which may coincide with nuclease sensitive sites found in chromatin from fat body but not in chromatin from embryos (13). The three genes have an intron in the same place close to the 5' end and the first 20 amino acids are hydrophobic as expected of secreted proteins (12).

P element mediated transformation has been carried out and the alpha transformant shows correct temporal and spatial expression (13).

In a search for electrophoretic variants of the three LSP-1 polypeptides to map their coding sequences we found a homozygous viable γ-null allele. By chance some time later we found a homozygous viable β-null allele and finally after a search we found a homozygous and hemizygous viable α-null allele. These three alleles were combined to give an LSP-1 null stock which is viable and which we have maintained in our laboratory for nearly two years.

This observation poses the question: what selection pressure(s) maintains the abundance and conserves the sequence of this apparently dispensible protein? In other words what is the function of LSP-1?

RESULTS AND DISCUSSION

Three functions have been proposed for the larval serum proteins:

i) that they are storage proteins. If labelled calliphorin is injected into unlabelled larvae the label is distributed among adult proteins roughly in proportion to their abundance (14). But, if it is a general storage protein why is there an unusual amino acid composition? (i.e. rich in aromatic amino acids)

ii) that they are tanned into adult cuticle. Antibodies raised against the *Calliphora* protein were shown to decorate sections of the adult cuticle (15). We were unable to repeat this observation in *Drosophila* using either monoclonal or polyclonal antibodies. Further, we could find no difference, after examination with an electronmicroscope, between sections of adult cuticle from flies with zero, six or ten doses of LSP-1 genes.

iii) ecdysteriods in the larva of *Calliphora vicina* are bound to macromolecules including calliphorin (16). The protein may be a carrier protein for hormones.

If any or all of these represent functions of LSP-1 then they are not indispensible functions as LSP-1 null flies survive!

LSP-2 is related to LSP-1. They share 50% of their methionine containing peptides (17); their size, amino acid composition and regulation are very similar. Perhaps LSP-2 or some other haemolymph protein is over synthesised to compensate for the lack of LSP-1. We analysed, quantitatively and qualitatively on gels, the haemolymph protein of larvae from stocks with 0, 2, 6 and 10 copies of LSP-1 genes and found no evidence for compensation for the lack of LSP-1 by LSP-2 or any other protein.

Natural selection acts on survival, fecundity or both.

Survival

I carried out a number of experiments to compare the survival of a wild type and an LSP-1 null stock.

i) With respect to longevity at 18°C, 25°C and 29°C both stocks showed the same lifespan.

ii) With respect to such basic parameters as larval and adult weight, duration of embryonic, larval and pupal development, % of larvae pupating and of pupae eclosing the two stocks were indistinguishable.

iii) In a direct comparison heterozygous γ-null flies were crossed *inter se* (these flies were α-null, β-null) and the three classes of adult progeny were counted. We would expect a 1:2:1 ratio of homozygotes to heterozygotes to homozygotes. In the experiment the γ-plus homozygotes were significantly under represented among the progeny. There was no evidence from this experiment to suggest that the presence of LSP-1 increased the chance of survival of *Drosophila*. The contrary result obtained is probably due to vagaries in the genetic background.

All of these studies were carried out under laboratory conditions and it is possible that the advantages in possessing LSP-1 are only manifest under the more stressful conditions in the wild. I stressed wild type and LSP-1 null stocks by starvation, dessication, increasing larval density, the presence of heavy metal ions in the diet, high and low pH and increasing concentrations of urea in the medium. In no case did these stress conditions discriminate between stocks with and without LSP-1.

Fecundity

The first suggestion that the fecundity of the null stock might be disturbed came from the observation that the null flies laid but few eggs and that these looked peculiar. Some null eggs were almost transparent and scanning electronmicrographs showed that others had abnormal exochorions with no chorionic ridges and flaccid as opposed to tumescent chorionic appendages. We call these curates eggs (18). This result may be a coincidence as there is no obvious relationship between the observed quality of the eggs and their developmental fate. However, the number of curates eggs was inversely related to the number of LSP-1 genes present in the mother.

At 25°C 95% of eggs laid by our wild type Oregon-R stock hatch whereas only 37% of eggs laid by the null stock hatch. Nearly 1000 eggs of each stock were analysed. By this criterion the null stock was about 40% as fit as the wild type stock.

A possibility for this decrease in the number of eggs hatching is that either males or females of the null stock or both are sterile or partially sterile. I set up crosses between pairs of virgin flies and examined the vials for progeny one week later. Only vials in which both parents were alive after five days were counted. The results were surprising (Table 1).

TABLE 1
FERTILITY

Cross	%sterility	number of pairs
Oregon-R x Oregon-R	8	177
LSP-1 null x LSP-1 null	54	184
Oregon-R x LSP-1 null[a]	51	136
LSP-1 null x Oregon-R	54	142

[a]female parent first

If you assume that the Oregon-R flies are 100% fertile then the fertility of the null males and null females can be estimated and the product of these two estimates gives the fertility for the LSP-1 null stock as 23% compared with the observed 46%. If you take the worst case and distribute the 8% sterility equally between Oregon-R males and females then the product of the estimates of fertility of null males and females is 27% compared with the observed 46%. This difference is significant and was observed in two separate experiments. It is difficult to explain except by assuming some incompatability between Oregon-R and the null stock. Analysis of the genital apparatus of sterile male and female flies did not reveal any obvious difference from the wild type. Developing eggs and motile sperm tails were observed in all cases examined.

One source of incompatability would be mating behaviour. We have compared the mating behaviour in different crosses (Table 2).

TABLE 2
MATING BEHAVIOUR

Cross	of pairs courting % mating in 2h	number of pairs
Oregon-R x Oregon-R	50	40
Oregon-R x LSP-1 null[a]	53	40
LSP-1 null x Oregon-R	33	40
LSP-1 null x LSP-1 null	24	40

[a]female parent first

Inbred stocks, such as Oregon-R, are known to show reduced mating behaviour compared with outbred stocks and in this case LSP-1 null is an outbred stock. This means that the experiments are not strictly comparable and the % of Oregon-R mating is a depressed value for an outbred wild type stock. Nevertheless the lowest % of flies mating were the null flies crossed *inter se*. The highest % was any cross with Oregon-R females. Detailed examination of the courtship behaviour explains this result. Null females are reluctant to mate and reject the overtures of null males. Wild type males are more aggressive or persistent and so have a greater success with null females. With the more receptive Oregon-R females the null males perform as well as wild type males.

Although there is a mating behaviour difference which would be important in the wild where flies meet but fleetingly, the difference does not explain the previous result in any obvious way. But, if the difference is indeed due to the presence or absence of LSP-1 these results lend weight to the argument that the relevance of LSP-1 is to the fecundity of flies.

The reduced mating of null flies was not due to lethargy as the amount of time spent moving and the speed of movement of wild type and null flies was the same.

The number of eggs laid by the null stock was obviously less than the number laid by the wild type stock. Single virgin male and female flies were mated for four days and the number of eggs laid on the 5th day after eclosion was counted. The null stock laid 20% of the number of eggs laid by the wild type stock. Wild type females lay almost as many eggs when crossed with null males (but fewer are fertile see below) as when crossed with wild type males. Null females lay more eggs when crossed with wild type males than when crossed with null males. This last result is probably attributable to the fact that mated females lay more eggs than unmated ones and more null females will have been mated when crossed to wild type males, echoing the observation made on mating behaviour above.

The null stock lays only 20% of the eggs laid by the wild type stock and only 40% of these are fertile. The fitness of the null stock compared with the wild type stock is then 8%. If this were due to a single factor difference it would readily account for the elimination of null flies from the population. With three genes and the possibility of gene doses between 0 and 6 we must assume that stocks with intermediate gene copy number have intermediate fitness values. This assumption although not tested by counting eggs and fertility of eggs is probably correct, see below.

In the end what really matters is how many fertile progeny are left by a particular genotype. We set up single pairs of virgin male and female flies on the day they eclosed. Every day for the next 24 days these were transferred to a new vial. The vials were kept and the number of progeny emerging in each vial was counted. Only pairs which survived the full 25 days of the experiment were considered. Over this period the null flies left 10% of the adult progeny left by the wild type stock and stocks of flies with 2 or 4 LSP-1 genes left intermediate progeny numbers.

Crosses between null females and wild type males behaved like the null stock while crosses between wild type females and null males, if fertile, behaved like the wild type stock. This shows that if the null males are fertile then they are fully fertile otherwise even in fertile crosses fewer progeny would have been observed in the cross between wild type females and null males.

The fitness of the null stock with respect to the number of adult progeny left (10%) is in good agreement with the calculated fitness from the product of the number of eggs laid and the number of eggs which hatch (8%). Although it is fair to say that with respect to the number of progeny left during the first week of the experiment, which would approximate more realistically to the life span of the flies in the wild, the null stock is 15% as fit as the wild type stock.

However, not all the null progeny are fertile. Assuming equal amounts of male and female sterility the lowest estimate of the sterility of null flies comes from the observation that 50% of null x null crosses are sterile which gives a minimum estimate of 30% sterility. By a similar calculation only 4% of wild type flies are sterile. With respect to fertility the null flies are only 73% as fit as wild type. This gives a final fitness for the null flies of 10% progeny x 73% fertile = 7% of wild type.

While LSP-2 is similar to LSP-1 it does not fulfill the same function. We have no LSP-2 null alleles but we do have a heterozygous deficiency stock. When this is put into an LSP-1 null background it does not exacerbate the LSP-1 null phenotypes in the few experiments we have so far carried out.

Summary

The absence of LSP-1 appears to have a serious effect on the fertility of Drosophila. This effect is sufficient to eliminate null flies from the population. We are left with two problems:

i) to what extent are the differences observed due to the presence or absence of LSP-1 rather than differences in the genetic background?

ii) what is the primary effect of LSP-1? The phenotypes described above are presumably due to the adult fly developing in the absence of the protein. The continuing presence of the protein cannot be essential as its beneficial effects are observed long after it has disappeared from wild type flies.

ACKNOWLEDGEMENTS

I wish to thank my colleagues for permission to cite our joint unpublished work.

REFERENCES

1. Roberts DB, Brock HW (1981). The major serum proteins of Dipteran larvae. Experientia 37:103.
2. Roberts DB, Wolfe J, Akam ME (1977). The developmental profiles of two major haemolymph proteins from Drosophila melanogaster. J.Insect Physiol. 23:871.
3. Powell D, Sato JD, Brock HW, Roberts DB (1984) Regulation of synthesis of the larval serum proteins of Drosophila melanogaster. Dev. Biol. 102:206
4. Sato JD, Roberts DB (1983). Synthesis of Larval serum proteins 1 and 2 of Drosophila melanogaster by third instar fat body. Insect Biochem. 13:1.
5. Roberts DB, Evans-Roberts SM (1979) The genetic and cytogenetic localization of the three structural genes coding for the major protein of Drosohila larval serum. Genetics 93:663.
6. Schenkel H, Kejzlarova-Lepesant J, Berreur P, Moreau J, Scheller K, Beegegere F, Lepesant J-A (1985) Identification and molecular analysis of a multigene family encoding calliphorin, the major larval serum protein of Calliphora vicina. EMBO Journal 4:2983

7. Thomson JA (1975) Major patterns of gene activity during development in holometabolous insects. Adv. Insect Physiol. 11:321
8. Tahara T, Kuroiwa A, Obinata M, Natoria S (1984) Multi-gene structure of the storage protein genes of Sarcophaga peregrina. J.Mol.Biol. 174:19
9. Brock HW, Roberts DB (1983) Location of the LSP-1 genes in Drosophila species by in situ hybridization. Genetics 103:75
10. Roberts DB, Evans-Roberts SM (1979) The X-linked -chain gene of Drosophila LSP-1 does not show dosage compensation. Nature 280:691.
11. Smith DF, McClelland A, White BN, Addison CF, Glover DM (1981) The molecular cloning of a dispersed set of developmentally regulated genes which encode the major larval serum protein of Drosophila melanogaster. Cell 23:441.
12. Delaney SJ, Smith DF, McClelland A, Sunkel C, Glover DM (1986) Sequence conservation around the 5' ends of the larval serum protein 1 genes of Drosophila melanogaster. J.Mol Biol. in press.
13. Jowett T (1985) The regulatory domain of a larval serum protein gene in Drosophila melanogaster. EMBO Journal 4:3789.
14. Levenbook L (1984) Insect storage proteins. In Kerkut GA, Gilbert LI (eds): "Comprehensive Insect Physiology Biochemistry and Pharmacology". Pergamon Press Chapter 9.
15. Scheller K, Zimmerman H-P, Sekeris CE (1980) Calliphorin, a protein involved in cuticle formation of the blowfly Calliphora vicina. Z. Naturf. 35c:387.
16. Enderle U, Kauser G, Reum L, Scheller K, Koolman J (1983) Ecdysteroids in the haemolymph of blowfly larvae are bound to calliphorin. In Scheller K (ed): "The Larval Serum Proteins of Insects", Georg Thiome Verlag p 40.
17. Brock HW, Roberts DB (1980) Comparison of the Larval serum proteins of Drosophila melanogaster 106:129.
18. True Humiliation (1895) Punch 109:222.

Molecular Entomology, pages 295–304

INHIBITORY AND STIMULATORY CONTROL OF DEVELOPMENTALLY REGULATED HEMOLYMPH PROTEINS IN *Trichoplusia ni*[1]

Grace Jones, Shivanand T. Hiremath*, Gary M. Hellmann*, Mietek Wozniak and Robert E. Rhoads*

Departments of Entomology and Biochemistry*, University of Kentucky, Lexington, Kentucky 40506

ABSTRACT In the *Trichoplusia ni* system, two groups of hemolymph proteins associated with metamorphic development are regulated in opposite senses by juvenile hormone. The expression of one group, consisting of several storage proteins, is suppressed by juvenile hormone, while that of the other group, juvenile hormone esterases, is stimulated. The storage proteins are shown to be regulated at the mRNA level.

INTRODUCTION

During the last decade there has been substantial progress in the study of hormonal regulation of gene expression. A number of model systems have been developed, each with its own advantages for the study of a given aspect of hormone-gene interaction. A number of invertebrate model systems have been developed, and insect systems have figured prominently in these efforts.

The wealth of genetic information on the genus *Drosophila* and the availability of mutants have caused this insect to be a popular subject for study. For example, DOPA decarboxylase has received a great deal of attention as an enzyme inducible by ecdysteroids (3). Several systems involving the calypterate Diptera have been used for the study of fat body protein synthesis and uptake, as influenced by ecdysteroids (15).

[1]This work was supported by NIH grants GM 33995 (GJ) and GM 20818 (RER).

Several systems for studying aspects of juvenile hormone (JH)-gene interaction have also been developed. Questions concerning JH regulation of 'adult' genes have focused on production of proteins during reproductive cycles of several cockroach species (16,23).

Hormonal regulation of expression of genes involved in larval metamorphosis of Lepidoptera has been investigated most extensively in Manduca sexta, Galleria mellonella, and Bombyx mori. In M. sexta, each region of the cuticle has its own complement of proteins, and the proteins of each region are deposited as gene expression of these proteins proceeds. The effect of JH is to determine whether the suppressive action of ecdysteroid on the expression of these cuticular proteins is permanent. A number of hemolymph storage proteins have been reported to be suppressed in these species following JH treatment, and the developmental appearance of translatable mRNA for these proteins has been examined (4,12,18,19). However, no evidence has presented demonstrating that JH acts directly on the fat body to reduce mRNA levels for these proteins.

We report here a system based on Trichoplusia ni hemolymph proteins. In this system, we are able to examine the ability of JH to both stimulate and inhibit hemolymph protein gene expression.

MATERIALS AND METHODS

Larvae of T. ni were reared and staged as described previously (5,20). Hemolymph proteins and translation products were analyzed by sodium dodecyl sulfate-containing polyacrylamide gel electrophoresis (SDS-PAGE) (13). Hemolymph proteins were also analyzed by electrophoresis on nondenaturing 4-16% polyacrylamide gels as described by the supplier (Pharmacia), and by isoelectric focusing (IEF) over a pH range of 3.5-10 on an LKB Multiphor unit (6,22). Poly(A)-containing RNA was prepared from ten grams of larvae as described elsewhere (2,14). The RNA was then translated in a rabbit reticulocyte lysate system (1). For hormonal manipulation experiments, larvae were treated with 100 nmol of a potent JH analog, fenoxycarb, at head capsule slippage to the final larval instar, and bled two days later. In fat body culture experiments, the material was excised and incubated as described elsewhere (11).

RESULTS

JH Regulation of Acidic Hemolymph Storage Protein

There are several acidic proteins which increase to very high abundance during the final larval instar, especially towards the end of the feeding stage (fifth instar day 2, designated L5D2). One protein of isoelectric point 5.8 and molecular weight 76,000 essentially disappears when larvae are treated with 100 nmol of the JH analog and bled two days later on L5D2 (Fig. 1).

The direct action of JH on the fat body to suppress synthesis of this protein was demonstrated in vitro. The presence of the JH analog in the incubation medium effectively decreased the appearance of the protein in the medium (Fig. 2).

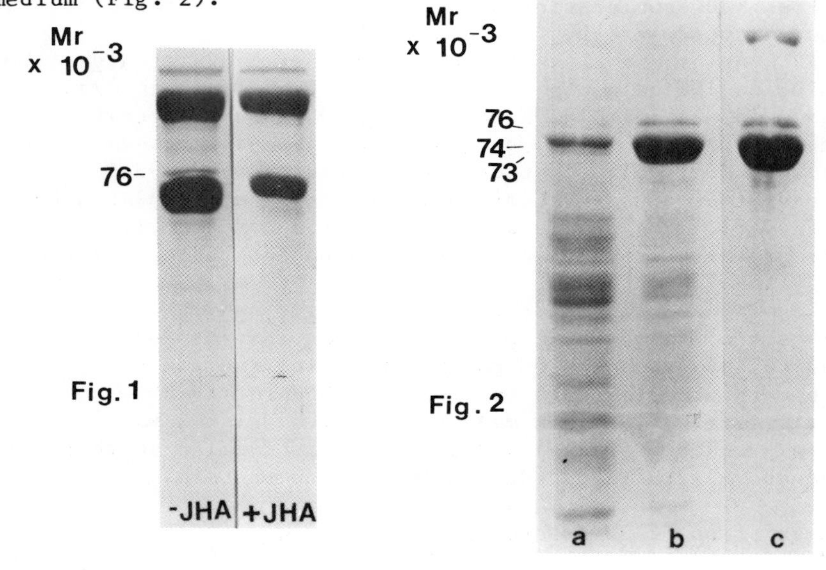

FIGURE 1 (left). Reduction of 76,000 dalton protein in hemolymph of L5D2 larvae two days after topical treatment with 100 nmoles of the JH analog fenoxycarb. Proteins were separated by 10% SDS-PAGE.

FIGURE 2 (right). Reduced expression of 76,000, 74,000 and 73,000 dalton proteins in organ culture of fat bodies. Proteins secreted into the medium were subjected to 10% SDS-PAGE. (a) Medium containing 100 uM fenoxycarb. (b) Control medium containing no fenoxycarb. (c) L5D2 hemolymph (without fenoxycarb treatment).

In order to test the level at which JH acts to regulate expression of this protein, mRNA from control and treated larvae was translated in a rabbit reticulocyte lysate cell-free system. A translation product with a molecular weight of 76,000 was markedly decreased, suggesting that JH acts to regulate this protein at the mRNA level (Fig. 3). In other experiments (not presented) we have shown that an mRNA of a size sufficient to code for a 76,000 dalton protein is decreased in JH analog treated larvae (9).

JH Regulation of Basic Hemolymph Storage Proteins

In addition to the acidic protein described above, a group of basic proteins dramatically increased in abundance in the hemolymph at the time of larval metamorphosis. These proteins had isoelectric points near 8.3 and were also strongly suppressed by the JH analog (Fig. 4).

Although these proteins focused as 4-6 bands on high resolution IEF gels, their molecular weights fell into two classes, approximately 73,000 and 74,000 when IEF-purified fractions were separated by SDS-PAGE (Fig. 5). However, they were difficult to study experimentally by SDS-PAGE because of an abundant comigrating protein not regulated by JH (Fig. 6). The suppressive action of JH on the levels of these proteins is more easily seen on nondenaturing gradient gels (Fig. 7).

JH appears to act directly on the fat body to suppress synthesis of these basic proteins. When the fat body in organ culture was exposed to the JH analog and the proteins which were secreted into the medium subjected to SDS-PAGE, the size of the protein mass near 74,000 daltons was strongly reduced (Fig. 1). This effect is similar to that noted above with the acidic 76,000-dalton protein.

JH Regulation of JH Esterases

In contrast to the storage proteins, JH esterases are stimulated by JH. This stimulatory action of JH has been shown with both *in vivo* and *in vitro* approaches. When L5D3 larvae are treated with 10 nmol of fenoxycarb and bled 12 hrs later, an increase in hemolymph JH esterase activity is observed (8,21). *In vitro* experiments show that JH appears to act directly on the fat body to increase JH esterase production, and this increase can be prevented with actinomycin D (11).

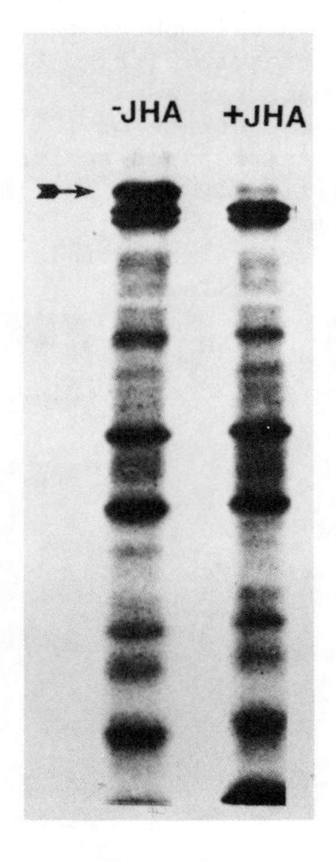

Fig. 3

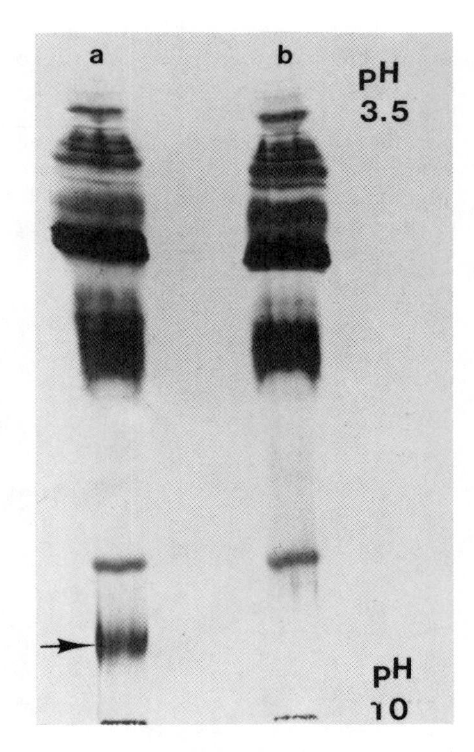

Fig. 4

FIGURE 3. SDS-PAGE and fluorogram of products obtained after cell-free translation of mRNA from L5D2 larvae treated two days previously with ethanol (-JHA) or 100 nmols of the JH analog fenoxycarb (+JHA).

FIGURE 4. Protein profiles obtained after isoelectric focusing of hemolymph proteins from L5D2 larvae treated two days earlier with (a) ethanol or (b) 100 nmoles of fenoxycarb. The arrow indicates a group of JH-suppressible basic proteins.

JH esterase activity occurs in the hemolymph as two major (6) and two minor electrophoretic forms (7; Fig. 8). At least the two major forms can be induced to a higher than normal level in vitro by the presence of 100 nM of the JH II homolog (11). Antiserum was prepared against total JH esterase. The antiserum specifically reacted with JH esterase in preparations of total hemolymph protein from either L5D2 or L5D4 larvae (Fig. 9). Pretreatment of larvae with fenoxycarb increased the amount of JH esterase.

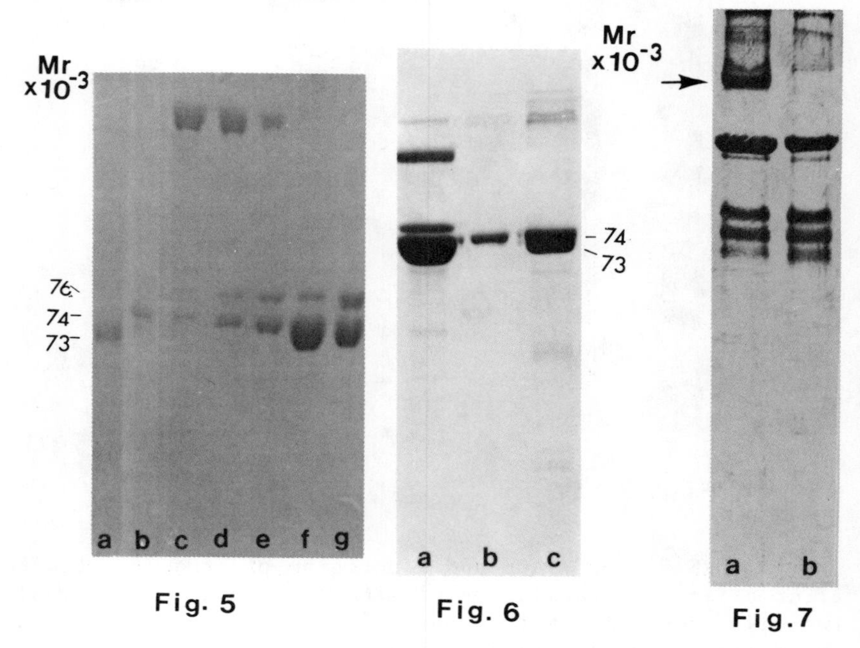

FIGURE 5. 7% SDS-PAGE of protein eluted from gel slices of a wide range (pH 3.5-10) IEF gel. The proteins eluted from each slice of the IEF gel were loaded onto individual lanes of the SDS gel. From left to right, the samples loaded were eluted from the basic to the acidic regions.

FIGURE 6. 10% SDS-PAGE of (a) L5D2 hemolymph, (b) same protein sample as in lane b, Fig. 5, (c) mixture of the two basic proteins abundant in lanes a and b of Fig. 5.

FIGURE 7. 4-16% nondenaturing gradient gel of (a) L5D2 hemolymph protein and (b) L5D2 hemolymph protein of larvae topically treated two days earlier with fenoxycarb (100 nmol). Arrow indicates protein(s) suppressed by fenoxycarb.

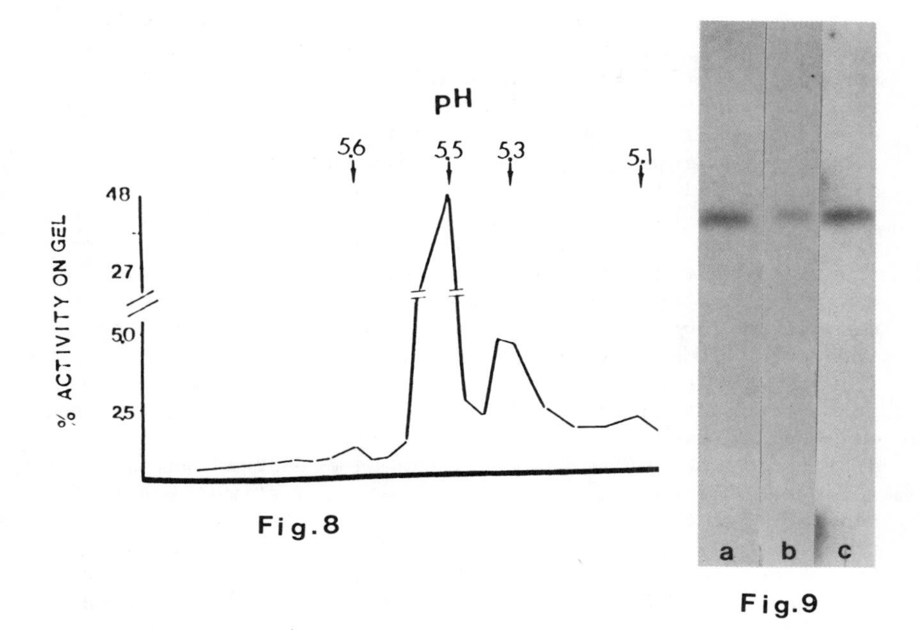

FIGURE 8. JH esterase activity eluted from narrow range (pH 4-6.5) isoelectric focusing gels. Peaks of activity were observed at pH 5.6 (minor peak), 5.5 (major peak), 5.3 (major peak) and 5.1 (minor peak). These forms all occur at several times during larval development (7).

FIGURE 9. Immunoblot (10% SDS-PAGE) using antiserum prepared against a mixture of all four electrophoretic forms of JH esterase. (a) purified JH esterase, (b) total hemolymph protein of L5D4 larvae, (c) L5D4 larvae treated the previous day with fenoxycarb.

DISCUSSION

Current Insect Model Systems

The frequency with which new model systems of insect gene regulation have been proposed might lead one to believe that few important aspects of gene regulation in insects are being left unscrutinized. Many of the model systems being used are very simple systems. In the systems where direct hormone action on gene activity has been demonstrated, the direct stimulatory action of the one hormone, and the absence of simultaneous stimulatory or inhibitory action of other hormones, have facilitated elucidation of the level at which the single hormone is acting. Consequently, these systems have greatly increased our understanding of insect hormone action.

However, the simplicity of these leading insect systems also prevents their direct application toward some common scenerios in endocrine regulation of gene expression. In most of these systems, only one stimulatory or inhibitory hormone is involved, or only one family of proteins is affected. Thus they are not designed for study of situations where there is 1) inhibition of one protein and stimulation of another by the same hormone, 2) simultaneous and qualitatively similar hormonal effects on distinctly different protein families, 3) the same protein being stimulated at one time by one hormone and stimulated at another time by another hormone (of a different chemical class) and 4) the same protein being stimulated at one time by one hormone and inhibited at the same time by another hormone. Examples of these situations are being studied at the endocrine and molecular levels in vertebrates, but such forms of hormone-gene interaction are not being as aggressively analyzed in invertebrate systems.

The *Trichoplusia ni* system reported here is especially suitable for the study of multiple hormone-multiple protein family interactions because 1) the storage proteins and JH esterase are closely associated with or are critical for insect metamorphosis, 2) the storage proteins are in high abundance and the JH esterases in low abundance, enabling a comparison of mechanisms by which the same hormone regulates two proteins which dramatically differ in the needed level of expression, 3) two distinctly different storage protein families are involved, and 4) in addition to regulation by JH, JH esterases are regulated by stimulatory and inhibitory neurohormonal factors (10). That several hormones impinge on the regulation of these proteins does not make the system intractable, because the temporal organization of the interaction with the protein genes enables each form of hormone(s)-gene(s) interaction to be probed directly. With approproate molecular tools we will be able to answer a number of important and interesting questions on mechanisms of hormone regulation of gene expression in insects.

ACKNOWLEDGEMENTS

We thank Dr. Davy Jones and Anita Click for their interaction on many aspects of this study. This paper (86-7-74) is published with the approval of the Director of the Kentucky Agricultural Experiment Station.

REFERENCES

1. Chu L-Y, Rhoads RE (1978). Biochemistry 17:2450-2455.
2. Desrosiers R, Friderici K, Rottman F (1974). Proc Natl Acad Sci USA 71:3971-3975.
3. Gietz RD, Hodgetts RB (1985). Develop Biol 107:142-155.
4. Izumi S, Tojo S, Tomino S (1980) Insect Biochem 10:429-434.
5. Jones D, Jones G, Hammock BD (1981) J Insect Physiol 27:779-788.
6. Jones D, Jones G, Rudnicka M, Click AJ, Sreekrishna, SC (1986). Experientia 42:5-47.
7. Jones G, Click A (1986) J. Insect Physiol (in press).
8. Jones G, Hammock BD (1983) J Insect Physiol 29:471-475.
9. Jones G, Hiremath S, Hellman GM, Rhoads RE (1986) (in preparation).
10. Jones G, Wing KD, Jones D, Hammock BD (1981) J Insect Physiol 27:85-89.
11. Jones G, Jones D, Click A (1986) Insect Biochem (submitted).
12. Kumaran AK, Memmel NA, Ray A (1985) Intl Symp Dev Biol Abstract 128
13. Laemmli UK, Favre M (1973) J Mol Biol 80:575-579.
14. LeMeur M, Glanville N, Mandel J L, Gerlinger P, Palmiter R, Chambon P (1981) Cell 23:561-571.
15. Munn EA, Feinstein A, Greville GD (1969). J Insect Physiol 15:1935-1950.
16. Pau R, Levenbook L, Bauer AC (1979). Experientia 35:1449-1451.
17. Rudnicka M, Jones D (1986) Insect Biochem (accepted)
18. Riddiford LM and Hice RH (1985) Insect Biochem 15:489-502.
19. Ryan, RO, Keim RS, Well MA, Law JH (1985) J Biol Chem 260:782-787.
20. Shorey HH, Hale RL (1965). J Econ Entomol 58:522-524.

21. Sparks TC, Hammock BD (1979) J Insect Physiol 26:551-560.
22. Winter A, Ek K, Anderson VB (1977). LKB Application Note 250, LKB Instruments, Inc., Gaithersbug, MD
23. Wyatt, GR, Dhadialla TS, Roberts PE (1984). Hoffmann JA, Porchet M (eds). In "Biosynthesis, metabolism and mode of action of invertebrate hormones." New York: Springer, p 444-453.

Molecular Entomology, pages 305–314

A STORAGE HEXAMER FROM HYALOPHORA THAT BINDS RIBOFLAVIN AND RESEMBLES THE APOPROTEIN OF HEMOCYANIN

William H. Telfer and Holman C. Massey, Jr.

Biology Department, University of Pennsylvania, Philadelphia, PA 19104-6018

ABSTRACT A storage hexamer from pupal hemolymph of *H. cecropia* contains after isolation 2-3 non-covalently bound riboflavin molecules per mole. In SDS-PAGE the subunits exhibit a variable, intrachain kink entailing at least one sulfhydryl and a heavy metal. The metal is apparently cupric ion, which is able to generate the kink when added to open-chain subunits. The amino acid and subunit compositions of this and other storage hexamers resemble those of the arthropod hemocyanins.

A newly isolated storage hexamer from the hemolymph of *Hyalophora cecropia* (1) exhibits properties that broaden our views on the evolution and functions of this class of insect proteins. Although its rise and fall during metamorphosis suggest a primary function in amino acid and carbohydrate storage, the new protein has a supplementary ability to store riboflavin and copper. Whether, armed with these two ligands, the protein is also an enzyme, or the phylogenetic descendant of an enzyme, remains to be determined.

We also show here that amino acid and subunit compositions of the hemolymph flavoprotein (HFP), and to a lesser extent all other insect storage proteins, resemble those of hemocyanin, the major hemolymph protein and oxygen carrier of many of

(1) Supported by NIH grant, GM-32909

the arthropods that lack tracheal systems. These similarities suggest that the hexameric storage proteins either evolved from an early arthropodan hemocyanin, or more recently from fortuitously similar cellular proteins. In either case it is clear that these proteins have been shaped and maintained by very similar adaptive constraints.

RELATIONS TO OTHER STORAGE PROTEINS

Two other hexamers that we have isolated from *Hyalophora* pupal hemolymph have recognizable homologues among the storage proteins that have been described from other lepidoptera. The most concentrated protein of pupal hemolymph in this species is arylphorin (2), whose aromatic amino acid composition and antigenic reactions indicate a close resemblance to the arylphorins of *Manduca* (3) and *Bombyx* (4). We have also isolated a much less concentrated hemolymph hexamer with the high methionine content of the fat body storage protein designated as SP 1 in *Hyalophora* (5), *Bombyx* (4), and Manduca (6).

By contrast there is as yet no evidence for a homologue of HFP in insects other than saturniids. For instance, antibodies against *Hyalophora* HFP do not react with hemolymph proteins of *Manduca* pupae or larvae. And since the riboflavin conjugates give HFP a color that is too bright to be overlooked when it has been separated from the carotenoid-bearing lipoproteins of the hemolymph, it is unlikely that the hexamers isolated in other laboratories carry significant amounts of this ligand.

The amino acid composition of HFP is compared in the first three columns of Table 1 with those of the other two storage proteins that we have isolated. It lacks the high methionine content of the fat body storage protein, and has less tyrosine and phenylalanine than arylphorin. It is also set apart from these two proteins by an extremely low value for cysteine--at the most two residues in each chain of circa 700 amino acids, and an unusually high histidine content--about 48 residues per chain. The average amino acid contents for 314 polypeptide chains published by Dayhoff et al (Table 1, fifth column) include 14 times more cys and less than a third of the his found in HFP.

TABLE 1
AMINO ACID COMPOSITIONS OF ARTHROPOD HEXAMERS
(moles %)

	FBSP	Aryl	HFP	Hemoc	314 Seq
Asp	13.5	10.9	10.7	14.5	9.8
Thr	5.2	4.9	4.5	4.6	6.1
Ser	5.4	4.8	6.9	4.1	7.4
Glu	8.2	10.9	10.6	11.6	9.9
Pro	4.1	7.2	3.4	3.9	5.2
Gly	(4.2)	4.8	4.7	(3.6)	8.4
Ala	(3.6)	(4.3)	(4.4)	(3.3)	8.6
Cys	(0.9)	(0.5)	(0.2)	(0.7)	2.9
Val	7.7	6.6	7.0	5.0	6.6
Met	(6.8)	1.5	2.4	2.8	1.7
Iso	4.7	4.3	6.7	5.4	4.5
Leu	10.0	6.8	8.6	8.0	7.4
Tyr	6.1	(9.1)	(7.7)	6.2	3.4
Phe	5.3	(8.6)	4.7	(7.6)	3.6
His	1.1	3.6	(6.1)	(7.3)	2.0
Lys	8.6	6.4	5.5	5.1	6.6
Arg	5.5	3.8	4.2	6.9	4.9
Try		1.1	1.7	1.8	1.3

Hemoc: Panulirus (7)
314 Seq: (8)
(___): Value is more than twice or less than half of the average for 314 seq.

HEAVY METAL LIGAND

The possibility of heavy metal binding was suggested by analysis of an apparent intrachain kink revealed by SDS-PAGE (1). After dissociation by SDS in the presence of dithiothreitol (DTT) or mercaptoethanol, HFP forms a single band with an Mr of 85 kd. Without the reducing agent varying amounts of a second, faster band with an Mr of about 80 kd appear (Fig. 1, lane 1). Magee et al (1) showed that the faster band is due to 85 kd monomer that cannot unfold completely in SDS.

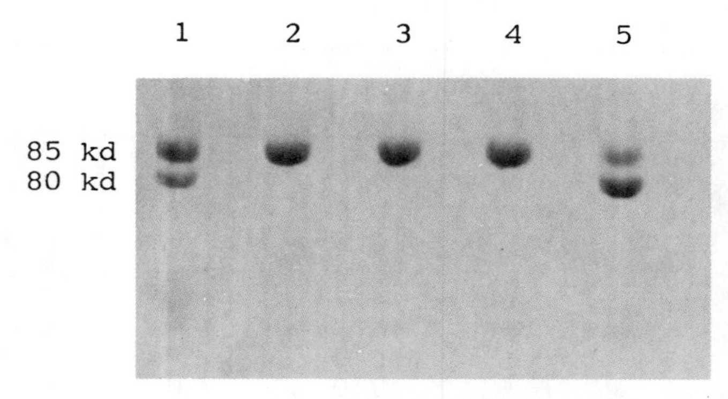

Figure 1. SDS-PAGE of HFP in the absence of disulfide reducing agent. Lane 1, without additives; 2, with 8 mM NEM; 3, 8 mM NEM followed by 5 mM $CuCl_2$; 4, 1.6 mM $CuCl_2$ followed by NEM; 5, 1.6 mM $CuCl_2$.

The salient feature of the DTT effect proved, to our surprise, not to be its ability to reduce disulfide bonds, for the kink in the monomer could also be opened by N-ethylmaleimide (NEM) (Fig. 1, lane 2), which implies that the sulfhydryl group is already in a reduced state.

The key to understanding the SDS-resistant intrachain kink is the finding that cupric chloride can greatly increase the fast to slow band ratio (Fig. 1, lane 5). We therefore propose that DTT opens the intrachain bond by chelating a copper ion, rather than by disulfide reduction. The copper would form a bridge between a sulfhydryl and a second binding site located elsewhere in the amino acid sequence (Fig. 2). NEM by contrast opens the chain by alkylating the sulfhydryl that bonds the cupric ion. The other binding site could be either the second sulfhydryl group detected by amino acid analysis in HFP or, as in copper binding to hemocyanin, one or more imidizole rings of histidine. Lanes 3 and 4 of Fig. 1 confirm this interpretation by showing that the addition of NEM prior to copper prevents kinking, and that kinking generated by copper addition can be reversed by NEM. Of 9 heavy metal ions that were tested, only cupric ions were able to generate the intrachain kink.

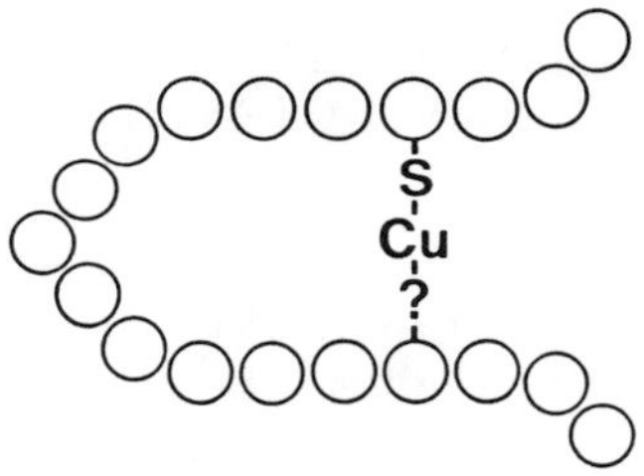

Figure 2. Schematic model of the intrachain kink generated by cupric ions in HFP monomers.

RIBOFLAVIN BINDING

Riboflavin binding to pupal hemolymph proteins with the solubility of albumins was first reported in Hyalophora by Chefurka (9). About half of the riboflavin in the hemolymph remains bound to protein during the ion exchange procedures used in our fractionations, and over 90% of this is associated with a single hexamer (Magee it al, in prep). In addition to these 2-3 molecules of riboflavin per mole of protein, the isolate can bind 7 or more exogenous molecules of ligand, for a total of over 10.

Binding involves aromatic groups of the protein and the isoalloxazine component of the riboflavin, for it quenches the fluorescence of both. Thus, denaturing the protein, which releases the riboflavin, results in a 50-fold increase in flavin fluorescence (1). And adding 10 equivalents of flavin to isolated HFP causes up to 40% quenching of aromatic amino acid fluorescence in the protein. Fluorescence quenching of the protein is elicited to an equal degree by equimolar amounts of FMN, FAD, lumiflavin, and riboflavin. Binding therefore appears not to be affected by the strongly hydrophilic ribityl group, or by the phosphate or adenine nucleotide that are added distally to form intracellular co-enzymes.

APPARENT INDEPENDENCE OF THE BINDING OF COPPER AND RIBOFLAVIN

Thus far we have been unable to detect an effect of copper on riboflavin binding. The results of an experiment designed to detect this possibility are shown in Table 2. The sample of HFP in this case increased its riboflavin fluorescence nearly 70-fold upon boiling, indicating over 98% quenching of ligand fluorescence. When treated with 10 mM DTT, there was a slight but insignificant increase in fluorescence. This is over 4 times the concentration of DTT required to convert kinked HFP subunits to the open form, and thus to chelate sulfhydryl-complexed metals.

TABLE 2
RIBOFLAVIN FLUORESCENCE IN HFP

Treatment	Relative Fluorescence
Untreated	9
Boiled	690
10 mM DTT	11
", 25 mM NEM	305
", ", Dialyzed 24 hr	60
", ", ", Boiled	665

Excitation wavelength 455 nM
Emission wavelength 520 nM

Amounts of NEM sufficient to alkylate the sulfhydryl involved in metal complexing also had little effect on riboflavin binding. (Data not shown.)

By contrast, higher concentrations of NEM caused up to a 40-fold increase in fluorescence (Table 2), apparently due to competition for the riboflavin binding sites. This effect was reversible, for lowering the NEM concentration by dialysis against a 50X volume of buffer resulted in a substantial return of quenching, without significant loss of riboflavin ligand (Table 2, last two lines). The result suggests a rapid method for screening potential ligands, which should allow us to define more precisely the features of the isoalloxazine complex responsible for riboflavin binding, and to determine the range of other ligands that might be carried at this binding site.

THE QUESTION OF LOW CYSTEINE AND DISULFIDE CONTENTS OF INSECT HEMOLYMPH PROTEINS

Low cys contents have been consistently reported for hexameric storage proteins of insects (10), although none are lower than the value of 2 per subunit found in HFP. Levenbook's review also indicates that the lack of disulfide involvement in subunit association is a general characteristic of this class of proteins. Among the highest recorded values for cys is that of the fat body storage protein of *Hyalophora* listed in Table 2, and even that is less than a third of the average cys content of 314 polypeptide sequences summarized by Dayhoff (8). By contrast, disulfide bonds play a much more prominent role in the serum proteins of vertebrates. Bovine serum albumin, for instance, has 17 disulfide bonds per 64 kd molecule, with each bond playing a discrete role in the folding of domains within the native protein (reviewed in 11).

There are indications that other insect hemolymph proteins also lack structurally significant disulfide bonds. A comparison of SDS-PAGE patterns in the presence and absence of disulfide reducing agents indicated that arylphorin, vitellogenin, and lipophorin all dissociate and unfold completely in SDS without the assistance of disulfide reduction (1). What are the physiological and adaptive bases of this remarkably consistent difference from the serum proteins of vertebrates? Whether they relate to oxidation versus reducing environments, closed versus open circulatory systems, or some other difference between the bloods of these groups cannot really be determined without parallel information on other groups of invertebrate animals.

STORAGE PROTEINS AND HEMOCYANIN

Arthropod hemocyanins, wherever they occur among the Xiphosura, Arachnida, Crustacea, or Chilopoda, consist of 500 kd hexamers that may aggregate to even larger sized molecules (reviewed by Van Holde and Miller, 12). A number of hemocyanins have been crystallized and analyzed by X-ray diffraction, so that understanding of their structure has advanced well beyond that of the storage proteins. (Molluscan hemocyanins, which have been even more intensively studied, are sufficiently different from those of arthropods to indicate a different phylogenetic origin (12)).

There are also many similarities in amino acid composition between arthropod hemocyanins and the storage proteins (e.g. Table 2), including a tendency toward contents that are high in aromatic amino acids, and low in glycine, alanine, and cys. HFP, with its high histidine and ability to sequester cupric ions, has additional similarities. While many of these compositional similarities may be matters of coincidence, the molecular weights and hexameric structures in particular indicate, if not common origins for these two families of proteins, at least evolutionary histories that have adaptated them to common features of the circulatory system.

Among other possibilities, the salient adaptive features may include the following. The high hemolymph concentrations required of these proteins may dictate their high molecular weights in order to minimize effects of viscosity and osmotic pressure. On the other hand they must be able to diffuse freely between the hemolymph and the cells that synthesize and consume them, and their size may thus be limited by the porosity of the basement laminae which coat most arthropodan tissues. And in the absence of disulfide bonds, stabilization of the tertiary and quaternary structures of these proteins may depend in some measure on the maximized number of intrasubunit bonds possible in hexamers.

Whatever this list of adaptive forces includes, it appears to have been powerful enough to have generated and maintained the 500 kd hexameric configuration through much of the phylogenetic history of the arthropods. If the hemocyanins and storage proteins are a monophyletic family, then the configuration has been maintained throughout a very ancient history indeed. Even the gene duplications giving rise to the three or four hexamers coexisting in many contemporary species occurred long enough ago to permit complete antigenic individuality to have arisen. If the origins are polyphyletic, then the 500 kd hexamer is sufficiently advantageous to have been reinstituted at least four times in the lepidopteran lineage.

REFERENCES

1. Magee J, Massey HC Jr, Telfer WH (1986). In preparation.
2. Telfer WH, Keim PS, Law JH (1983). Arylphorin, a new protein from *Hyalophora cecropia*: comparisons with calliphorin and manducin. Insect Biochem 13:601.
3. Kramer SJ, Mundall EC, Law JH (1980). Purification and properties of manducin, an amino acid storage protein of the haemolymph of larval and pupal *Manduca sexta*. Insect Biochem 10:279.
4. Tojo S, Nagata M, Kobayashi M (1980). Storage proteins in the silkworm, *Bombyx mori*. Insect Biochem 10:289.

5. Tojo S, Betchaku T, Ziccardi VJ, Wyatt GR (1978). Fat body protein granules and storage proteins in the silkmoth, Hyalophora cecropia. J Cell Biology 78:823.
6. Ryan RO, Keim PS, Wells MA, Law JH (1985). Purification and properties of a predominantly female-specific protein from the hemolymph of the larva of the tobacco hornworm, Manduca sexta. J Biol Chem 260:782.
7. Keiper HA, Gaastra W, Beintema EF, van Bruggen EFJ, Schepman AMH, Drenth J (1975). Subunit composition, X-ray diffraction, amino acid analysis and oxygen binding of Panulirus interruptis hemocyanin. J Molec Biol 99:619.
8. Dayhoff M (1972). Atlas of protein sequence and structure. 5:D355.
9. Chefurka W (1953) PhD thesis, Harvard University.
10. Levenbook L (1985). Insect storage proteins. In Kerkut GH, Gilbert LI (eds):"Comprehensive Insect Physiology, Biochemistry, and Pharmacology" Oxford: Pergamon Press 8:307.
11. Peters TJr (1985). Serum Albumin. Adv Protein Chem 37:161.
12. Van Holde K, Miller KI (1982). Hemocyanins. Quart Rev Biophys 15:1.

Molecular Entomology, pages 315–328

AFFINITY PURIFICATION AND CHARACTERISTICS OF JUVENILE HORMONE ESTERASE FROM LEPIDOPTERA[1]

Bruce D. Hammock, Yehia A. I. Abdel-Aal, Terry N. Hanzlik, Glenn E. Croston, R. Michael Roe[2]

Departments of Entomology and Environmental Toxicology
University of California, Davis, California 95616

ABSTRACT At several times during the development of lepidopterous larvae, the methyl ester of juvenile hormone is hydrolyzed by a highly specific group of esterases known as juvenile hormone esterase (JHE). Studies with inhibitors have demonstrated that JHE activity is essential for normal development. These inhibitors have included both organophosphates and 3-substituted thiotrifluoropropanones. The later compounds appear to act as "transition state mimic inhibitors" of JHE and yield slow tight binding kinetics. JHE is extraordinarily stable to extremes of pH, redox conditions and organic solvents. The enzyme also can be purified from a variety of species by affinity chromatography using the above transition state ligands. Experimental evidence indicates that the enzyme's catalytic site is involved in binding to the column.

INTRODUCTION

The elucidation of the mechanism by which larval insects initiate metamorphosis into pupae and then into adults remains one of the most exciting mysteries in developmental biology. Among those insects studied in the order Lepidoptera, a

[1]This work was supported by grants from NIH Grant Ro 1 ES02710-06, USDA 85-CRCR-1, NSF DCB-8518697, and The Herman Frasch Foundation.

[2]Present address: Department of Entomology, North Carolina State University, Raleigh, NC 27650.

lowering of the titer of juvenile hormone (JH) and a subsequent release of ecdysone in response to the prothoracicotropic hormone certainly are key endocrine events in metamorphosis. The role of JH in this process certainly is more complex than was suggested by the "high-low-no" hypothesis (1). For instance JH has been detected in a variety of insects immediately before pupation. In Manduca sexta this burst of JH appears to retard precocious development of adult like structures (2) while in Trichoplusia ni the prepupal burst of JH is essential for successful pupation (3). However, the basic arguments summarized by Gilbert (1), that a profound reduction in JH titer is associated with the initiation of metamorphosis and that JH must be absent for pupation to occur, still hold.

It is generally accepted in endocrinology that regulation of hormone titers occurs by modulation of biosynthesis against a rather constant background of catabolism. Certainly a reduction in the rate of JH biosynthesis is a key factor in the decline of the hormone in the early part of the last larval instar. However, Weirich et. al., (4) reported that radioactive JH was metabolized to the corresponding JH acid by diluted hemolymph from M. sexta. Whether this JH esterase (JHE) activity was simply associated with the decline in JH or whether it played an active role in the decline remained in question.

The most direct answer to this question was provided by treating larvae of T. ni with a potent inhibitor of JHE and observing that the insects remained in the feeding stage and delayed their pupation (5). This experiment has been repeated with various insect species and JHE inhibitors from several different chemical classes (6-8). Several arguments support the presence of this regulatory system. First, the insect must reduce its JH titer dramatically over a very short period, thus a dual system of reducing biosynthesis while increasing catabolism would be of survival benefit. The concept that regulation of hormone titers was entirely by biosynthesis was developed in rather long lived species where such rapid changes in hormone titer were not critical. Possibly a better vertebrate system for comparison with larval Lepidoptera would be the nervous system where rapid changes are needed and where catabolism and reuptake are as if not more important than biosynthesis and release.

Second, JH is a very lipophilic material and can be anticipated to partition into lipophilic depots making precise regulation over a brief time scale difficult. To address this problem the insect has evolved a JH carrier system which helps to keep the hormone in solution (9). However, both the lipophilic nature of JH and the presence of a carrier protein for the hormone reduce turnover making it difficult for the insect to effect dramatic reductions in JH titer over a short period of time. Thus, an enzyme system which can extract the hormone by mass action from these niches is critical for JH reduction. One can further argue that when the JHE is present, the carrier protein speeds degradation of JH by keeping it in solution, while when enzymes with a high affinity for JH are absent, that the carrier protein prolongs the life of JH (10).

Third, there is evidence that during wandering when the titer of JH is reduced dramatically, that the activity of the corpora allata is reduced but not eliminated. Since the JH titers in *T. ni* larvae just before pupation were the highest detected during the last two instars of larval life, the corpora allata must increase their biosynthetic capability again in the prepupa. Thus, it is reasonable to assume that it is efficient for the insect to reduce JH production but not to dismantle its biosynthetic machinery for the hormone. Since even trace amounts of JH will disrupt development during the change of commitment from larva to pupa, it seems critical that the insect degrade existing hormone as well as small amounts of JH that continue to be produced.

Thus, presence of JHE is critical for the normal metamorphosis of lepidopterous larvae. In addition, its appearance in high levels in the hemolymph is one of the earliest events in the developmental sequence leading to pupation. The molecular probes developed to investigate its appearance may lead to the discovery of still earlier events in development. Preliminary studies indicate that its initial appearance is under the control of factor(s) from the brain and subesophageal ganglion with juvenile hormone playing a minor role. However, its subsequent appearance clearly is influenced directly by the prewandering peak of JH in *M. sexta* and *T. ni* (11-13). Aspects of JH degradation and the regulation of JHE have been covered in previous reviews (9,10,14-17). This manuscript will describe the biochemical properties of the enzyme itself.

INHIBITION OF JHE

It was early noted that O,O-di-isopropyl phosphofluoridate (DFP) was a poor inhibitor of the enzymes metabolizing JH in *M. sexta* while it is a powerful inhibitor of most serine esterases and proteases (18). It subsequently has proven to be very useful in selective inhibition of other esterases in insect hemolymph while preserving JHE. Yet, it is important that this tool is not overused. Hammock (10) discussed various definitions of JHE with the simplest being any esterase which hydrolyzes JH. DFP appears to be a powerful tool in some insects for distinguishing the esterase with a low K_M for JH from other esterases probably due to the steric hindrance of the isopropyl groups. However, this apparent selective inhibition cannot be applied without testing each system individually. Older samples of DFP were almost inactive as inhibitors of JHE while being extraordinarily powerful inhibitors of general carboxylesterases of insect hemolymph and mammalian liver. Some recent samples of DFP, however, seem to contain trace impurities which inhibit JHE (Roe, unpublished).

DFP has proven useful in inhibiting esterases not involved in JH hydrolysis in Lepidoptera, while O-ethyl S-phenyl phosphoramidothioate (EPPAT) has proven to be a powerful, irreversible inhibitor of JHE *in vivo* and *in vitro* (5,19-21). Other, commercially available compounds such as paraoxon are very useful for inhibiting JHE *in vitro*, but their high toxicity limits their use *in vivo* (5).

Two series of reversible inhibitors that have proven exceptionally useful include substituted trifluoroketones ($RC(O)CF_3$) and 3-substituted thio-1,1,1-trifluoro-2-propanones ($RSCH_2C(O)CF_3$). The former compounds appear to be classical reversible inhibitors with K_i's as low as 3.2×10^{-9}M (22). It is interesting that the structure activity relationships among these compounds suggest that the properties needed for binding to JHE are in some ways similar to the properties of JH which the enzyme recognizes. The trifluoroketones appear to act as "transition state analogs" yet the compounds appeared to be inactive in *in vivo* bioassays designed to detect their ability to dramatically reduce the rate of JH catabolism.

The more potent thiopropanones yielded complex kinetics when they were tested as inhibitors of JHE (7,8,23-25). They can be classified as reversible inhibitors since it is likely that the compound which dissociates from the enzyme has the same chemical structure as the compound which bound to the enzyme. However, as is common with many powerful, reversible inhibitors, these compounds appeared to give slow tight binding kinetics. Possibly the slow dissociation of the inhibitor from the catalytic site of JHE as well as its high affinity for the site allowed one to detect JH-like effects when the compounds were applied *in vivo*. As will be discussed later, the affinity of these compounds for JHE and their unique kinetic behavior have proven to be very useful.

TABLE 1
KINETIC PARAMETERS OF JHE AND JH BINDING PROTEIN(S) FROM *T. ni* USING JH II AS SUBSTRATE OR LIGAND

Kinetic Parameter	JHE	JH Binding Protein
Molar concentration in the plasma at near maximum levels	1.49×10^{-6}	8.1×10^{-6}
k_{cat}	31.8 min^{-1}	Not Applicable
K_m(M)	7.06×10^{-8}	1.75×10^{-7}
V_{max} nmoles/min/ml	65.0	Not Applicable
k_d min^{-1}	Not Applicable	9.32×10^{-2}
$k_a M^{-1} min^{-1}$	Not Applicable	5.33×10^{5}
k_{cat}/K_M $M^{-1}min^{-1}$	4.50×10^{8}	Not Applicable
	$[k_{cat}/K_M] \times E_t$	$k_a \times [B.P.]$
	$6.72 \times 10^{2} (min^{-1})$	$4.32 (min^{-1})$
$t_{0.5}$ Sec.	0.062 (For hydrolysis)	9.62 (For association)
Relative $t_{0.5}$	1.00	155

KINETIC CONSTANTS OF JHE

A variety of factors, including the low solubility of JH, have made it difficult to obtain clean estimates of the kinetic constants of JHE in insect hemolymph. Recently by using the almost stoichiometric binding of some trifluoromethylketone inhibitors of JHE, it has been possible to titrate the catalytic sites of JHE in crude hemolymph using Ackerman-Potter plots (23). The concentration of JHE in the hemolymph determined by this method is surprisingly close to that estimated by affinity purification of the enzyme (25,27). Careful kinetic studies on both the crude and the affinity purified enzyme have given rise to kinetic constants such as those shown in Table 1.

In most physiological systems one assumes that the substrate concentration approaches the K_M of the enzyme which metabolizes it. However, JHE appears to function as a scavenger enzyme and the titer of JH II in _T. ni_ is far below the K_M of JHE. Thus, the ratio of k_{cat} to K_M is an appropriate kinetic parameter to use in estimating the capacity of the enzyme to hydrolyze JH under _in vivo_ conditions. Since JH II is by far the major JH in the ultimate larval instar of _T. ni_, it is the only substrate treated here. The k_{cat}/K_M ratio of JHE indicates that degradation of the hormone by the enzyme approaches the diffusion controlled encounter of enzyme and substrate.

It appears clear from previous studies that there is adequate JH binding protein in the hemolymph to complex the majority of the JH present. The relatively low K_M of JHE and its high turnover indicate that it is likely to hydrolyze JH molecules as soon as they are released by the JH binding protein in the hemolymph. A numerical argument for this can be made by noting that the $t_{0.5}$ for hydrolysis of JH II by JHE is approximately 150 times faster than the $t_{0.5}$ for association of the JH II with the hemolymph binding protein (Table 1). These data provide support for the argument presented in the introduction for the dual role of the JH binding protein in both stabilizing JH when the JHE is at low titers and enhancing clearance when JHE is present. These data also indicate that insect control strategies based upon inhibition of JHE would be very unlikely to succeed. However, it is likely that even limited expression of JHE at inappropriate times during development could overpower the biosynthetic capacity of the corpora allata to produce JH.

STABILITY OF JHE

While examining the influence of a variety of reagents on JHE activity, it was noted that JH is exceptionally stable to a variety of treatments which denature many enzymes. These treatments include high levels of oxidants, inorganic ions, many classical esterase inhibitors, and extremes of pH. For instance when JHE from T. ni was incubated in H_2O_2 for 10 minutes at 30.C, it retained 94 percent of its catalytic activity. As reported earlier by several workers, JHE is catalytically active over a wide range of pH's (FIG. 1). This figure also shows that the enzyme is very stable to extremes of pH's for extended periods when analyzed at pH 7.4. It was observed that some organic solvents dramatically increased the activity of JHE of M. sexta both in the crude hemolymph and as a pure enzyme. Such activation was noted earlier for other carboxylesterases by a variety of workers including Barker and Jencks (26) who attributed the behavior to a modifier site in porcine hepatic esterases. However, the activation of JHE from M. sexta was much higher than that observed in vertebrate systems with activation as high as 1200 percent with acetone at a concentration of 2 M. Activation was also found with low concentrations of ethanol with 147 percent of control enzyme activity found with an ethanol concentration of 0.17 M (1 percent ethanol). This certainly indicates an effect at solvent concentrations commonly used in enzyme assays.

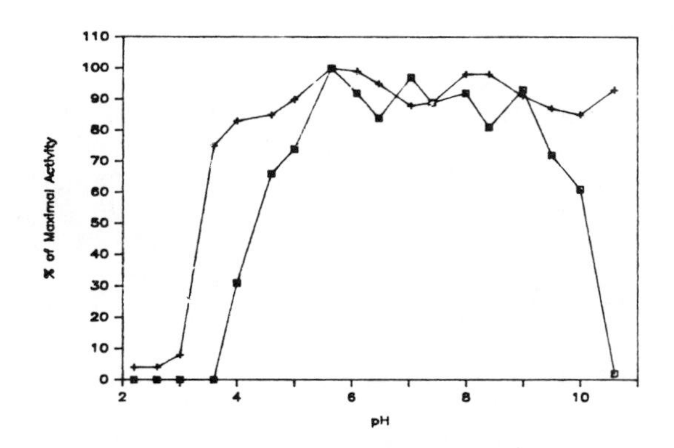

FIGURE 1. Relative activity of JHE partially purified from prepupal homogenates of T. ni when incubated with JH III at different pH's (■) or held at 4.C at varying pH's then diluted with sodium phosphate buffer to pH 7.4 for analysis (+).

It is possible that such a regulator site could be of biological significance in influencing the activity of JHE *in vivo*. However, the JHE's from other Lepidoptera examined do not seem to be influenced as dramatically by the presence of solvents as does the enzyme from *M. sexta*.

AFFINITY PURIFICATION OF JHE

Because of its interesting catalytic properties and its apparent biological role there has been interest in purifying JHE. However, the low concentration of the enzyme in the hemolymph and the difficulty in obtaining large amounts of biological material has made classical purifications very difficult (10). Since some trifluoropropanones appeared to bind to JHE with a high degree of selectivity, these compounds were used as affinity ligands in an effort to purify the enzyme from several species.

$$HS(CH_2)_4SH + BrCH_2C(O)CF_3$$

$$HS(CH_2)_4SCH_2C(O)CF_3$$

FIGURE 2. Synthesis of a ligand for the affinity purification of JHE.

Several ligands have been examined, but the simplest results from the reaction of 1,4-butane dithiol with 3-bromo-1,1,1-trifluoroacetone as shown in FIGURE 2. The resulting thiol is reacted in turn with epoxy activated Sepharose. When 10 ml of diluted hemolymph was passed through 25 microliters of this gel no decrease in protein content was noted in the effluent but over 99 percent of the JHE activity was absent. Since JHE is such a minor protein that it cannot be seen as a discrete band on a SDS-PAGE gel of crude hemolymph, proof that the enzyme was actually bound to the column was not straight forward. Mixing of the effluent with inhibited enzyme indicated that the column was not bleeding an inhibitory substance. However, removal of the JHE activity bound to the column proved to be a difficult task.

Detergent, ionic, pH and other gradients alone or in combination failed to remove significant levels of JHE activity or protein. Even when gradients containing powerful inhibitors of JHE were run through the column, the bound esterase was not recovered. The solution to this dilemma came from recalling that the thio trifluoropropanones were slow tight binding inhibitors. Thus, when an inhibitor solution was allowed to incubate with the affinity gel for an extended period, excellent and sometimes quantitative recovery of JHE was obtained (25,27).

Recovery of enzyme activity also was difficult since very powerful inhibitors such as 3-octylthio-1,1,1-trifluoro-2-propanone (OTFP) were used to elute the enzyme from the column. Fortunately, JHE is very stable in both the crude and purified states and long dialysis times were successful in removing the inhibitor and in recovering the catalytic activity. As shown in FIGURE 3, recovery of enzyme activity is more rapid at acidic pH's. This increased rate of recovery probably results from a combination of two reasons. First, reduced catalytic activity of JHE was noted at low pH (FIG. 1), and this may translate to reduced binding of the enzyme to the inhibitor.

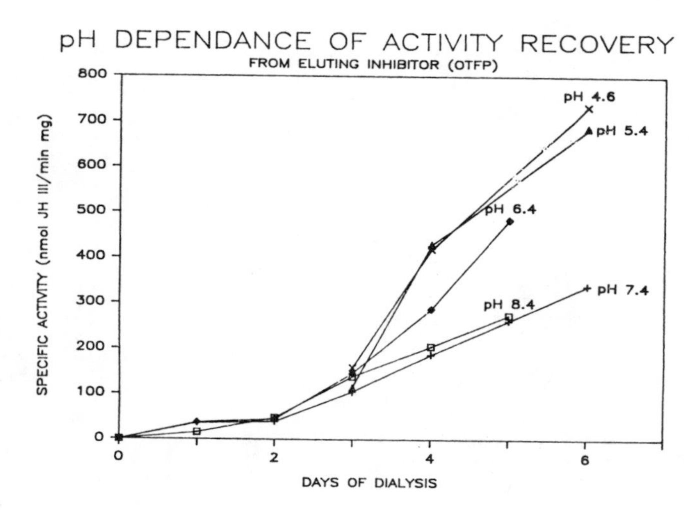

FIGURE 3. Plot of the pH dependant rate of recovery of JHE activity from T. ni during dialysis of the the enzyme eluted from the affinity column by OTFP. Equal amounts of eluted enzyme containing 200 μg/ml BSA were dialyzed against buffer containing 0.2M sodium phosphate or acetate buffers with 0.01% phenyl thiourea, 5% sucrose and 0.02% sodium azide at room temperature. All samples were assayed at pH 7.4. Full recovery of this enzyme takes approximately 2 weeks.

Second, acidic conditions greatly accelerate the chemical formation and decomposition of hemiketals which could accelerate dissociation of the enzyme and inhibitor. In fact this experiment provides circumstantial evidence that the enzyme inhibitor complex resembles a hemiketal.

As mentioned above, JHE is thought to bind to the affinity column by interaction of the trifluoroketone with the catalytic site. FIGURE 4 provides direct evidence for this hypothesis by showing that inhibitors of JHE block its binding to the column. These data are from whole body homogenates of day 2 last instar larvae of *T. ni* which contain several proteins binding to the affinity column. The inhibitors were $2x10^{-4}$M DFP, $1x10^{-5}$M EPPAT, and $1x10^{-5}$M OTFP. Enzyme activity was monitored during the batch loading procedure and showed that little JHE activity was lost from the control and DFP treated homogenate prior to addition of the affinity gel. However, the activity dropped sharply once the gel was added. In the homogenates treated with OTFP and EPPAT, no JHE activity was present upon addition of the gel. Following loading, the

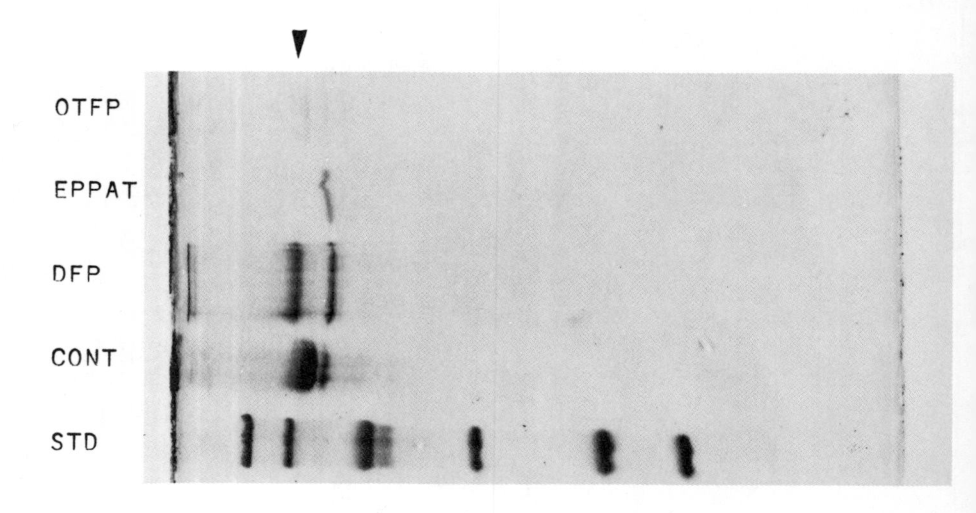

FIGURE 4. SDS-PAGE gel of proteins recovered from affinity purification of homogenates of T. ni treated with various inhibitors. A more complete washing of the gel resulted in a 1 step purification of JHE migrating with an estimated Mr of 65,000 shown by an arrow. Standards of molecular weights 92.5, 66.2, 45, 31, 25, and 14.5 KD are in the left (bottom) lane of the gel.

gel was washed and eluted according to standard procedures. As shown in FIGURE 4, no JHE was detected in the samples treated with EPPAT or OTFP while both protein and activity were recovered from the control and DFP treated gels.

The trifluoroketone affinity column has proven to be very useful for a variety of purposes. For instance, it is directly applicable to the purification of JHE from a variety of species as shown in FIGURE 5.

Although classical purification procedures for JHE are conceptually feasible, small biomass makes it difficult to purify sufficient quantities of a low abundance protein for subsequent studies. The high yield of the affinity chromatography procedure has made it possible to purify sufficient material to radiolabel the catalytic site of JHE from the above species, raise antibodies and obtain sequence information (25,27). The latter data indicate the similarity among JHE isozymes in that the N-terminal sequence for the major component of JHE from _H. virescens_ is TRP-GLN-R while the minor component has two additional amino acids giving a sequence of SER-ALA-TRP-GLN-R. The affinity column has also made it possible to compare JHE isolated from different tissues and from different stages of larvae.

The chemical and molecular probes developed for JHE should be of great value in expanding our understanding of this regulatory enzyme. Hopefully, this technology can be applied to biologically interesting enzymes in other systems.

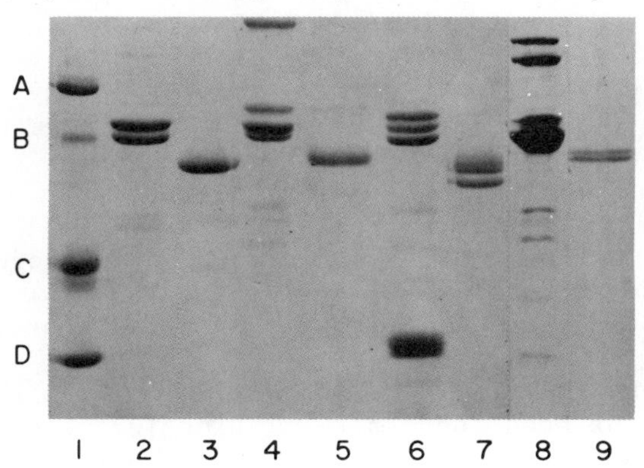

FIGURE 5. SDS-PAGE of the crude hemolymph (lanes 2,4,6,8) and affinity purified JHE (3,5,7,9) from _M. sexta_, _H. virescens_, _B. mori_, and _H. zea_, respectively. Protein standards from 92.5 to 31KD are shown in lane 1. Adapted from Abdel-Aal and Hammock (26).

REFERENCES

1. Schneiderman HA, Gilbert LI (1964). Control of growth and development of insects. Science 143:325.
2. Kiguchi K, Riddiford LM (1978). A role of juvenile hormone in pupal development of the tobacco hornworm, Manduca sexta. J Insect Physiol 24:673.
3. Jones G, Hammock BD (1985). Critical roles for juvenile hormone and its esterase. Arch Insect Biochem Physiol 2:397.
4. Weirich G, Wren J, Siddall JB (1973). Developmental changes of the juvenile hormone esterase activity in haemolymph of the tobacco hornworm, Manduca sexta. Insect Biochem 3:397.
5. Sparks TC, Hammock BD (1981). Comparative inhibition of the juvenile hormone esterases from Trichoplusia ni, Musca domestica, and Tenebrio molitor. Pestic Biochem 14:290.
6. Sparks TC, Hammock BD, Riddiford LM (1983). The haemolymph juvenile hormone esterase of Manduca sexta (L.) - inhibition and regulation. Insect Biochem 13:529.
7. Prestwich GD, Wai-Si E, Roe RM, Hammock BD (1984). Synthesis and bioassay of isoprenoid 3-alkylthio-1,1,1-trifluoro-2-propanones: potent, selective inhibitors of juvenile hormone esterase. Arch Biochem Biophys 228(2):639.
8. Hammock BD, Abdel-Aal YAI, Mullin CA, Hanzlik TN, Roe RM (1984). Substituted thiotrifluoropropanones as potent inhibitors of juvenile hormone esterase. Pestic Biochem Physiol 22:209.
9. Goodman WG, Chang ES (1985). Juvenile hormone cellular and hemolymph binding proteins. In Kerkut GA, Gilbert LI (eds): "Comprehensive Insect Physiology Biochemistry and Pharmacology," New York: Pergamon Press, p 491.
10. Hammock BD (1985). Regulation of juvenile hormone titer: degradation. In Kerkut GA, Gilbert LI (eds): "Comprehensive Insect Physiology Biochemistry and Pharmacology," New York: Pergamon Press, p 431.
11. Sparks TC, Hammock BD (1979). Induction and regulation of juvenile hormone esterases during the last larval instar of the cabbage looper, Trichoplusia ni. J Insect Physiol 25:551.
12. Jones G, Wing KD, Jones D, Hammock BD (1981). The source and action of head factors regulating juvenile hormone esterase in the larvae of the cabbage looper, Trichoplusia ni. J Insect Physiol 27:85.
13. Jones G, Hammock BD (1983). Prepupal regulation of juvenile hormone esterase through direct induction by juvenile hormone. J Insect Physiol 29:471.

14. Hammock BD, Jones D, Jones G, Rudnicka M, Sparks TC, Wing KD (1981). Regulation of juvenile hormone esterase in the cabbage looper, Trichoplusia ni. In Sehnal F, Zabaz A, Menn JJ, Cymborowski B (eds).: "Regulation of Insect Development and Behavior," Wroclaw, Poland: Wroclaw Technical University Press, p 219.
15. Hammock BD, Quistad GB (1981). Metabolism and environmental fate of juvenile hormone, juvenoids and benzoylphenylureas. In Hutson DH, Roberts TR (eds): "Advances in Pesticide Biochemistry," Sussex, England: John Wiley & Sons, Ltd., p 1.
16. Sparks TC, Hammock BD (1983). Insect growth regulators: resistance and the future. In Georghiou GP, Satio T (eds): "Pest Resistance to Pesticides: Challenges and Prospects," New York: Plenum Press, p 615.
17. Hammock BD, Abdel-Aal YAI, Hanzlik TN, Jones D, Jones G, Roe RM, Rudnicka M, Sparks TC, Wing KD (1984). The role of juvenile hormone metabolism in the metamorphosis of selected Lepidoptera. In Hoffman J, Porchet M (eds): "Biosynthesis, Metabolism and Mode of Action of Invertebrate Hormones," Berlin, Heidelberg: Springer-Verlag, p 416.
18. Kramer KJ, Sanburg LL, Kezdy FJ, Law, JH (1974). The juvenile hormone binding protein in the hemolymph of Manduca sexta Johannson (Lepidoptera: Sphingidae). Proc Natl Acad Sci USA 71:493.
19. Sparks TC, Hammock BD (1979). A comparison of the induced and naturally occurring juvenile hormone esterases from last instar larvae of Trichoplusia ni. Life Sci 25:445.
20. Sparks TC, Rose RL (1983). Inhibition and substrate specificity of the haemolymph juvenile hormone esterase of the cabbage looper, Trichoplusia ni. Insect Biochem 13:633.
21. Roe RM, Hammond AM, Sparks TC (1983). Characterization of the plasma juvenile hormone esterase in synchronous last instar female larvae of the sugarcane borer, Diatraea saccharalis. Insect Biochem 13:163.
22. Hammock BD, Wing KD, McLaughlin J, Lovell VM, Sparks TC (1982). Trifluoromethylketones as possible transition state analog inhibitors of juvenile hormone esterase. Pestic Biochem Physiol 17:76.
23. Abdel-Aal YAI, Hammock BD (1985). 3-Octylthio-1,1,1-trifluoro-2-propanone, a high affinity and slow binding inhibitor of juvenile hormone esterase from Trichoplusia ni (Hubner). Insect Biochem 15(1):111.

24. Abdel-Aal YAI, Hammock BD (1985). Use of transition state theory in the development of bioactive molecules. In Hedin PA (ed): "Bioregulators for Pest Control," Washington DC, ACS Symposium Series 276, p 293.
25. Abdel-Aal YAI, Hammock BD (1985). Apparent multiple catalytic sites involved in the ester hydrolysis of juvenile hormones by the hemolymph and by an affinity-purified esterase from Manduca sexta (Johannson) (Lepidoptera: Sphingidae). Arch Biochem Biophys 243(1):206.
26. Barker DL, Jencks WP (1969). Pig liver esterase. Some kinetic properties. Biochem 8:3890.
27. Abdel-Aal YAI, Hammock BD (1986). Transition state analogs as ligands for affinity purification of juvenile hormone esterase. Science. Accepted for publication.

IV. INSECT-SPECIFIC PROCESSES

Molecular Entomology, pages 331–355

MECHANISMS OF INSECT CUTICLE STABILIZATION. HOW DO TOBACCO HORNWORMS DO IT?[1]

Karl J. Kramer[2,3], Theodore L. Hopkins[4], Jacob Schaefer[5], Thomas D. Morgan,[4] Joel R. Garbow[5], Gary S. Jacob[5], Edward O. Stejskal[5] and Roy D. Speirs[2]

[2]U.S. Grain Marketing Research Laboratory, Agricultural Research Service, U.S. Department Agriculture and [3]Dept. of Biochemistry, Kansas State University, Manhattan, KS 66502, [4]Dept of Entomology, Kansas State University, Manhattan, KS 66506, [5]Monsanto Co., St. Louis, MO 63167

ABSTRACT Many mechanistic schemes for the sclerotization or hardening of the insect cuticle (exoskeleton) have been proposed with most involving the formation of crosslinks between cuticular proteins, chitin and/or catechols. We have used ^{13}C- and ^{15}N-cross polarization magic angle spinning nuclear magnetic resonance (CPMAS-NMR) to study the chemical changes that occurred when cuticle from *Manduca sexta* was sclerotized. ^{13}C- and ^{15}N-Labeled amino acids and glucose were introduced into hornworms by injection or ingestion, and the subsequent metabolism of these compounds determined in lyophilized powders of pupal and adult cuticles. Untanned pupal cuticle exhibits a spectrum generally dominated by protein carbon resonances while tanned cuticle shows addition of primarily chitin and catechol carbons. Our studies indicate that carbon-nitrogen bonds are formed in the process of sclerotization with the side chain of histidine playing a prominent role in a link between protein and the catecholamine ring moiety and perhaps in a direct link between protein and chitin chains. Chemical studies have shown that N-β-alanyldopamine (NBAD) and its β-hydroxylated derivative, N-β-alanylnorepinephrine (NBANE), are important

---

[1] This work was supported by National Science Foundation grant PCM-8411408 and U.S. Department Agriculture competitive research grant 85-CRCR-1-1667.

metabolites for sclerotization with the latter catecholamine produced from the former by an insoluble enzyme in pupal cuticle. NBANE released by mild acid hydrolysis of pupal cuticle may be derived from an adduct of NBAD bonded through the β-carbon with the chitin-protein matrix or through other weak interactions since NBANE is more difficult to extract from the chitin-protein matrix than NBAD, its precursor. However, a substantial amount of free NBANE is present in cuticle and it may be a precursor for quinonoid tanning agents because its extractability decreases as the cuticle tans.

INTRODUCTION

The insect cuticle is a composite supramolecular assembly of protein, chitin, catechols, minerals, lipids, water, pigments and other components that has contributed greatly to the success of insects in becoming one of the largest and most diverse groups of animals (Hepburn 1985). The cuticle determines body shape, allows the insect to grow, move, communicate, reproduce and cope with environmental hazards such as predators, pathogens and toxic substances, serves as an anchorage to skeletal muscles and other organs, and protects against physical damage and dessication. The cuticular linings of the tracheal system and the fore and hindguts allow for exchanges of respiratory gases, nutrients, water and waste products.

As a consequence of the rigidity or physical constraints of the exoskeleton, insects must form new cuticles and shed old confining ones periodically in order to continue growth and development. The stabilization of a new cuticle, together with destabilization and recycling of certain components of the old one are critical events in the molting process. Extracellular assembly of cuticle involves many complex processes including organization of the chitin-protein matrix into laminated procuticle probably involving crosslinking by phenolic or quinonoid residues (sclerotization), mineralization, pigmentation, dehydration and water-proofing that cause changes in mechanical properties and appearance (Andersen 1985, Kramer et al. 1985, Blomquist and Dillwith 1985). Although the relative importance of the individual cuticular components in imparting physical and chemical properties is quantitatively unknown, it is generally believed that crosslinked protein (sclerotin) and the polysaccharide chitin [β(1 4) linked 2-acetamido-2-deoxy-D-glucopyranoside] play the major roles in stabilizing the exoskeleton, except in those species which heavily mineralize their cuticle.

To date there is no definitive work published that establishes the covalent interactions between cuticular components as well as the precise molecular architecture of the exoskeleton. Except for bityrosine and trityrosine links between proteins in wing hinges and dipteran cuticle (Andersen 1964, 1966, Lipke et al. 1983), we do not know the chemical structure of any crosslinks. In order to obtain a better understanding of the structure of insect cuticle, we have conducted solid state nuclear magnetic resonance (NMR) analysis of *Manduca sexta* cuticle and chitin prepared therefrom.

The techniques of single and double cross-polarization and magic angle spinning have been developed in recent years to enhance sensitivity and sharpen signals, yielding high resolution NMR spectra of dilute nuclei (e.g. ^{13}C or ^{15}N) in the solid state (Schaefer et al. 1979,1981). This noninvasive technique, cross polarization magic angle spinning (CPMAS) NMR, has already been used to measure carbohydrate, lignin and protein ratios in grass species as well as peptidoglycan crosslinking (isopeptide) in bacteria and is generally applicable to investigations where determination of covalent interactions in intractable material is needed (Himmelsbach et al. 1983, Jacob et al. 1983). CPMAS-NMR involves cross polarization to enhance the ^{13}C or ^{15}N signal, higher power dipolar decoupling to eliminate dipolar line broadening due to protons, and spinning of the sample about an axis at a particular angle to the static field to eliminate chemical shift anisotropy. The chemical shift values of solid state spectra are isotropic values which are similar to those obtained in solution and may be used for structural elucidation in terms of both molecular and crystal structures. We report here some results of a CPMAS-NMR study of tobacco hornworm cuticle assembled from natural abundance sources and ^{13}C- or ^{15}N-enriched amino acids and glucose to determine molecular composition and specific atoms that undergo covalent modification during cuticle sclerotization.

We have also determined the identity and concentration of catecholamines in *M. sexta* pupal cuticle during tanning using liquid chromatography with electrochemical detection (Hopkins et al. 1984). Catecholamines are oxidized to electrophilic intermediates that may covalently bond with the chitin-protein matrix of insect cuticle during sclerotization

(Brunet 1980, Lipke et al. 1983, Andersen 1985). The diphenol moiety of the catecholamine is readily oxidized to a benzoquinone by cuticular phenoloxidases, but the side chain also undergoes important reactions that apparently lead to sclerotizing agents (Peter 1980, Sugumaran and Lipke 1983). N-β-Alanyldopamine (NBAD) is a major catecholamine in developmental stages of insects that form brown cuticles (Hopkins et al. 1982, 1984, Kramer et al. 1984, Roseland et al. 1985). We now report that N-β-alanylnorepinephrine (NBANE) is another major catecholamine in brown pupal cuticle of *M. sexta* and that it is synthesized from NBAD by an insoluble cuticular enzyme. The significance of NBAD and NBANE in relation to cuticle sclerotization is discussed.

MATERIALS AND METHODS

NMR Analysis.

Cuticle was dissected at 4°C from larvae, pupae and adults of *Manduca sexta* (L.) and scraped or cut free of epidermis, muscle, head, scales and appendages. After rinsing with cold deionized water, it was lyophilized and subsequently ground into a coarse powder using a Wiley micromill. Prior to dissection some larvae were fed or injected with isotopically labeled (^{13}C or ^{15}N) amino acids or glucose (Merck Stable Isotopes, Montreal) and pupal cuticle was obtained at ecdysis (soft and untanned) and 7 days postecdysis (hard and tanned). Pupal exuviae (exocuticle) were also collected as well as adult cuticle three days after adult ecdysis. Magic angle spinning ^{13}C-NMR spectra were obtained at 50.3 MHz using 2 ms cross polarization transfers from protons and 50 kHz radio frequency fields with the dried samples spinning a 3.2 kHz in a boron nitride double bearing 0.7 ml hollow rotor (Schaefer et al. 1979, 1981, Yannoni 1982). The chemical shifts are in parts per million (ppm) downfield from external TMS. Spectra were collected under spinning sideband suppression conditions. ^{15}N-NMR spectra were obtained at 20.2 MHz using matched spin-lock cross polarization transfers with 5 ms contacts and 35 kHz H_1's. The chemical shifts are in ppm downfield from external ammonium ion. Assignment of resonances were made by comparison to solid state and solution spectra of model compounds (Allerhand 1978, Levy and Lichter 1979, Levy et al. 1980, Peter et al. 1984).

Chitin Preparation

Chitin was prepared by boiling finely ground freeze dried *M. sexta* fifth instar larval or pupal cuticles (5 $mg \cdot ml^{-1}$) in 1 M sodium hydroxide for four to 20 hr.

Amino Acid and Hexosamine Analysis

Hydrolysis was carried out for 20 hr in 6 N HCl at 110°C *in vacuo*. Amino acids and glucosamine were separated by cation exchange chromatography and detected by post column ninhydrin derivatization.

Catechol Analysis

Cuticle was homogenized and extracted in 1.2 M HCl containing 5 mM ascorbic acid and an internal standard α-methyl dopa (6 $\mu g\ ml^{-1}$) (Hopkins et al. 1984). Free catechols were recovered from alumina and quantitated by reverse-phase high performance liquid chromatography with electrochemical detection (Hopkins et al. 1984, Morgan et al. unpublished data).

RESULTS AND DISCUSSION

Solid State ^{13}C-NMR of *M. sexta* Cuticles

As a first step in the study of *M. sexta* cuticle structure, the natural abundance ^{13}C-CPMAS NMR spectra of sclerotized larval, pupal and adult cuticles were obtained (Fig. 1). Although the spectrum of each cuticle is unique, there are general features that can be used to estimate the relative abundance of the major cuticular components such as protein, chitin, catechols and lipid. There are many different individual proteins in cuticle that become progressively unextractable as the cuticle tans and together their protein carbons contribute substantially to resonances 1, 2, 4, 5 and 10 to 15 (Table 1). In particular the peptide backbone α-carbon signals at 55 ppm and 60 ppm are diagnostic of general protein levels especially in those cuticles of low chitin content. Resonances 1, 6 to 11 and 14 are due to chitin carbons with the highly resolved signals at 74 ppm, 75 ppm, 82 ppm and 104 ppm reflective of GlcNAc carbons 3, 5, 4 and 1, respectively. The signals at 104 ppm (chitin carbon 1) and 82 ppm (chitin carbon 2) are the best representative

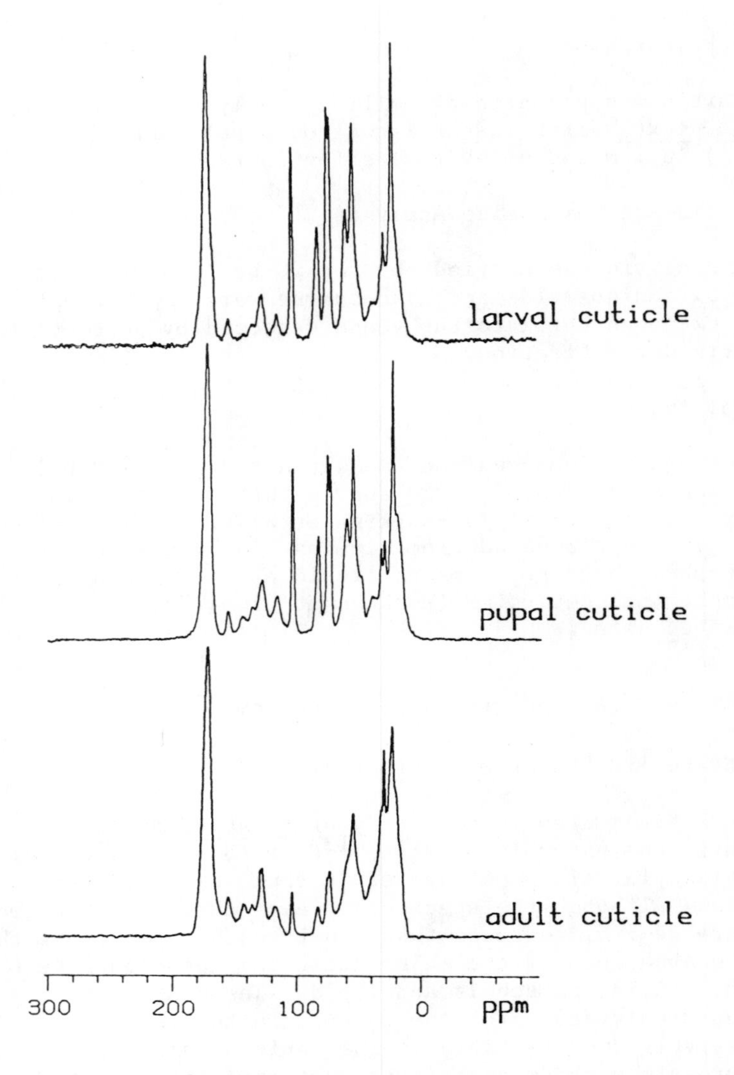

FIGURE 1. ^{13}C-CPMAS NMR spectra (natural abundance) of M. sexta larval, pupal and adult cuticles. The larval and adult cuticles are devoid of cuticle from the head and appendages. Scales were removed adult cuticle.

of chitin content because they are due to single carbon types, are well resolved from other resonances, and appear as prominent well-defined signals in the spectra of whole cuticular material. The aromatic carbon region of the spectra which includes resonances 2 to 5 is also of diagnostic value. The catecholic ring carbon 3 at 144 ppm can be used to estimate catechol levels in tissues. The signal at 33 ppm is representative of methylene carbons, most of which probably come from surface lipids.

On a relative concentration basis the adult cuticle exhibits the highest protein concentration while the larval cuticle has the most chitin. The adult cuticle is remarkably low in chitin content. Catechols are abundant in both pupal

Table 1. Chemical assignments of resonances in the CPMAS-^{13}C-NMR spectra of *M. sexta* cuticles and chitin

Resonance	δ-Values (ppm) *		Assignment
	Cuticle	Chitin	
1	172	172	Carbonyl carbon in chitin, protein, lipid and catechol acyl groups
2	155	--	Phenoxy carbon in tyrosine, guanidino carbon in arginine
3	144	--	Phenoxy carbon in catechols
4	129	--	Aromatic carbons
5	116	--	Tyrosine carbons 3 and 5, imidazole carbon 4, catechol carbons 2 and 5
6	104	102	GlcNAc carbon 1
7	82	82	GlcNAc carbon 4
8	75	74	GlcNAc carbon 5
9	74	72	GlcNAc carbon 3
10	60	60	GlcNAc carbon 6, amino acid α-carbon
11	55	54	GlcNAc carbon 2, amino acid α-carbon
12	44	--	Amino acid, catechol and lipid aliphatic carbons
13	33	--	Amino acid, catechol and lipid aliphatic carbons
14	23	21	Methyl carbons in chitin, protein, lipid and catechol acetyl groups, amino acid methyne carbons
15	19	--	Amino acid and lipid methyl carbons

* δ-Values relative to external TMS reference.

and adult cuticles with larval cuticle containing substantially less. The major chemical difference detected by solid state NMR analysis between stiff (pupal and adult) and flexible (larval) cuticles of M. sexta is the presence of higher levels of catechols in the former type of cuticle.

Changes in ^{13}C-NMR During Hardening of M. sexta Pupal Cuticle

The pupal cuticle of M. sexta undergoes major chemical changes as the process of sclerotization takes place. The untanned pupal cuticle exhibits a spectrum composed primarily of protein and lipid carbon signals (Fig. 2, left spectrum). Cuticular protein appears to be rich in aromatic amino acid residues as evidenced by signals between 120 ppm and 160 ppm. The sharp signal at 33 ppm indicates that a substantial quantity of lipid is present in untanned cuticle perhaps representing the newly secreted waxy layer of the epicuticle. The untanned pupal cuticle spectrum resembles that of tanned adult cuticle (Fig. 1) except for the presence of substantially more catechols in the latter. After tanning has occurred in pupal cuticle, the presence of increased amounts

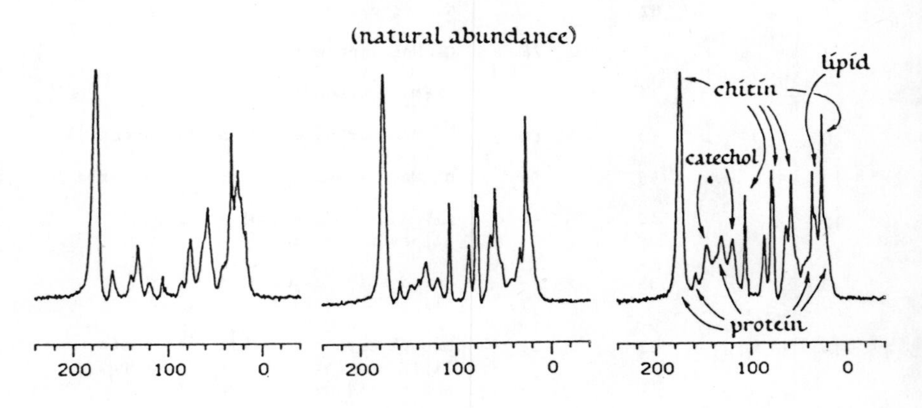

FIGURE 2. ^{13}C-CPMAS NMR spectra (natural abundance) of M. sexta untanned (left) and tanned (middle) pupal cuticles and pupal exuvium (right).

of chitin and catechols is apparent (Fig. 2, middle spectrum). Lipid is diminished considerably. The ^{13}C-spectrum of pupal exuvium, which is the outermost and most heavily sclerotized part of the cuticle (exocuticle) that remains after molting fluid enzymes such as proteases and chitinases have digested away major portions of the endocuticle, is similar to that of tanned cuticle except that there are substantially increased levels of catechols and lipid carbons (Fig. 2, right spectrum). The exocuticle is enriched in the type of chemical bonding (covalent and noncovalent) responsible for cuticular sclerotization. These results support the hypothesis that the pupal cuticle of *M. sexta* is sclerotized by incorporation of catechols into the protein-lipid composite of newly ecdysed soft cuticle. Chitin content is also substantially greater in tanned cuticle, suggesting an increased amount is laid down in the endocuticle after ecdysis.

Incorporation of ^{15}N-Labeled Amino Acids into *M. sexta* Pupal Cuticle During Tanning.

The natural abundance NMR spectra do not show any evidence for the presence of crosslinks or adducts between the cuticular components probably because of the low natural abundance of ^{13}C and ^{15}N and the low levels of such bonding. In order to search for covalent interactions that occur as a result of sclerotization, isotopically enriched cuticle samples assembled from ^{15}N- and ^{13}C-labeled precursors were subjected to solid state NMR analysis. This technique has the potential to determine whether specific atoms undergo covalent modifications that give rise to new resonances in the spectra.

Since amino acid side chain nitrogen atoms are potential nucleophiles that participate in the formation of carbon-nitrogen adducts, *M. sexta* pupal cuticle was labeled with 1,3-^{15}N-histidine or ε-^{15}N-lysine by injection into late stage larvae and ^{15}N-CPMAS NMR spectra of untanned and tanned cuticles were recorded (Fig. 3). For reference, chemical assignments of the nitrogen resonances are listed in Table 2. For the 1,3-^{15}N-histidine labeled samples, untanned cuticle shows that, in addition to the protonated ring nitrogen at 140 ppm and the aromatic nitrogen at 225 ppm, some of the nitrogen label has been metabolized to amide nitrogen at 100 ppm, probably into the peptide backbone and glucosamine nitrogen (Fig. 3a, upper spectrum). In tanned cuticle, a new

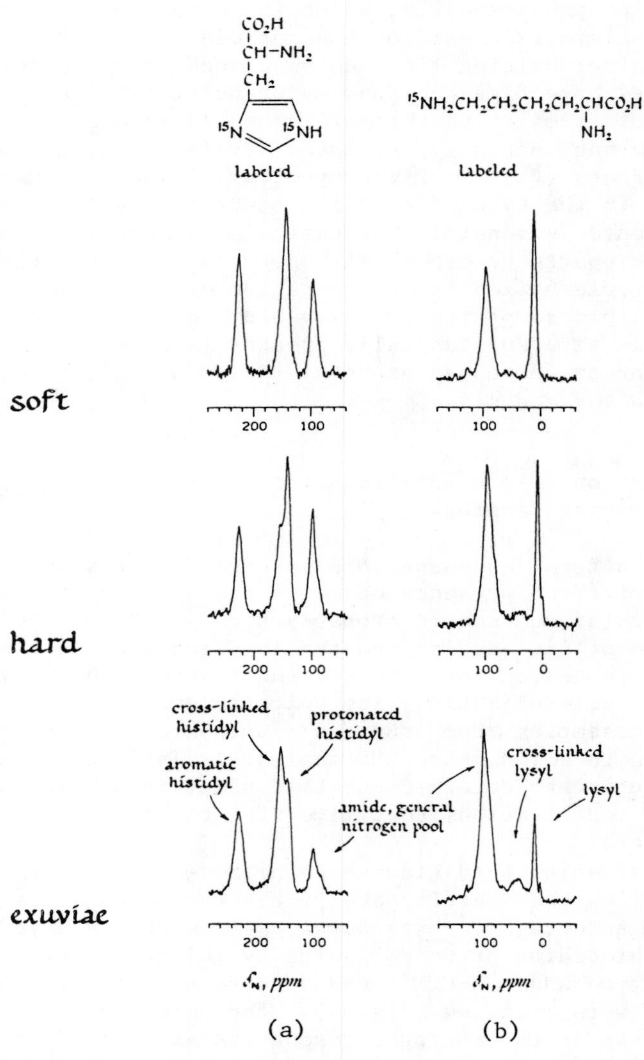

FIGURE 3. ^{15}N-CPMAS NMR spectra of M. sexta untanned (top) and tanned (middle) pupal cuticle and pupal exuvium (bottom) labeled with 1,3-^{15}N-histidine (a) and ε-^{15}N-lysine (b).

Table 2. Chemical assignments of resonances in the CPMAS-^{15}N-NMR spectra of *M. sexta* cuticle labeled with ^{15}N-enriched histidine or lysine.

Resonance	δ-Values (ppm)*	Assignment
1	0	Ammonium ion
2	15	Primary amine
3	35-50	Secondary amine (crosslinks?)
4	100	Amide
5	140	Imidazole protonated nitrogen
6	160	Imidazole alkylated nitrogen (crosslink?)
7	225	Imidazole aromatic nitrogen

*δ-Values relative to external ammonium ion reference.

signal is observed at 160 ppm (Fig. 3a, middle spectra) that becomes the major resonance in pupal exuvium (Fig. 3a, bottom spectrum). The new nitrogen structure is a nonprotonated probably alkylated atom because delayed proton decoupling had a minor effect on the 160 ppm signal while the signal at 140 ppm became substantially diminished. The results suggest that histidyl residues in cuticular protein become alkylated at the 1-nitrogen position during cuticle sclerotization.

In the ε-^{15}N-lysine labeled cuticle spectra, the ε-nitrogen is observed at 15 ppm in untanned cuticle together with a significant amount of amide nitrogen at 100 ppm (Fig. 3b, upper spectra). The possibility of an isopeptide linkage between lysyl and either aspartyl or glutamyl side chains was considered, but pupal cuticle double-labeled with ε-^{15}N, δ-^{13}C-lysine incorporated ^{15}N but not ^{13}C into the amide structure. There was no double cross polarization (DCP) in the amide nitrogen signal whereas there was a DCP component in the amino nitrogen signal. Apparently some of the ε-amino nitrogen label of lysine is catabolized apart from the rest of the lysyl side chain and metabolized for utilization into amide nitrogen in cuticular protein or chitin. With tanned pupal cuticle, a small broad signal centered at 40 ppm appears (Fig. 3b, middle spectrum) and increases in relative intensity in the pupal exuvium (Fig. 3b, bottom spectrum). This signal occurs in the secondary amine nitrogen region and may be attributed to an alkylation of the ε-nitrogen of lysine in cuticular protein. The results of the ^{15}N-labeling experiments indicate that protein side chain nitrogens from histidyl and lysyl residues participate in the formation of carbon-nitrogen adducts during M. sexta pupal cuticle tanning with the former amino acid participating to a greater degree.

Incorporation of ^{13}C-Labeled Amino Acids and Glucose into M. sexta Pupal Cuticle During Tanning.

^{13}C-NMR analysis of M. sexta pupal cuticle labeled in methyl or methylene carbons of glycine, alanine or lysine revealed that isotope was incorporated into the corresponding residue in cuticular proteins during cuticle morphogenesis. For example, when pupal or adult cuticle was labeled with ε-^{13}C-lysine, ^{13}C was substantially enriched over natural abundance primarily in the signal at 44 ppm where the ε-methylene carbon is resolved, a result indicating that there are no aldol or Shiff base type crosslinks involving lysine in cuticle that are found in connective tissue proteins such as collagen or resilin. When ^{13}C-C_9 uniformly labeled tyrosine was fed to fifth instar larvae (the low solubility of tyrosine limited its adminstration by injection), ^{13}C was enriched in nearly all pupal cuticle carbon signals with relatively more label found in the catechol aromatic carbon 3 signal at 144 ppm, the carbonyl carbon signal at 172 ppm, and unexpectedly in a methyl carbon signal at 23 ppm. This result suggested that although catechols were derived from tyrosine, substantial catabolism and scrambling of the carboxyl, α- and β-carbons occurred when ^{13}C-tyrosine was admixed in the diet, fed and taken up through the gut, such that the phenolic ring is split apart from the aliphatic side chain. Results from DCP experiments using pupal cuticle from hornworms fed C_9-^{13}C-tyrosine and injected with 1,3-^{15}N-histidine suggested that there is no link between a tyrosyl ring carbon and a histidyl ring nitrogen.

Chitin in pupal cuticle was labeled with ^{13}C by feeding fifth instar larvae uniformly labeled ^{13}C-glucose. Rather surprisingly only chitin carbon signals were enhanced over natural abundance carbon levels by about a factor of 3. The presence of a chitin carbon that was covalently bonded to another cuticular component was not observed but may have been masked by other natural abundance carbon signals in pupal cuticle.

Association of Protein Histidine with Chitin in *M. sexta* Pupal Cuticle.

In 1960 Hackman presented evidence to support the hypothesis that chitin was covalently bonded to protein in cuticle from the sheep blowfly and several other invertebrate species with the linkage to protein through histidyl and aspartyl residues. We have also obtained chemical and spectroscopic evidence for a chitin-histidyl protein adduct in *M. sexta* cuticle. Chitin was extracted from larval and pupal cuticles by repeated hot alkali treatment (1 M NaOH at 95°C for 1 hour repeated four times). After acid hydrolysis amino acid and hexosamine analyses revealed that histidine is the major residual amino acid in the chitin preparations with substantially more associated with the more highly sclerotized pupal cuticle (Table 3). The so-called pupal

Table 3. Glucosamine and amino acid composition (μmole/g) of chitin prepared from *M. sexta* larval and pupal cuticles.*

Residue	Larval chitin	Pupal chitin
Glucosamine	1581.4±428.5 (93)	680.4±226.1 (52)
Aspartic acid/amide	4.6±0.8 (<1)	51.5±2.7 (4)
Threonine	3.4±0.9 (<1)	9.5±1.1 (<1)
Serine	5.6±1.0 (<1)	9.6±1.5 (<1)
Glutamic acid/amide	4.8±0.3 (<1)	83.3±2.9 (6)
Glycine	6.5±0.6 (<1)	49.3±0.4 (4)
Alanine	12.6±2.4 (<1)	57.9±0.4 (4)
Valine	5.3±0.4 (<1)	105.8±1.5 (8)
Isoleucine	2.3±0.7 (<1)	27.8±5.2 (2)
Leucine	4.3±0.7 (<1)	25.2±4.1 (2)
Histidine	67.0±8.2 (4)	182.9±8.9 (14)
Lysine	<2 (<1)	16.3±6.2 (1)
Glucosamine:histidine	23.6	3.7

*Mean values ±S. E. from two analyses hydrolyzed for 20 hr *in vacuo* in 6M HCl. Proline, methionine, cysteine, tyrosine, phenylalanine, arginine and β-alanine were present in trace amounts (<2 μmole/g) or could not be quantitated due to interfering substances. Mole percent given in parentheses.

chitin preparation was actually more of a chitinoprotein material where protein accounted for nearly 50% of the sample dry weight with one histidine present per four 2-amino-2-deoxy-D-glucopyranoside residues. When chitin was prepared from M. sexta pupal cuticle labeled with 1,3-^{15}N-histidine, subjected to varying times of exposure to 1N NaOH and analyzed by both ^{13}C- and ^{15}N-CPMAS NMR, additional evidence for a covalent association between a chitin chain and a histidyl residue was obtained (Fig. 4). Pupal chitin can be prepared in an almost homogeneous state by alkali treatment

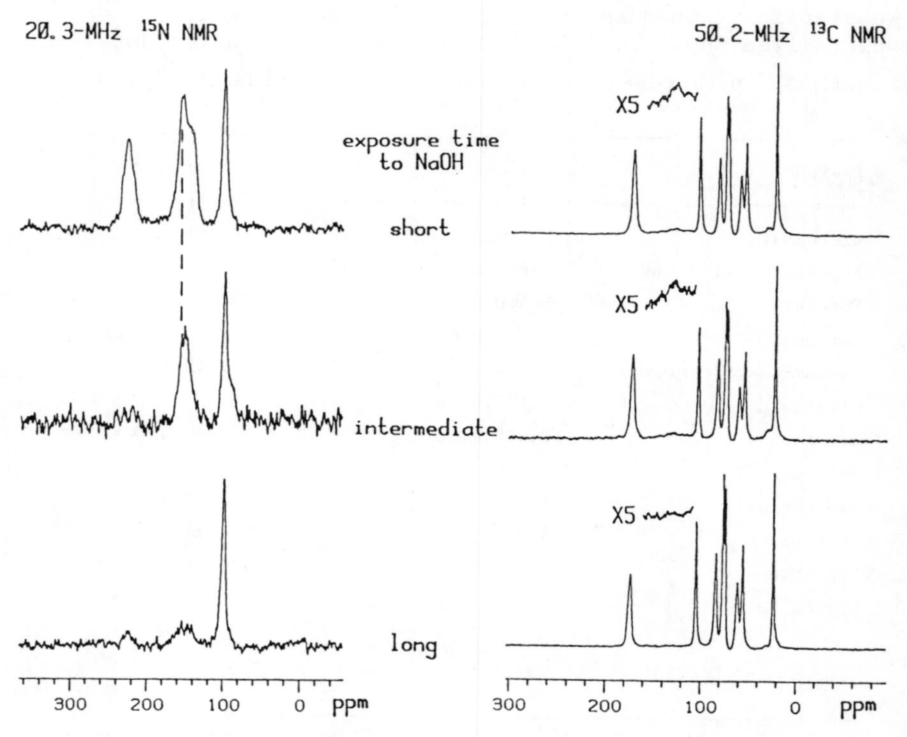

FIGURE 4. CPMAS NMR spectra of chitin extracted from 1,3-^{15}N-histidine labeled M. sexta cuticle by alkali treatment. ^{15}N spectrum, left; ^{13}C spectrum, right. Exposure time to alkali was for approximately 8 (short), 15 (intermediate) and 20 hours (long).

with only chitin carbon resonances appearing in the NMR spectrum except for some rather minor signals at 30 ppm and 130 ppm where histidyl and catecholamine carbon signals are known to occur. With longer incubation periods in alkali, the minor carbon signals are diminished. ^{15}N-NMR revealed that histidyl nitrogens are present in the pupal chitin preparation with the crosslinked or alkylated histidyl nitrogen signal at 160 ppm being very prominent. The histidyl nitrogen resonances were also diminished in the chitin preparation by additional alkali treatment. The ^{13}C-histidyl abundance in the various preparations matched closely the ^{15}N-histidyl abundance. The exact structure of the histidyl-chitin adduct is unknown at the present time, but it may resemble a direct Maillard condensation product or consist of an indirect crosslink mediated by a catecholamine derivative.

Catechols in Pupal Cuticle.

The pathway for tyrosine being converted to N-acylated catecholamines for incorporation into *M. sexta* cuticle is shown in Fig. 5 (Hopkins et al. 1984). The catecholamines may be oxidized to quinonoid derivatives which can electrophilically bond to cuticular components or polymerize into pigments such as melanin. The catechols may also be impregnated directly into the chitin-protein matrix. N-Acylation of dopamine decreases the nucleophilicity of the nitrogen atom such that if oxidation does occur, the rate of subsequent intramolecular cyclization of the quinonoid is greatly reduced (Kramer et al. 1983). The quinonoid itself thus becomes relatively long-lived and available for intermolecular condensation reactions with nucleophilic groups from other cuticular components. The major catechols that can be extracted from *M. sexta* pupal cuticle during the process of sclerotization are N-β-alanyldopamine (NBAD) and N-β-alanylnorepinephrine (NBANE). These compounds were extracted in cold 1.2 M HCl from the outer cuticle of the forewing and the dorsal cuticle of the third and fourth abdominal segments and their concentrations measured by liquid chromatography with electrochemical detection. During the first three hours following pupal ecdysis, acid-extractable NBANE accumulated more rapidly than NBAD in both types of cuticle, reaching peak levels of approximately 2 μmole g^{-1} (Fig. 6, Morgan et al. 1986). The outer cuticle of the forewing is still essentially colorless and soft at

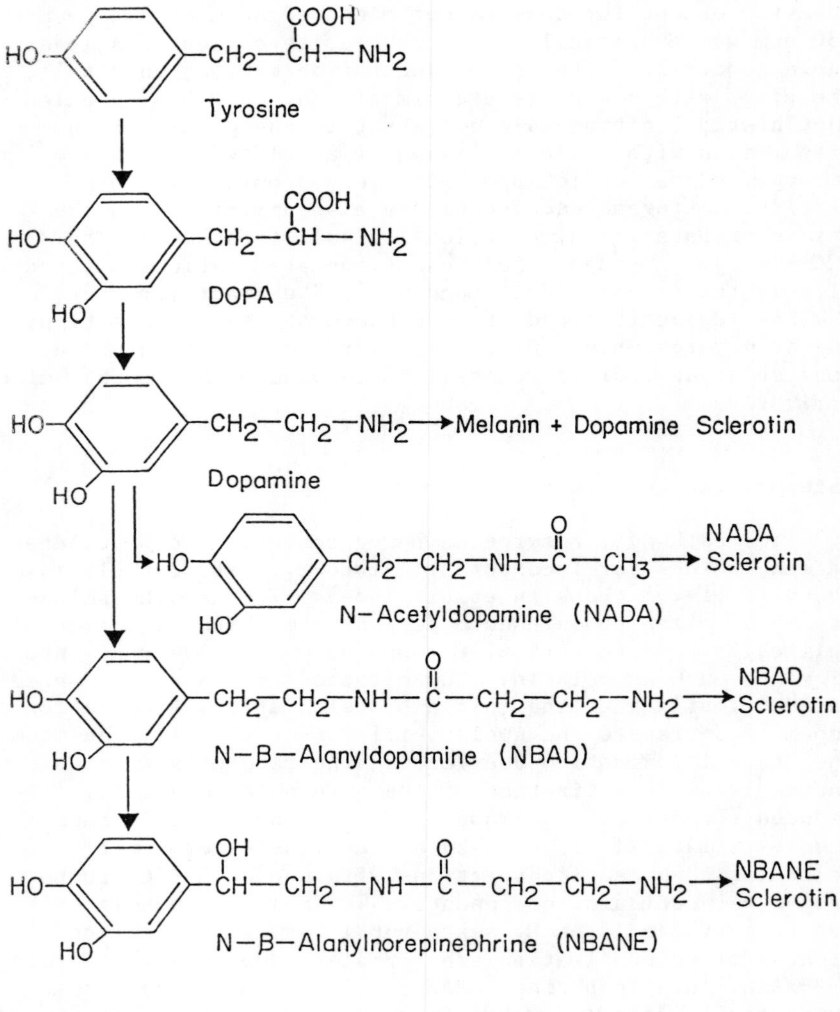

HO–C$_6$H$_4$–CH$_2$–CH(COOH)–NH$_2$
Tyrosine

↓

HO, HO–C$_6$H$_3$–CH$_2$–CH(COOH)–NH$_2$
DOPA

↓

HO, HO–C$_6$H$_3$–CH$_2$–CH$_2$–NH$_2$ → Melanin + Dopamine Sclerotin
Dopamine

→ HO, HO–C$_6$H$_3$–CH$_2$–CH$_2$–NH–C(=O)–CH$_3$ → NADA Sclerotin
N-Acetyldopamine (NADA)

↓

HO, HO–C$_6$H$_3$–CH$_2$–CH$_2$–NH–C(=O)–CH$_2$–CH$_2$–NH$_2$ → NBAD Sclerotin
N-β-Alanyldopamine (NBAD)

↓

HO, HO–C$_6$H$_3$–CH(OH)–CH$_2$–NH–C(=O)–CH$_2$–CH$_2$–NH$_2$ → NBANE Sclerotin
N-β-Alanylnorepinephrine (NBANE)

FIGURE 5. Scheme for the metabolism of tyrosine to N-acylcatecholamines for sclerotization and melanization of *M. sexta* cuticle.

two hours postecdysis, and only faint brown is visible at three hours. In contrast abdominal cuticle shows extensive tanning at ecdysis, and has about two-fold higher NBANE

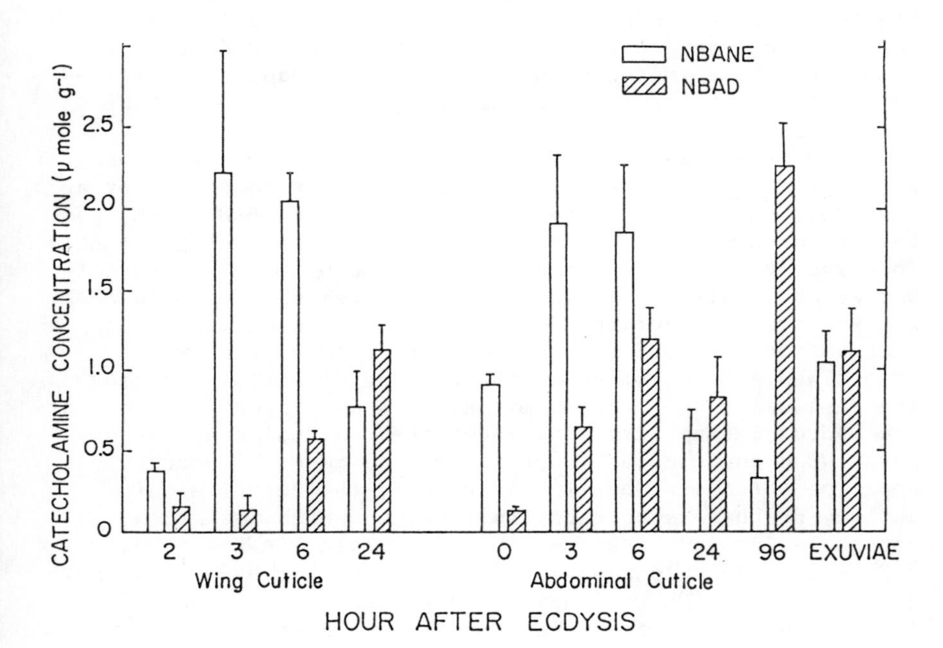

FIGURE 6. N-β-Alanyldopamine (NBAD) and N-β-alanylnorepinephrine (NBANE) concentrations in the pupal cuticle of M. sexta during sclerotization. The cuticle samples were homogenized and extracted with cold 1.2 MHCl. Means ±S.E.M., n=3.

concentrations at zero hours than wing cuticle had at two hours. NBAD concentrations were slower to increase and did not reach an equivalent level to that of NBANE until 24 hours after ecdysis. NBANE decreased by more than 50% by that time while NBAD was near a concentration of 1 μmole g^{-1}. By 96 hours after ecdysis NBANE had fallen to 0.3 μmole g^{-1} in abdominal cuticle, while NBAD was at the highest concentration recorded (2.3 μmole g^{-1}). Substantial levels of both compounds (~1 μmole g^{-1}) were present in acidic extracts of pupal exuviae after adult eclosion. Other extracting solvents such as 80% methanol or 7M urea-0.1% sodium dodecyl sulfate were able to release less than 10% of the NBANE extracted with cold acid, whereas more than 50% of the NBAD could be recovered with nonacidic solvents. This differential extractability suggests stronger interactions of NBANE with the chitin-protein matrix, perhaps even covalent binding. The temporal pattern of catecholamine levels in pupal cuticle is consistent with NBANE being produced from NBAD with the former compound becoming subsequently oxidized to a quinonoid intermediate that crosslinks the cuticular biopolymers.

The synthesis of β-hydroxylated catecholamines appears to occur mainly in the cuticle itself and is catalyzed by an insoluble enzyme (Morgan et al. 1986). When NBAD (5 mM) was incubated at pH 6 with pupal integumental homogenates, NBANE increased to 40 ± 4 μmole g^{-1} after 10 min. based on cuticle wet weight, while in cuticular homogenates 49 ± 9 μmole NBANE g^{-1} was produced during the same period. Essentially the same amount of NBANE accumulated during incubation with either whole homogenates or the residue fraction. Heating the residue at 80°C for 10 min at pH 6 destroyed about 85% of the hydroxylating activity. When the residual fraction of pupal wing cuticle was incubated with 1 mM NBAD, N-acetyldopamine (NADA) or dopamine to compare the relative suitability of these substrates for enzymes which modify the catecholamine side chain, 22.0 ± 3.0 μmole NBANE g^{-1}, 3.8 ± 0.2 μmole N-acetylnorepinephrine (NANE)g^{-1} or 0.9 ± 0.1 μmole norepinephrine (NE) g^{-1} was produced after 10 min at 27°C. These results indicated that NBAD is the best catecholamine substrate for side chain modification catalyzed by an insoluble cuticular enzyme leading to hydroxylation or reaction with nucleophiles during cuticle sclerotization.

After extraction with cold 1.2 MHCl, large amounts of NBANE (1.7 ± 0.2 μmole g^{-1}) were released as NE equivalents, from pupal cuticle after heating the cuticular homogenate in 1.2 M HCl. Some of the NBANE released by hot and cold dilute acid may be a hydrolysis product derived from a covalent adduct of NBAD bonded through the β-carbon with the chitin-protein matrix of cuticle.

One hypothesis that may explain how ring or β-carbon adducts of catecholamines occur in insect cuticle invokes a catecholamine oxidation mechanism where cuticle becomes sclerotized by quinone tautomeric structures that contain electrophilic aromatic and aliphatic carbons. Nucleophilic nitrogen, oxygen or sulfur atoms in a cuticular biopolymer may attack intermediates such as the _o_-quinone, quinone methide or α,β-dehydrocatechol that might be generated by cuticular enzymes such as tyrosinase or laccase (Fig. 7). The N-acylation of catecholamines with acetate or β-alanine

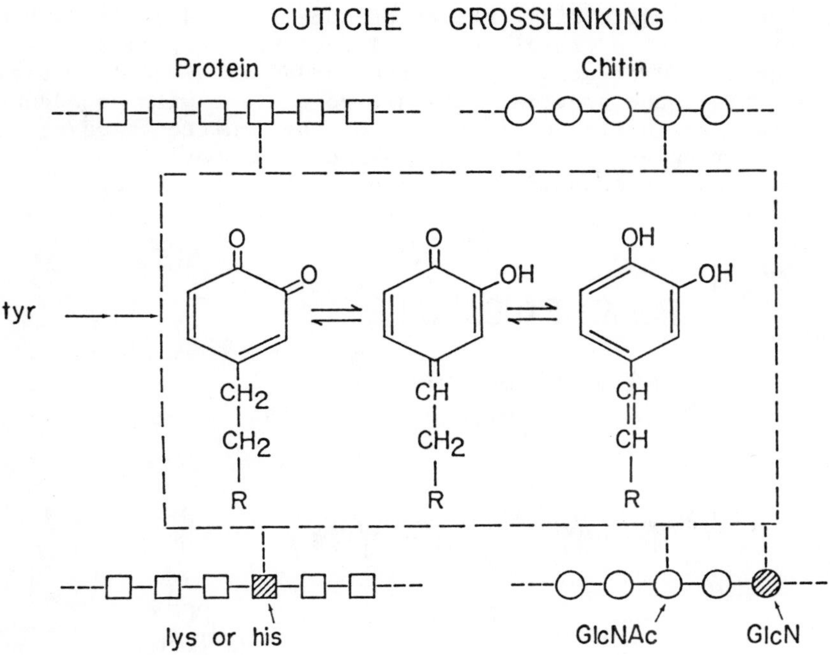

FIGURE 7. Proposed intermediates for covalent modification of *M. sexta* cuticle by oxidized catechol metabolites.

may facilitate electron delocalization from the ring carbons of the oxidized products to aliphatic side chain carbons and promote side chain modification.

Crosslinks Between Histidine and Catecholamine in *M. sexta* Pupal Cuticle.

The possibility of the existence of a crosslink between histidine and a catecholamine in *M. sexta* pupal cuticle was examined after labeling the cuticle with both 1,3-^{15}N-histidine and ring C_6-^{13}C-dopamine. Preliminary results show that sufficient incorporation of both compounds occurred to allow detection of a ^{13}C-^{15}N adduct in the exuvium by solid state NMR analysis. The dominant signals in the ^{13}C-spectrum were those in the 120-160 ppm region where the catecholic aromatic carbons occur (Fig. 8). A double cross component was also observed at 135 ppm indicating that a bond between one of the aromatic ring carbons of dopamine and an imidazole ring-nitrogen of a protein-histidyl residue had formed. Although the exact structure of the adduct is unknown at the present time, the spectral data are consistent with an adduct resulting from nucleophilic attack of the 1-nitrogen of the imidazole ring of histidine on the 2 or 6-aromatic ring carbon of an o-quinonoid derivative of NBAD, the major catecholamine found in tanning pupal cuticle.

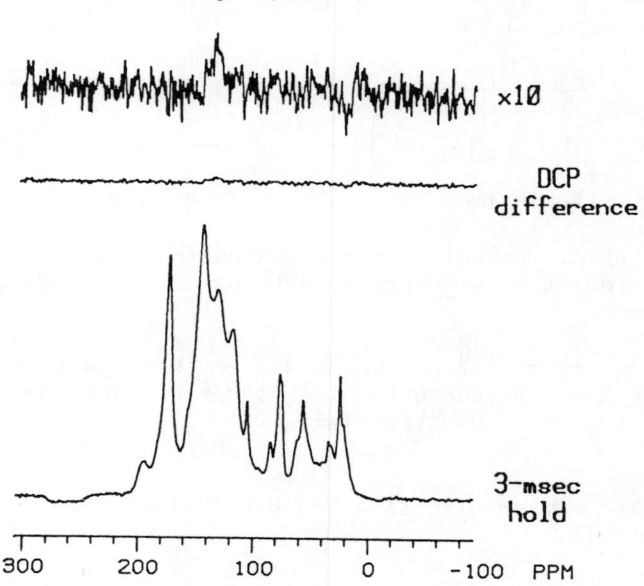

FIGURE 8. ^{13}C-CPMAS NMR spectrum of *M. sexta* pupal exuvium labeled with ring C_6-^{13}C-dopamine and L-1,3-^{15}N-histidine. Double cross polarization difference signals shown in upper traces.

Concluding Remarks

In summary there appears to be a general sequence of events that insects use to assemble and sclerotize their cuticles. First, varying amounts of protein and chitin are secreted in a layered matrix or procuticle into which are transported catecholic sclerotizing precursors, minerals and lipids. Second, crosslinks appear to be established between chitin and protein or protein and protein either directly or through a crosslinking agent such as a catecholic or quinonoid derivative. Third, free catechols accumulate, some of which may displace water and dehydrate the cuticle. Minerals are incorporated in all cuticles, but in specialized cases are the principal stabilizing agents. Although we have spectroscopic and chemical evidence for the involvement of specific protein side chains, catechols and chitin residues, we still do not know the precise structure of any adduct or crosslink in *M. sexta* cuticle. There appears to be a link between chitin and a histidyl residue in cuticular protein, perhaps in the form of a Maillard addition product where a chitin aldehyde function condenses with the protonated histidyl nitrogen. Other possibilities include an indirect crosslink between chitin and protein chains via either a quinone or a quinone methide intermediate where histidine is the nucleophilic group (Fig. 9, group B). The participation of a quinone methide intermediate in cuticle stabilization helps to explain the analytical results obtained from catechol extraction experiments. The addition of water to a quinone methide intermediate would generate β-hydroxylated catecholamines such as NBANE that would be extracted from cuticle. Also hot acid treatment of cuticle would hydrolyze a putative chitin/protein matrix-catechol β-carbon adduct or crosslink structure and release compounds such as NBANE.

For future study we are continuing to prepare more isotopically enriched single and double labeled *M. sexta* cuticles with ^{13}C-dopamine to label catechols, ^{13}C-glucose to label chitin and both ^{13}C- and ^{15}N-amino acids to label protein. If sufficient enrichment of cuticular components takes place, then the probability of identifying by solid state NMR carbon-nitrogen or carbon-carbon bonds that form during sclerotization should be much improved. Studies are also continuing on the catecholic or quinonoid metabolites generated by cuticular enzymes and the isolation of adducts formed from their reaction with available nucleophiles in the chitin-protein cuticular matrix. Although the study of insect cuticle stabilization mechanisms is far from complete, some progress has been made and with future applications of techniques such as solid state NMR, there is promise of substantially more information to come.

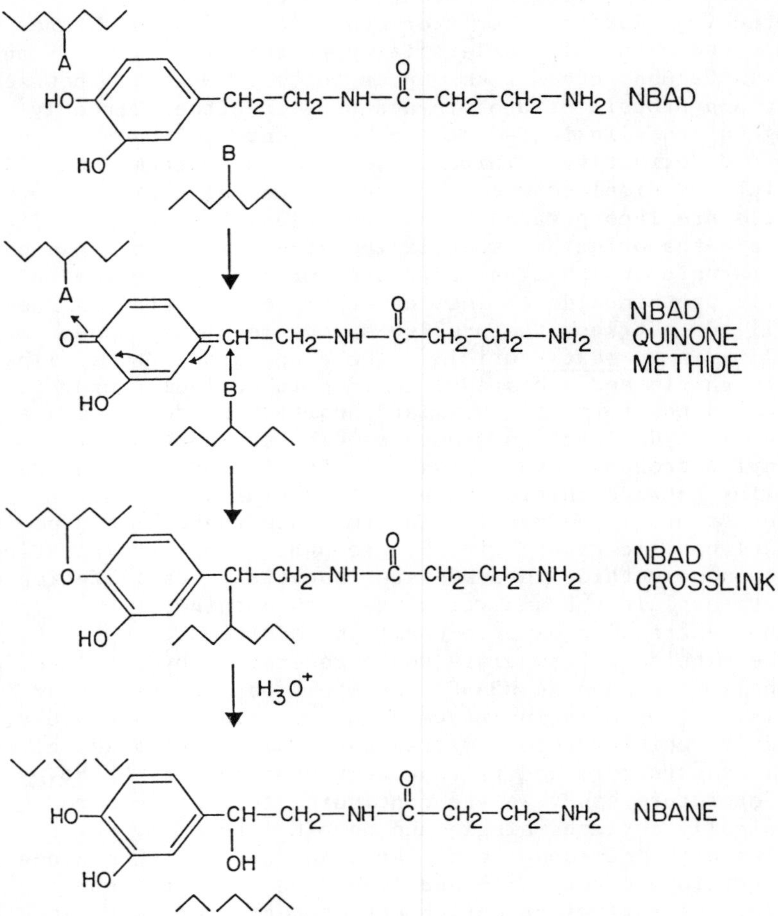

FIGURE 9. Crosslinking of compounds A and B by NBAD quinone methide via β-carbon and catecholic oxygen.

ACKNOWLEDGMENTS

Contribution no. 86-412-A. Department of Biochemistry and Entomology, Kansas Agricultural Experiment Station,

Manhattan, KS 66506. Cooperative investigation between Agricultural Research Service, U.S. Department of Agriculture, the Kansas Agricultural Experiment Station and Monsanto Co. Mention of a proprietary product does not imply approval by the USDA to the exclusion of other products that may also be suitable. We are grateful to Dr. Tamo Fukamizo, Renee Elonen, Tom Czapla and Leon Hendricks for discussion and assistance with these studies.

REFERENCES

1. Allerhand A (1978). Natural abundance carbon-13 nuclear magnetic resonance spectroscopy of proteins. Observation and uses of nonprotonated aromatic carbon resonances. Accounts Chem Res 11:469.
2. Andersen SO (1964). The cross-links of resilin identified as dityrosine and trityrosine. Biochem Biophys Acta 93:213.
3. Andersen SO (1966). Covalent cross-links in a structural protein, resilin. Acta Physiol Scand 66, Suppl 263: 1-18.
4. Andersen SO (1985). Sclerotization and tanning in cuticle. In Kerkut GA and Gilbert LI (eds): "Compre Insect Physiol Biochem Pharmacol," Vol 3 Oxford: Pergamon, p 59.
5. Blomquist GJ, Dillwith JW (1985). Cuticular lipids. In Kerkut GA and Gilbert LI (eds): "Compre Insect Physiol Biochem Pharmacol," Vol 3 Oxford: Pergamon, p 17.
6. Brunet PCJ (1980). The metabolism of the aromatic amino acids concerned in the cross-linking of insect cuticle. Insect Biochem. 10:467.
7. Grun L, Peter MG (1983). Selective crosslinking of tyrosine-rich larval serum proteins and of soluble _Manduca sexta_ cuticle proteins by nascent N-acetyl-dopamine quinone and N-β-alanyldopamine quinone. In Scheller K (ed) "The Larval Serum Proteins of Insects," Stuttart: Georg Thieme Verlag, p 102.
8. Grun L, Peter MG (1984). Incorporation of radiolabeled tyrosine, N-acetyldopamine, N-β-alanyldopamine, and the arylphorin manducin into sclerotized cuticle of tobacco hornworm (_Manduca sexta_) pupae. Z. Naturforsch C 39:1066.
9. Hackman RH (1960). Studies on chitin IV. The occurrence of complexes in which chitin and protein are covalently linked. Aust J Biol Sci 13:568.

10. Hepburn HR (1985). Structure of the integument. In Kerkut GA and Gilbert LI (ed): "Compre Insect Physiol Biochem Pharmacol" Vol 3, Oxford: Pergamon, p 1.
11. Himmelsbach DS, Barton FE, Windham WR (1983). Comparison of carbohydrate, lignin and protein ratios between grass species by cross polarization-magic angle spinning carbon-13 nuclear magnetic resonance. J Agri Food Chem 31:401.
12. Hopkins TL, Morgan TD, Aso Y, Kramer KJ (1982). N-β-Alanyldopamine: major role in insect cuticle tanning. Science 217:364.
13. Hopkins TL, Morgan TD, Kramer KJ (1984). Catecholamines in haemolymph and cuticle during larval, pupal and adult development of *Manduca sexta* (L). Insect Biochem 14:533.
14. Jacob GS, Schaefer J, Wilson GE (1983). Direct measurement of peptidoglycan cross-linking in bacteria by ^{15}N nuclear magnetic resonance. J Biol Chem 258:10824.
15. Kramer KJ, Nuntnarumit C, Aso Y, Hawley MD, Hopkins TL (1983). Electrochemical and enzymatic oxidation of catecholamines involved in sclerotization and melanization of insect cuticle. Insect Biochem 13:475.
16. Kramer KJ, Dziadik-Turner C, Koga D (1985). Chitin metabolism in insects. In Kerkut GA and Gilbert LI (eds) "Compre Insect Physiol Biochem Pharmacol" Vol 3, Oxford: Pergamon, p 75.
17. Levy GC, Lichter RL (1979). "Nitrogen-15 nuclear magnetic resonance spectroscopy." New York: Wiley and Sons, p 221.
18. Levy GC, Lichter RL, Nelson GL (1980). "Carbon-13 nuclear magnetic resonance spectroscopy." Second ed, New York: Wiley and Sons, p 338.
19. Lipke J, Sugumaran M, Henzel W (1983). Mechanisms of sclerotization in dipterans. Adv Insect Physiol 17:1.
20. Morgan TD, Hopkins TL, Kramer KJ, Roseland CR, Czapla TH, Tomer KB, Crow FW (1986). N-β-Alanylnorepinephrine: biosynthesis in insect cuticle and possible role in sclerotization. Insect Biochem. In press.
21. Peter MG (1980). Products of *in vitro* oxidation of N-acetyldopamine as possible components in the sclerotization of insect cuticle. Insect Biochem 10:221.
22. Peter MG, Grun L, Forster H (1984). CP/MAS-^{13}C-NMR spectra of sclerotized insect cuticle and of chitin. Angew Chem Int Ed Engl 23:638.

23. Roseland CR, Grodowitz MJ, Kramer KJ, Hopkins TL, Broce AB (1985). Stabilization of mineralized and sclerotized puparial cuticle of muscid flies. Insect Biochem 15:521.
24. Schaefer J, Stejskal EO, McKay RA (1979). Double cross polarization NMR of solids. J Magn Reson 34:443.
25. Schaefer J, Stejskal EO, Sefiek MO, McKay RA (1981). Application of high-resolution ^{13}C and ^{15}N NMR of solids. Phil Trans R Soc Lond A299:593.
26. Sugumaran M, Lipke H (1983). Quinone methide formation from 4-alkylcatechols: A novel reaction catalyzed by cuticular polyphenol oxidase. FEBS Lett 155:65.
27. Yannoni CJ (1982). High resolution NMR in solids: the CPMAS experiment. Acc Chem Res 15:201.

Molecular Entomology, pages 357–367

QUINONE METHIDE SCLEROTIZATION[1]

M. Sugumaran

Department of Biology
University of Massachusetts at Boston
Harbor Campus, Boston, MA 02125

ABSTRACT Two molecular mechanisms have been proposed for the sclerotization reaction involving o-diphenols as cross linkers. The quinone tanning mechanism calls for the generation and subsequent reactions of highly reactive o-benzoquinones with cuticular components. The alternate mechanism known as β-sclerotization requires the participation of unknown intermediates and is characterized by the presence of covalently bound o-diphenols in cuticle. Recently, 1,2-dehydro-N-acetyl-dopamine has been isolated from β-sclerotized cuticle and suggested to be a key intermediate in β-tanning. However, we proposed the participation of a quinone methide derivative in this process. In order to test the involvement of 1,2-dehydro-N-acetyldopamine in β-sclerotization, we synthesized this compound from 3,4-dimethoxycinnamic acid. Radioactive trapping experiments reveal that this compound may not be involved in tanning. Our studies support the contention that quinone methides are the reactive intermediates in β-sclerotization. Analagous to quinone tanning where quinones participate, we wish to rename β-sclerotization as quinone methide sclerotization.

INTRODUCTION

One of the most intriguing and yet unresolved problems in cuticle biochemistry is the mechanism by which the exoskeleton of various insects is hardened and tanned for pro-

[1]This work was supported by N.I.H. grant 2R01-Al-14753.

tection of their soft bodies. Studies from several laboratories indicate the operation of two chemical mechanisms, viz., quinone tanning and β-sclerotization for stabilization of insect cuticle, both acknowledging the participation of catecholamine derivatives as sclerotizing agents (1-9). Quinone tanning invokes the reactions of enzymatically generated quinones with structural proteins as originally proposed by Pryor (1). Quinone protein adducts thus formed, upon further oxidation and coupling, produce crosslinks with two different proteins. The crosslinking in this case occurs via bridging through the aromatic ring of the sclerotizing agent. Recently, we have provided some evidence for the presence of such crosslinks in dipteran cuticle (10,11). The alternate mechanism, known as β-sclerotization, is based on the presence of unmodified catechol units in cuticle (12). In this case, the crosslinking seems to occur via the side chain of the catecholamine derivatives (12). Release of a number of ketocatechols modified at the β-position upon acid hydrolysis of sclerotized cuticle and the enzymatic release of tritium from side chain tritiated N-acetyldopamine during incubation with cuticular enzyme(s) attest to the presence of such crosslinks in cuticle (12-14). However, neither the sclerotizing agent nor the mechanism of crosslinking has been identified for this mode of tanning.

In 1982, Andersen's group extracted 1,2-dehydro-N-acetyldopamine from the sclerotized tissues of <u>Locusta migratoria</u> using a hot alkali treatment and suggested that its corresponding quinone could be the sclerotizing agent for β-tanning (Figure 1) (9,15). On the other hand, we

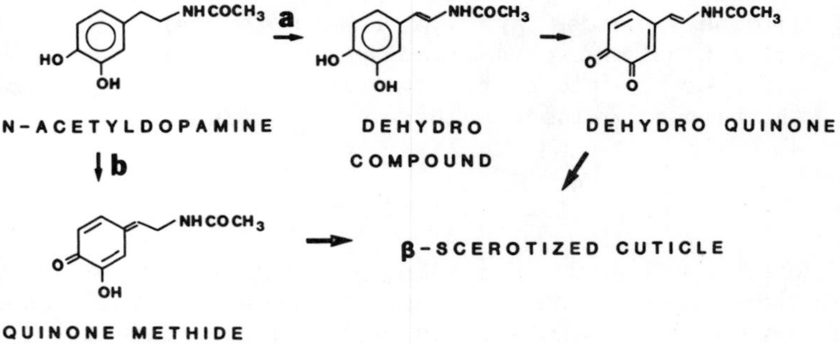

FIGURE 1. Alternate mechanisms for β-sclerotization. a. Andersen's mechanism; b. quinone methide mechanism.

proposed an alternate mechanism involving quinone methides as reactive intermediates (8,16). In order to resolve this controversy, we synthesized 1,2-dehydro-N-acetyldopamine and studied some of its properties. Using radioactive trapping experiments, we now show that this compound is not freely liberated in cuticle. Thus, our studies discount the participation of 1,2-dehydro-N-acetyldopamine in β-sclerotization.

METHODS

Synthesis of 1,2-dehydro-N-acetyldopamine was achieved by the following route: 3,4-dimethoxy cinnamic acid --> 3,4-dimethoxy cinnamyl chloride --> 3,4-dimethoxy cinnamyl azide --> 3,4-dimethoxy styrenyl isocyanate --> 3,4-dimethoxy-1,2-dehydro-N-acetyl phenethylamine --> 1,2-dehydro-N-acetyl dopamine. The details of the synthetic procedure will be published elsewhere.

Preparation of cuticular enzyme from *Sarcophaga bullata*, conditions for HPLC analysis of catechols and spectral studies on phenoloxidase mediated oxidation of catechols were carried out as outlined in an earlier publication (16).

Radioactive trapping experiments were carried out as follows: 10 mg of cuticular powder in 300 μl of 0.025 M sodium phosphate buffer, pH 6.0, was incubated with 10 μC_i of [Ring-^{3}H]-N-acetyldopamine (specific activity ~100 mC_i/mmole). After exactly 15 seconds, 1 mg each of N-acetyldopamine and 1,2-dehydro-N-acetyldopamine were added to the reaction mixture and the reaction was arrested by the addition of 300 μl of 10 M acetic acid. The contents were filtered and chromatographed on a Biogel P-2 column (30 cm x 1.5 cm) using 0.2 M acetic acid as eluant. Fractions of 3.5 ml were collected and an aliquot from each tube (1 ml) was used for radioactive measurements.

RESULTS

1,2-Dehydro-N-acetyl dopamine was synthesized employing the scheme of reactions outlined under 'methods'. The synthetic compound had the same properties as that of the isolated product (15). Repeated attempts to isolate this compound from *Sarcophaga bullata* cuticle ended in vain. Therefore, a radioactive trapping experiment was devised to evaluate its generation. Since this compound is claimed to be generated freely in cuticle, incubation of cuticular phenoloxidase with radioactive N-acetyl dopamine should produce the radioactive dehydro compound. The trace amount of radioactive dehydro compound thus formed can be diluted by exogenous addition of cold compound and its further oxidation can be prevented by arresting the reaction. Reisolation of the dehydro compound from the reaction mixture and radioactive analysis will confirm the intermediary formation of this compound if it carries radioactivity. Figure 2 gives the results of such a trapping experiment. It can be clearly seen that no radioactivity is trapped in 1,2-dehydro-N-acetyl dopamine, thus ruling out this compound as a freely generated intermediate. At the same time Figure 2 shows the generation of a radioactive product corresponding to N-acetyl norepinephrine. Further support for the production of N-acetylnorepinephrine comes from HPLC studies. HPLC analysis of a reaction mixture containing N-acetyldopamine

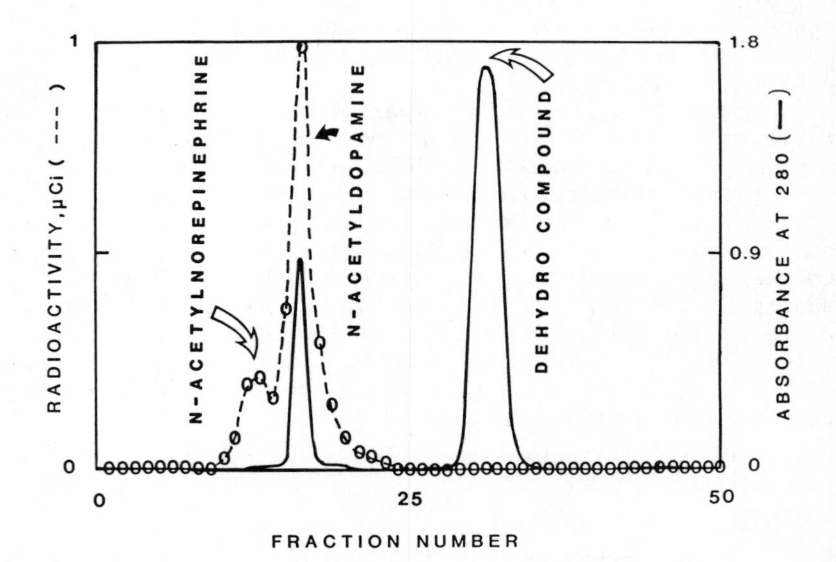

FIGURE 2. Biogel P-2 column chromatography of the reation mixture used for radioactive trapping experiment. For experimental details refer to 'methods'. Note the absence of radioactivity under the dehydrocompound.

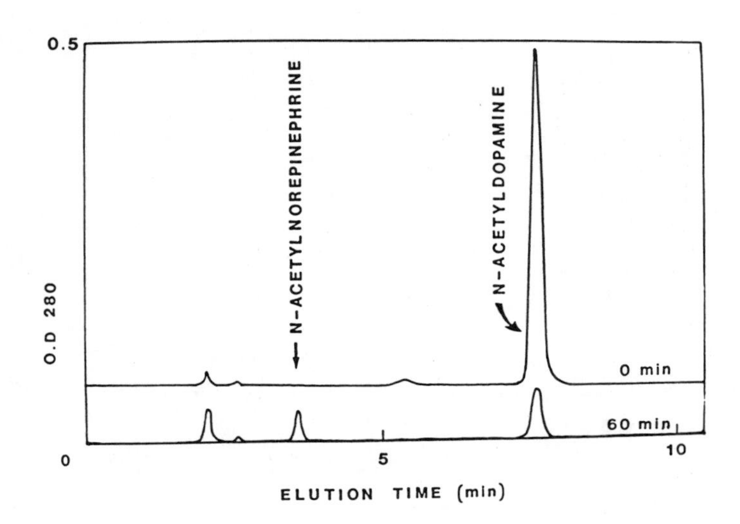

FIGURE 3. HPLC analysis of the reaction mixture containing N-acetyldopamine and cuticular phenoloxidase.

and cuticle at zero time and after 60 min incubation clearly illustrates the appearance of a peak at 3.6 minutes corresponding to N-acetylnorepinephrine (Figure 3). Cochromatography with authentic compound and spectral studies confirmed the identity of the material to be N-acetylnorepinephrine.

Since Andersen's mechanism calls for the oxidation of dehydro compound to the corresponding quinone (Figure 1), a study on the oxidation of 1,2-dehydro-N-acetyl dopamine by cuticular as well as mushroom polyphenoloxidase was carried out. During oxidation with cuticular enzyme, dehydro compound was rapidly oxidized and bound to the cuticle but no quinone formation could be witnessed. Oxidation of this compound by mushroom tyrosinase also failed to produce any quinones as evidenced by the spectral studies shown in Figure 4. For comparison, the spectra of the quinones generated from N-acetyldopamine and 3,4-dihydroxycinnamic acid are also given in Figure 4. These studies clearly indicate that 1,2-dehydro-N-acetyl dopamine is not converted to the corresponding quinone upon oxidation by either cuticular phenoloxidase or mushroom tyrosinase.

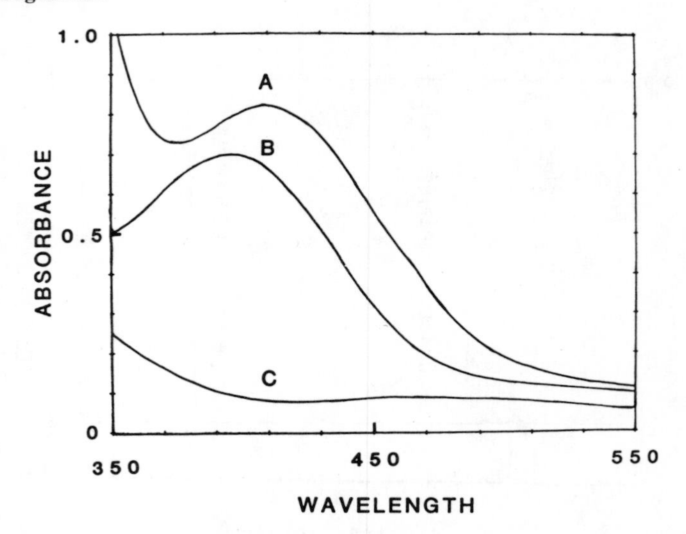

FIGURE 4. Visible spectral changes associated with the oxidation of A) caffeic acid, B) N-acetyldopamine, and C) 1,2-dehydro-N-acetyldopamine by mushroom tyrosinase.

DISCUSSION

The results of the radioactive trapping experiments shown in Figure 2 clearly illustrate that 1,2-dehydro-N-acetyl dopamine is not freely formed in cuticle during N-acetyl dopamine oxidation. The mechanism for β-sclerotization, involving 1,2-dehydro N-acetyldopamine, calls for its generation from N-acetyldopamine by a two electron oxidation process (Figure 1). The dehydrogenation reaction requires the presence of a desaturase similar to succinate dehydrogenase or fatty acyl CoA desaturase. However, evidence for the presence of such an enzyme system in cuticle is still lacking. Even if such an enzyme system exists in cuticle, it might require electron acceptors such as flavin or pyridine nucleotides which are not normally required for phenol oxidation. Moreover, this mechanism calls for further oxidation of the dehydro compound to dehydro quinone before crosslinking of structural protomers occurs. A dehydro quinone would produce a darker color cuticle as compared to a simple quinone because the dehydro quinone, due to the double bond conjugation to the quinone framework, is expected to absorb light strongly at a higher wavelength than the corresponding side chain saturated quinone. Andersen's mechanism requires the removal of at least four electrons from N-acetyldopamine to observe even a quinone-protein adduct. It would be less economical to the organism to use this pathway over the quinone methide mechanism, which requires only a two electron oxidation for the production of crosslinking agent. Apart from economy, it is not clear why a quinone would produce catechol type adducts when it can produce quinone-type adducts much more readily by Michael type 1,4-additions to available nucleophiles on protein side chains. Furthermore, our studies indicate that no such quinone is produced from 1,2-dehydro-N-acetyl dopamine upon oxidation with either cuticular phenoloxidase or mushroom tyrosinase (Figure 4). Thus it appears that neither 1,2-dehydro-N-acetyldopamine nor its corresponding quinone is formed from N-acetyldopamine. At present it is not clear why oxidation of dehydro compound by cuticle does not yield the corresponding quinone. Perhaps it might be converted a reactive quinone methide derivative before being incorporated into the cuticle.

In this context, it is interesting to note that mushroom tyrosinase, which usually catalyzes the conversion of diphenols to quinones, converts 3,4-dihydroxymandelic acid to 3,4-dihydroxybenzaldehyde by an unusual oxidative decarboxylation reaction (17). During this conversion, neither 3,4-dihydroxybenzoyl formate nor 3,4-dihydroxybenzyl alcohol is formed as a transient intermediate. Quinone methide formation accounts for this unusual reaction by a simultaneous oxidative decarboxylation to yield the "enone" form of the stable isomer 3,4-dihydroxybenzaldehyde. Cuticular, as well as hemolymph, phenoloxidase also catalyze this conversion. It, therefore, appears that quinone methide production is inherent with phenoloxidases, and appropriate modification at the active site can alter the course of this reaction in favor of either quinone or quinone methide production.

If, indeed, quinone methides are formed in cuticle, they can be easily trapped by a number of nucleophiles. Water, which is present in large excess in biological reactions, can add to quinone methide to produce side chain hydroxylated catechols. As expected, during N-acetyldopamine oxidation by cuticle, accumulation of N-acetylnorepinephrine can be easily witnessed.

The 1,2-dehydro compound does not account for the accumulation of N-acetylnorepinephrine in the reaction mixtures containing cuticle and N-acetyldopamine.

Spontaneous addition of water to N-acetyldopamine quinone methide will produce N-acetylnorepinephrine, while a separate enzyme is necessary to affect hydration of the double bond in the 1,2-dehydro compound. Such an enzyme catalyzed process would be normally stereo specific and produce either R or S forms of N-acetylnorepinephrine. However, nonenzymatic water addition to quinone methide does not require an enzyme and produces a racemic mixture. N-Acetylnorepinephrine formed in cuticle is found to be a racemic mixture, attesting to the formation of quinone methides in cuticle (18).

Both 4-methyl catechol and 3,4-dihydroxyphenylacetate, the former by lack of a β-carbon and the latter by virtue of a carboxyl group at the β-position, cannot form a dehydro compound similar to the one proposed for N-acetyldopamine, and yet they react with cuticle readily and produce lightly colored cuticle with the regeneration of the catecholic moiety (16).

Based on the above arguments the course of β-sclerotization can be explained as follows (Figure 5).

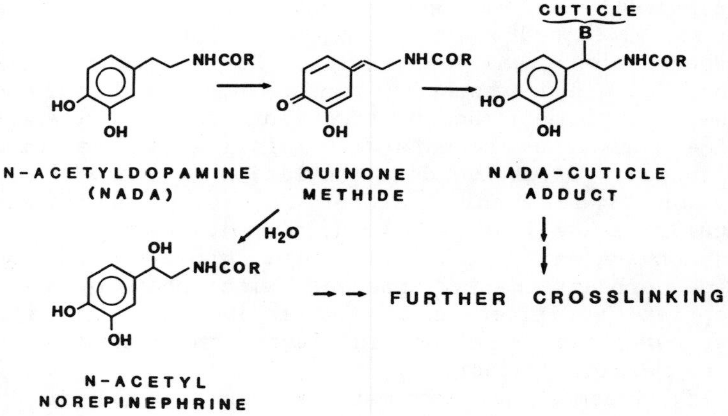

FIGURE 5. Proposed mechanism for sclerotization of insect cuticle involving quinone methides as reactive intermediates ($R=CH_3$). The above scheme is also applicable to N-β-alanyldopamine ($R=CH_2CH_2NH_2$). In this instance water addition to quinone methide leads to the production of N-β-alanylnorepinephrine.

N-Acetyldopamine is converted to highly reactive quinone methide intermediates which react with available nucleophiles in cuticle by a 1,6-Michael type addition reaction with the regeneration of a catecholic group. Since catechol-cuticle adducts are formed as final products as opposed to quinone cuticle adducts, they account for the light color of the β-sclerotized cuticle and acknowledge the presence of unmodified catechol groups in cuticle. N-Acetyldopamine cuticular adducts upon further oxidation and crosslinking, produce N-acetyldopamine cuticle crosslinks. The reaction of N-acetyldopamine quinone methide with water produces the observed racemic mixture of N-acetyl norepinephrine. Acid hydrolysis of cuticle produces ketocatechols modified primarily at the side chain. The above course of reaction is applicable not only to N-acetyldopamine or N-β-alanyldopamine, but to any compounds which can form a quinone methide.

Quinone methide sclerotization, apart from explaining the formation of N-acetylnorepinephrine and colourless cuticle, also accounts for the formation of 1,2-dehydro-N-acetyldopamine. 1,2-Dehydro-N-acetyldopamine was isolated from sclerotized cuticle by treatment with hot alkali, which means a base catalyzed β-elimination reaction was used to isolate this compound (15). Thus it does not appear to be a metabolite freely occurring in cuticle, but a chemically generated compound. The making of the 1,2-dehydro compound can be visualized by a β-elimination of quinone methide cuticle adduct (Figure 6).

Analogous to quinone tanning where quinones are produced and participate in crosslinking, quinone methide sclerotization produces isomeric quinone methides and uses

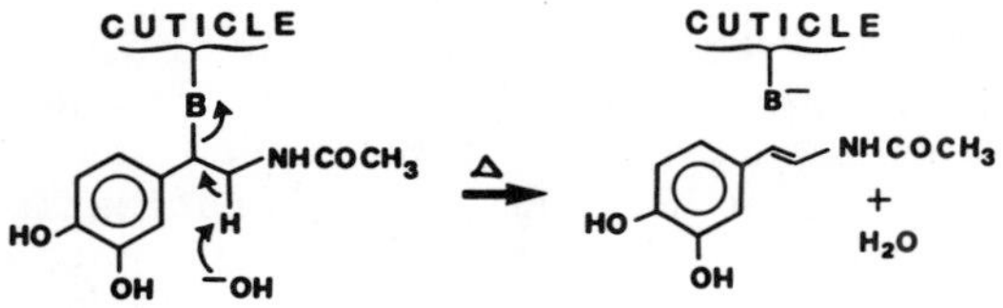

FIGURE 6. Generation of 1,2-dehydro-N-acetyl dopamine by β-elimination of quinone methide cuticle adduct [B = nucleophiles on cuticle].

them for crosslinking. Therefore we wish to rename the β-sclerotization process in which actually quinone methides participate, as "Quinone Methide Sclerotization".

ACKNOWLEDGMENTS

I thank Dr. B. Ramamurthy, Mr. B. Hennigan, Mr. V. Semensi, Ms. T. Rivera and Ms. D. Swanson for their help.

REFERENCES

1. Pryor MGM (1940). On the hardening of the cuticle of insects. Proc R Soc London Ser B 128:393.
2. Hepburn HR (1976). "The Insect Integument." Amsterdam: Elsevier.
3. Andersen SO (1977). Arthropod cuticles: their composition, properties and functions. Symp Zool Soc London 39:7.
4. Andersen SO (1979). Biochemistry of insect cuticle. Ann Rev Entomol 24:29.
5. Vincent JFV, Hillerton JE (1979). The tanning of insect cuticle - a critical review, and a revised mechanism. J Insect Physiol 25:653.
6. Sherald AF (1980). Sclerotization and coloration of the insect cuticle. Experientia 36:143.
7. Miller TA (1980). "Cuticle Techniques in Arthropods." New York: Springer-Verlag.
8. Lipke H, Sugumaran M, Henzel W (1983). Mechanisms of sclerotization in dipterans. Adv Insect Physiol 17:1.
9. Andersen SO (1985). Sclerotization and tanning of the cuticle. In Kerkut GA, Gilbert LI (eds): "Comparative Insect Physiology, Biochemistry and Pharmacology", Vol 3, Oxford: Pergamon Press, p 59.
10. Sugumaran M, Lipke H (1982). Crosslink precursors for the dipteran puparium. Proc Natl Acad Sci USA 79:2480.
11. Sugumaran M, Lipke H (1983). Sclerotization of insect cuticle: a new method for studying the ratio of quinone and β-sclerotization. Insect Biochem 13:307.
12. Andersen SO, Barrett FM (1971). The isolation of ketocatechols from insect cuticle and their possible role in sclerotization. J. Insect Physiol 17:69.
13. Anderson SO, Roepstorff P (1978). Phenolic compounds released by mild acid hydrolysis from sclerotized

cuticle. Purification, structure and possible origin from cross-links. Insect Biochem 8:99.
14. Andersen SO (1974). Evidence for two mechanisms of sclerotization in insect cuticle. Nature 251:507.
15. Andersen SO, Roepstorff P (1982). Sclerotization of insect cuticle. III. An unsaturated derivative of N-acetyldopamine and its role in sclerotization. Insect Biochem 12:269.
16. Sugumaran M, Lipke H (1983). Quinone methide formation from 4-alkylcatechols - a novel reaction catalyzed by cuticular polyphenoloxidase. FEBS Lett 155:65.
17. Sugumaran M, Lipke H (1984). Oxidative decarboxylation of 3,4-dihydroxymandelic acid - a novel reaction catalyzed by polyphenol oxidases. Fed Proc 43:1879.
18. Peter MG, Vaupel W (1985). Lack of stereoselectivity in the enzymatic conversion of N-acetyldopamine into N-acetylnorepinephrine in insect cuticle. JCS Chem Commun 848.

Molecular Entomology, pages 369–378

HEMOLYMPH PROTEINS PARTICIPATING IN THE DEFENCE SYSTEM OF *SARCOPHAGA PEREGRINA*

Shunji Natori

Faculty of Pharmaceutical Sciences, University of Tokyo, Bunkyo-ku, Tokyo 113, Japan

ABSTRACT Structure and function of two major proteins induced in the hemolymph of *Sarcophaga peregrina* larvae on injury of body wall were studied. One is a galactose-binding lectin termed *Sarcophaga* lectin and the other is a bactericidal protein termed sarcotoxin I. These proteins are likely to participate in the defence mechanism of this insect. Of these, *Sarcophaga* lectin was found to be synthesized at early embryonic stage and pupal stage, indicating that the same protein plays a role in the normal development as well as in the defence mechanism.

INTRODUCTION

Injection of dead or live bacteria into some insects is known to induce antibacterial proteins in the hemolymph (1). These proteins are thought to function in preventing infection by bacteria entering the body through the damaged epidermis. In 1977, I reported induction of antibacterial activity in the hemolymph of *Sarcophaga peregrina* larvae by injection of a light suspension of *E. coli* (2). Subsequent studies revealed that simple pricking of the body wall with hypodermic needle was sufficient for induction of antibacterial activity. We investigated the quantitative and qualitative changes in the hemolymph proteins of *Sarcophaga* larvae on injury of the body wall, and found that besides induction of antibacterial activity, many other striking changes of hemolymph proteins occurred (3). All these changes seemed likely to be related to enhancement of the defence system of the animal. Therefore, to obtain information on the defence system of this insect, a rational strategy was to isolate and characterize the hemolymph proteins that changed markedly in these conditions.

Here I will describe two of these proteins induced on injury of body wall of Sarcophaga larvae. One is Sarcophaga lectin and the other is sarcotoxin I. We found that Sarcophaga lectin has dual functions, in the defence system and in development. During development, unnecessary tissues and cells produced during ontogeny have to be eliminated. Therefore, there should be common processes in the defence system of mature larvae and in development in which Sarcophaga lectin is needed.

Sarcophaga Lectin.

Structure of Sarcophaga lectin. Sarcopahga lectin could be purified to homogeneity from the hemolymph of injured larvae by affinity chromatography on Sepharose 4B (4). When the purified protein was subjected to SDS polyacrylamide gel electrophoresis, two subunits were detected with molecular masses of 32,000 and 30,000, respectively. The molecular mass of intact lectin is about 190,000 and since the ratio of the intensities of the bands of the large and small subunits on SDS gel is about 2 : 1, we concluded that one molecule of Sarcophaga lectin consists of 4 large subunits and 2 small subunits. Tryptic peptide maps of the large subunit and small subunit were almost identical, so the small subunit may be produced from the large subunit by partial proteolysis or modification (5).

We isolated a cDNA clone, pLE10, for the large subunit of this lectin (6). Analysis of this clone showed that the large subunit consists of 260 amino acids with a putative signal sequence of 23 amino acids. Two possible glycosylation sites were found in the molecule. As shown in Fig. 1, this protein is extremely hydrophilic, except in the region of the signal sequence, and Asp, Asn, Glu and Gln residues constitute more than 32% of the total amino acid residues.

Developmental regulation of expression of the Sarcophaga lectin gene. Using pLE10 as a probe, we investigated the expression of the Sarcophaga lectin gene at various physiological and developmental stages of Sarcophaga by RNA blot hybridization (7). As is evident from Fig. 2(a), the Sarcophaga lectin gene is switched on in the fat body of third instar larvae when the body wall is injured, but is not normally expressed at this stage.

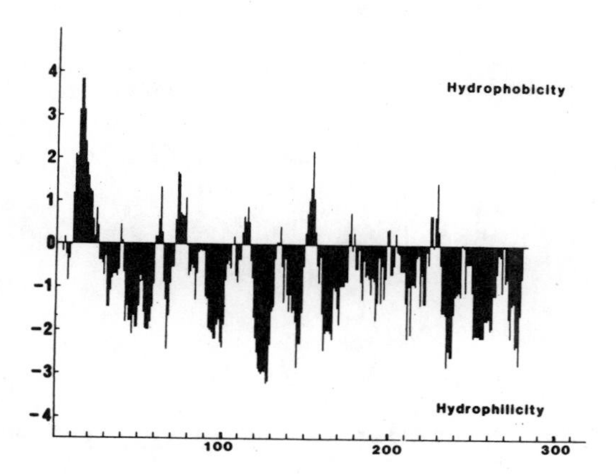

FIGURE 1. Hydropathy analysis of Sarcophaga lectin. Numbers of amino acid residues are shown at the bottom. Data presented as hydrophilic and hydrophobic portions are plotted above and below the vertical line, respectively.

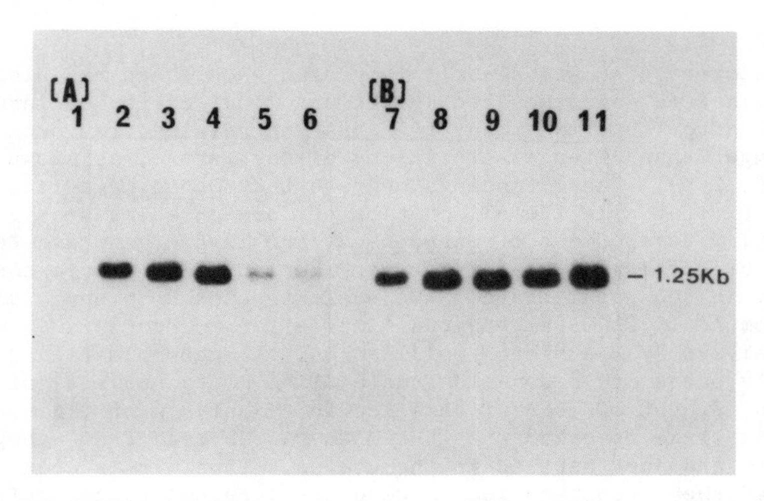

FIGURE 2. Expression of the Sarcophaga lectin gene with time after administration of insect saline A or sheep red blood cells B . Lane 1, RNA prepared from normal larvae; lanes 2 and 7, 3 and 8, 4 and 9, 5 and 10, and 6 and 11 indicate RNAs prepared 1.5, 3, 6, 12, and 24 hrs, respectively, after injections.

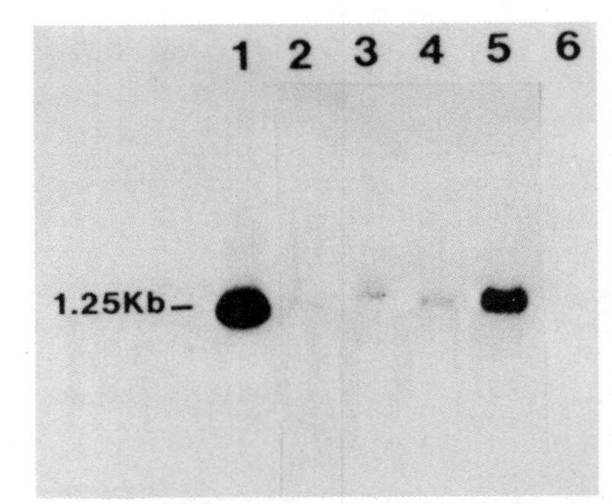

FIGURE 3. Expression of the Sarcophaga lectin gene in pupae and adults. RNA was extracted from pupae and adults, and examined by RNA blot hybridization. Lane 1, pupae harvested on day 1 after puparium formation; lanes 2 and 3, 4-day-old male and female adults, respectively; lanes 4 and 5, 8-day-old male and female adults, respectively; lane 6, normal third instar larvae.

Its expression was clearly increased when sheep red blood cells were injected into in the abdominal cavity of larvae, and after this treatment the mRNA was retained for much longer than after simple injury of the larvae, as shown in Fig. 2(b). These findings suggest that Sarcophaga lectin participates in the elimination of foreign substances. During development of Sarcophaga, the lectin gene was found to be expressed only at the very beginning of embryogenesis and in the pupal stage. RNA was extracted from pupae and from adult flies at various times after emergence and analyzed by RNA blot hybridization. As shown in Fig. 3, only pupae and 8-day-old female flies contained a significant amount of mRNA of this lectin. Analysis of flies at this stage revealed that the mRNA was derived from embryos that had just settled in the uterus. Thus, it is clear that the Sarcophaga lectin gene is expressed transiently in the early embryonic stage. From these results, we concluded that the Sarcophaga lectin gene is expressed in at least three different physiological conditions of this insect: in the early embryonic stage and the pupal stage during normal development, and in larvae whose body wall has been injured.

Function of Sarcophaga lectin. When the body wall of larvae is injured, the chance of invasion of foreign substances is expected to increase. Thus, *Sarcophaga* lectin probably participates in the elimination of foreign substances invading through the damaged body wall. To prove this possibility, we examined whether *Sarcophaga* lectin participates in the elimination of sheep red blood cells introduced into the abdominal cavity of larvae. When sheep red blood cells are injected, they are lysed and then disappear from the hemolymph. However, when antibody against *Sarcophaga* lectin or galactose, a hapten sugar of this lectin, was injected with the sheep red blood cells, hemolysis of the cells in the abdominal cavity was greatly subpressed, as shown in Fig. 4. Injection of preimmune serum or glucose with the cells had no effect on their lysis.

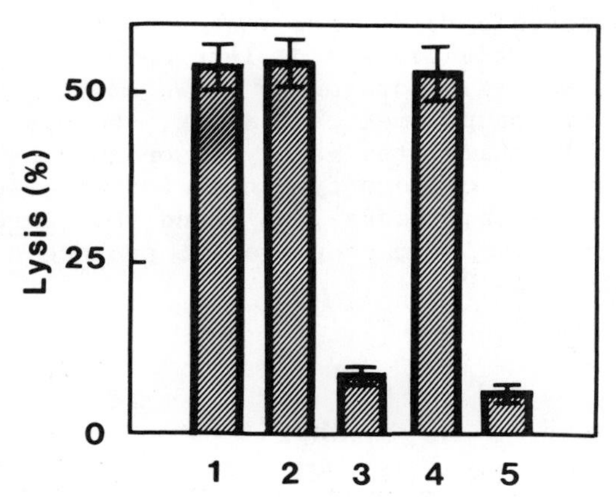

FIGURE 4. Inhibition of the lysis of sheep red blood cells by antibody against *Sarcophaga* lectin or its hapten sugar. Labeled sheep red blood cells were injected with antibody against *Sarcophaga* lectin or hapten sugar. After 60 min, hemolymph was collected and the percentage lysis of sheep red blood cells was determined. Larvae injected with 1, saline; 2, normal IgG; 3, antibody against *Sarcophaga* lectin; 4, glucose; 5, galactose.

Therefore, it is clear that *Sarcophaga* lectin participates in the lysis of red blood cells in the abdominal cavity of larvae (8).

However, this reaction is very complicated, because, Sarcophaga lectin itself has no hemolytic activity. When sheep red blood cells are injected, the Sarcophaga lectin gene is switched on and the fat body synthesizes and secretes Sarcophaga lectin into the hemolymph. Probably, the lectin stimulates hemocytes or some other cells to produce a hemolytic factor. The production of this factor would stop when lectin activity was blocked with antibody or galactose, resulting in inhibition of hemolysis. In any case, Sarcophaga lectin is essential for the elimination of sheep red blood cells introduced into the abdominal cavity, and thus it is a protein participating in the defence mechanism.

Then what is the function of this lectin during development? During embryogenesis and in the pupal stage, programmed death of certain cells occurs that is essential for subsequent development. For instance, in the pupal stage, most larval tissues are disintegrated and cells are killed and eliminated. Nothing is known about the mechanism of this programmed cell death. We assume that Sarcophaga lectin stimulates hemocytes or some other cells in these stages to produce a cytotoxic factor essential for programmed cell death, because, we found that Sarcophaga lectin stimulates murine macrophages to produce a cytotoxic protein, as shown in Fig. 5.

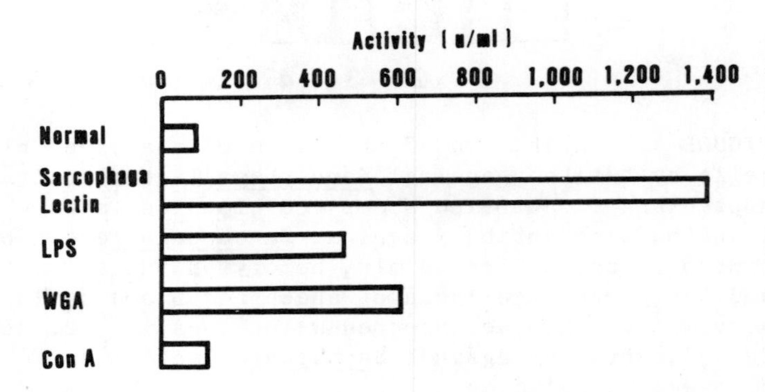

FIGURE 5. Effects of various stimulants on the production of a cytotoxic protein. J774.1 cells were cultured in the presence of various stimulants. After 24 hrs, cytotoxic activity in the medium was determined.

Of the various plant and animal lectins so far tested, Sarcophaga lectin showed especially high activity (9). We do not know anything about the situation in Sarcophaga cells, since insect hemocytes are difficult to culture in vitro. However, the results obtained with murine macrophages support the above idea.

Sarcotoxin I.

Structure of sarcotoxin I. Sarcotoxin I is an antibacterial protein induced on injury of the body wall of Sarcophaga larvae. We purified this protein to homogeneity from the hemolymph of injured larvae (10). Further analysis showed that it was a mixture of three proteins, named sarcotoxin IA, IB and IC, which could be separated by HPLC. We determined the amino acid sequence of each protein and found that all three proteins consist of 39 amino acid residues (11). These three proteins are very similar in primary structure and they differ in only 2 - 3 amino acid residues, as shown in Fig. 6.

1 5 10 15

H_3N-Gly-Trp-Leu-Lys-Lys-Ile-Gly-Lys-Lys-Ile-Glu-Arg-Val-Gly-Gln-His-Thr-Arg-Asp-

- - - - - - - - - - - - - - - - - - - -

- - - Arg - - - - - - - - - - - - - - -

20 25 30 35 39

Ala-Thr-Ile-Gln-Gly-Leu-Gly-Ile-Ala-Gln-Gln-Ala-Ala-Asn-Val-Ala-Ala-Thr-Ala-Arg-COO

- - - - Val-Ile - Val - - - - - - - - - - - - - -

- - - - Val - - - - - - - - - - - - - - - -

FIGURE 6. Amino acid sequences of sarcotoxin IA, IB and IC, respectively, from top to bottom. Charged amino acid residues are marked + or -, and hydrophobic amino acid residues are underlined. Identical amino acid residues to those of sarcotoxin IA are indicated by dashes.

The amino-terminal half of each molecule contains 9 charged residues, 7 of which are basic, while the carboxyl-terminal half of the molecule is rich in nonpolar amino acid residues. From these amino acid sequences, it is readily conceivable that the amino-terminal halves of the molecules are hydrophilic, whereas their carboxyl-terminal halves are hydrophobic. We assume that the hydrophobic regions of these molecules penetrates the bacterial membrane and that the basic amino acid residues in their hydrophylic reagions interact with acidic phospholipids of the bacterial membrane, resulting in perturbation of the membrane. Therefore, the primary target of sarcotoxin I is likely to be the bacterial membrane.

Function of sarcotoxin I. To determine the function of sarcotoxin I, we studied the effect of this substance on E. coli (12,13). As expected, sarcotoxin I has bactericidal activity, and results indicating that it interferes with the activity of the membrane are accumulating. Uptake of proline by E. coli cells stopped almost instantaneously when sarcotoxin I was added to the culture medium. Since proline is known to be transported via a membrane potential, it is likely that sarcotoxin I destroys the membrane potential. As shown in Fig. 7, the ATP content of E. coli cells decreased rapidly with time after addition of sarcotoxin I. Thus we conclude that when E. coli is treated with sarcotoxin I, both the membrane potential and the proton gradient disappear, resulting in loss of ATP generation, and that this is the main reason for the bactericidal effect of sarcotoxin I.

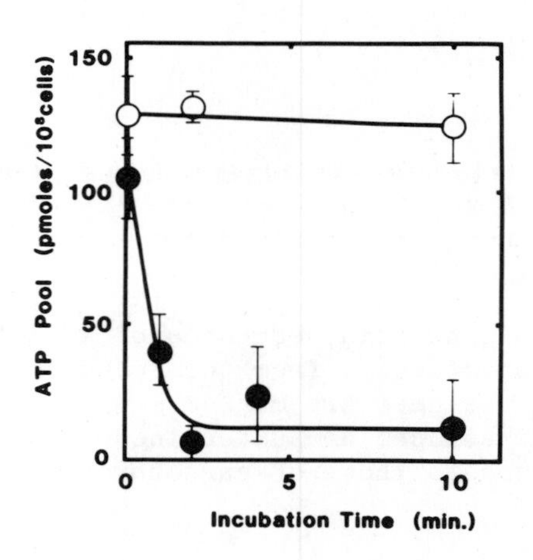

FIGURE 7. Effect of sarcotoxin I on the ATP pool of E. coli. After addition of sarcotoxin I, samples were

taken at intervals and ATP was measured. ○, Control cells treated with distilled water; ●, cells treated with sarcotoxin I.

DISCUSSION

Two proteins participating in the defence mechanism of Sarcophaga peregrina were studied. Of these, Sarcophaga lectin is especially interesting, because its appears not only when the defence mechanism is enhanced, but also at specific developmental stages: the early embryonic and pupal stages.

During normal development, unnecessary cells must be eliminated. For instance, in the pupal stage, larval cells suddenly become nonself and are ingested by hemocytes. As Sarcophaga lectin is induced at this stage, it is probably involved in the elimination of larval cells. I assume that Sarcophaga lectin stimulates hemocytes or some other cells to produce proteins similar to mammalian cytokines, which modulate the development of this insect. There is as yet no evidence to support this assumption, but on treatment with this lectin mouse macrophages produce a cytotoxic protein. It is noteworthy that there is a defence mechanism mediated by a humoral lectin such as Sarcophaga lectin in both vertebrates and invertebrates. In vertebrates, this mechanism may be active only in the early stage of ontogeny and become dormant during development, being replaced by the immune network. But probably even when the mechanism has become dormant, the receptor for this lectin on the surface of mammalian cells remains active and responds to Sarcophaga lectin (14). In contrast, in invertebrates such as insects this mechanism may persist throughout life, since invertebrates have no immune network.

REFERENCES

1. Qu X-M, Steiner H, Engström Å, Bennich H, Boman HG (1982). Insect immunity: Isolation and structure of cecropins B and S from pupae of the Chinese oak silk moth, Antheraea pernyi. Eur J Biochem. 127:219.
2. Natori S (1977). Bactericidal substance induced in the haemolymph of Sarcophaga peregrina larvae. J Insect Physiol 23:1169.
3. Takahashi H, Komano H, Kawaguchi N. Obinata M, Natori

S (1984). Activation of the secretion of specific proteins from fat body following injury to the body wall of Sarcophaga peregrina larvae. Insect Biochem 14:713.

4. Komano H, Mizuno D, Natori S (1980). Purification of lectin induced in the hemolymph of Sarcophaga peregrina larvae on injury. J Biol Chem 255:2919.
5. Komano H, Mizuno D, Natori S (1981). A possible mechanism of induction of insect lectin. J Biol Chem 256:7087.
6. Takahashi H, Komano H, Kawaguchi N, Kitamura N, Nakanishi S, Natori S (1985). Cloning and sequencing of cDNA of Sarcophaga peregrina humoral lectin induced on injury of the body wall. J Biol Chem 260:12228.
7. Takahashi H, Komano H, Natori S (1986). Expression of the lectin gene in Sarcophaga peregrina during normal development and under conditions where the defence mechanism is activated. J Insect Physiol in press.
8. Komano H, Natori S (1985). Participation of Sarcophaga peregrina humoral lectin in the lysis of sheep red blood cells injected into the abdominal cavity of larvae. Devel Comp Immunol 9:31.
9. Itoh A, Iizuka K, Natori S (1984). Induction of TNF-like factor by murine macrophage-like cell line J774.1 on treatment with Sarcophaga lectin. FEBS Lett 175: 59.
10. Okada M, Natori S (1983). Purification and characterization of an antibacterial protein from haemolymph of Sarcophaga peregrina (flesh fly) larvae. Biochem J 211:727.
11. Okada M, Natori S (1985). Primary structure of sarcotoxin I, an antibacterial protein induced in the hemolymph of Sarcophaga peregrina (flesh fly) larvae. J Biol Chem 260:7174.
12. Okada M, Natori S (1984). Mode of action of a bactericidal protein induced in the haemolymph of Sarcophaga peregrina (flesh fly) larvae. Biochem J 222:119.
13. Okada M, Natori S (1985). Ionophore activity of sarcotoxin I, a bactericidal protein of Sarcophaga peregrina. Biochem J 229:453.
14. Ohkuma Y, Komano H, Natori S (1985). Identification of target proteins participating in a lectin-dependent macrophage-mediated cytotoxic reaction. Cancer Res 45:288.

V. SEX SPECIFIC PROTEINS OF INSECTS

Molecular Entomology, pages 381–390

Increase in Serum Lysozyme Following Injection of Bacteria into Larvae of *Manduca sexta*[1]

P. E. Dunn, M. R. Kanost[2], and D. R. Drake[3]

Department of Entomology, Purdue University
W. Lafayette, In 47907

ABSTRACT Following a 6 hr lag, lysozyme activity increased in serum of insects treated with either viable or killed cells of *Pseudomonas aeruginosa*. The increase in lysozyme activity is not a response to wounding, since only a small, transient increase in enzyme activity was observed in saline-treated larvae. The level of lysozyme activity elicited by injection of bacteria was dose dependent. Lysozyme increase was not accompanied by a change in total serum protein concentration. Elevation of serum lysozyme elicited by bacteria was blocked by pretreatment of larvae with actinomycin D. The RNA synthesis required for increase in lysozyme activity occurred within 4 hr after injection of bacteria.

INTRODUCTION

The entry of bacteria into the body cavity of an insect elicits a series of active cellular and humoral defensive responses (1). These include immediate responses by circulating hemocytes and the synthesis of several antibacterial hemolymph proteins. The set of antibacterial proteins includes the bacteriolytic enzyme lysozyme and

[1]This work was supported in part by USDA-SEA Competitive Research Grant 59-2182-0-1-429-0. Purdue Univ. Agricultural Experiment Station Journal Paper No. 9236.
[2]Present address: Department of Biology, Queens University, Kingston, Canada K7L 3N6.
[3]Present address: Department of Microbiology, University of Tennessee, Knoxville, TN 37916.

families of bactericidal proteins (cecropins and attacins). Previous studies of the initial antibacterial responses of the tobacco hornworm, Manduca sexta, have demonstrated rapid phagocytosis and nodule formation by hemocytes following injection of viable or killed bacteria (2,3,4). Cell-free hemolymph (serum) of untreated (naive) larvae of M. sexta is not bactericidal, but becomes bactericidal after injection of bacteria (2,5). Hughes et al. (6) have described the synthesis of several hemolymph proteins in tobacco hornworm larvae injected with viable Enterobacter cloacae. One of these proteins (M23) has been identified as the enzyme lysozyme (6,7). While the role of lysozyme in defense against Gram negative bacteria is unknown, this enzyme is a ubiquitous, major component of insect antibacterial defensive responses. Studies of lysozyme synthesis may serve as a model to increase our knowledge of the regulation of humoral antibacterial responses in general.

In the present study, we have examined changes in the activity of lysozyme in serum of tobacco hornworm larvae after injection of viable and formalin-killed bacteria.

METHODS

Insects and bacteria

M. sexta eggs obtained from Carolina Biological Supply Co. were surface sterilized and newly hatched larvae were reared in the laboratory on artificial diet as described previously (3). Larvae in the second day of the fifth larval stadium were utilized for all experiments.

The origin of bacteria, culture conditions, and methods of quantitating viable cells were as previously described (3). Formalin killed bacteria were prepared by resuspending a pelleted (3,000xg, 20 min) broth culture in 0.37%(v/v) formaldehyde in 0.85%(w/v) NaCl (saline) and incubating the resuspended cells for 18 hr at 20-25°C. Formalin-killed cells were washed several times with saline and finally resuspended in the same medium to the desired concentration. A sample from the killed cells was plated on nutrient agar to check sterility. Total bacterial cell counts were obtained using a Petroff-Hausser counting chamber under phase-contrast optics.

Injection of Insects; Collection and Analysis of Hemolymph

All substances to be injected into larvae were suspended or dissolved in sterile, filtered (Millipore, 0.22 μm) saline and were delivered in a 5 μL volume as described by Dunn and Drake (3). Hemolymph was collected, serum was prepared, and lysozyme was assayed as described previously (3). Hemolymph protein concentration was measured by the biuret method (8) using bovine serum albumin (Miles) as standard.

Data Analysis

Prior to analysis of levels of significance of differences between mean serum protein concentrations or lysozyme activity levels, homogeneity of variances was determined using the Bartlett-Box F. Data exhibiting homogeneous variances were analyzed by one-way analysis of variance (ANOVA) and multiple comparisons between significantly different means were computed by the Student-Newman-Keuls (SNK) procedure. Data which exhibited heterogeneous variances were analyzed either by a t-test for populations with unequal variances (9) or by the non-parametric Kruskal-Wallis (K-W) test with multiple comparisons between significantly different means computed as described by Conover (10). Means different at alpha=0.05 were considered to be significantly different.

RESULTS

Serum Lysozyme Activity

Naive _M. sexta_ larvae contained a low, constitutive level of serum lysozyme (3). Injection of _P. aeruginosa_ into naive larvae elevated levels of serum lysozyme activity (Figs. 1 and 2). A small, transient increase (SNK test) in serum lysozyme activity was observed only at 17 hr post-inoculation (PI) in insects treated with sterile saline (Fig. 1A). In contrast, following an initial lag period, consistently elevated lysozyme activity was observed in serum from insects treated with either formalin-killed or viable bacterial cells.

In insects receiving 10^7 killed cells of strain 9027 (Fig. 1B) or of P11-1 (Fig. 1C), the initial 6 hr lag was followed by significant increases (K-W test) in lysozyme activity until 24 hr PI and this elevated level was maintained until 64 hr PI. When serum lysozyme activity at comparable times following treatment with strain 9027 and strain P11-1 were compared, no significant differences (t-test) in enzyme activity were detected.

In insects which received ca. 10^7 colony forming units (cfu) of strain 9027 (Fig. 2A), a lag period of 8 hr was followed by significant increases (K-W test) in serum lysozyme activity until 16 hr PI and this final level was maintained until 60 hr PI. When serum lysozyme activity in insects treated with 10^7 viable cells of strain 9027 was compared to enzyme activity at comparable times in insects

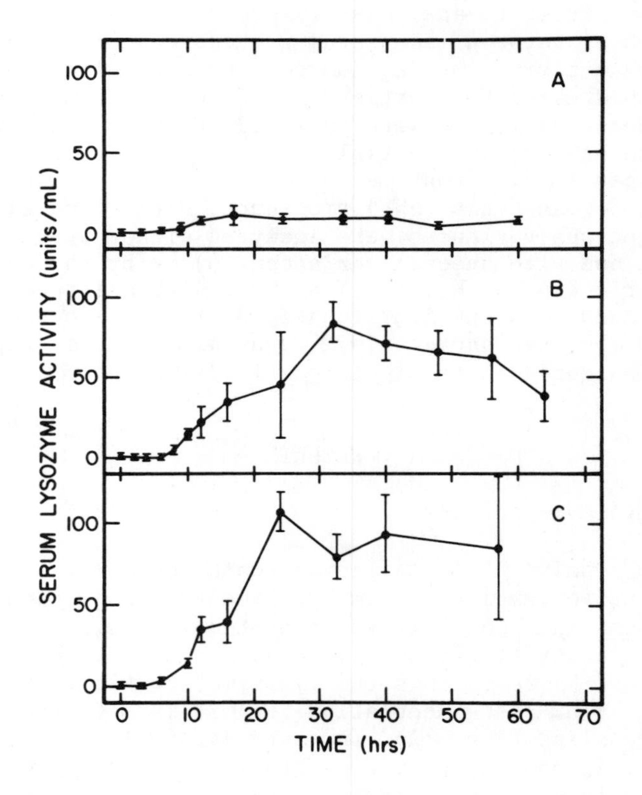

FIGURE 1. Serum lysozyme activity in M. sexta following injection of saline (A), formalin-killed P. aeruginosa 9027 (B), and formalin-killed P. aeruginosa P11-1 (C). Data represent the mean ±S.D. (n=3).

treated with 10^7 killed cells of the same strain, no significant differences (t-test) in response to treatment were detected.

A more complex time course of increase in serum lysozyme activity was observed in insects receiving a lower dose (8×10^3 cfu) of strain P11-1 (Fig. 2B). In this group of larvae, a 7 hr lag period was followed by a significant increase (K-W test) in enzyme activity until 10 hr PI. This level of serum lysozyme activity did not change until 31 hr PI when a second significant increase was observed.

It is noteworthy that the initial plateau in enzyme activity was low and that the second increase in lysozyme activity corresponded to the onset of bacterial multiplication observed previously when the fate of ca. 10^4 cfu of P. aeruginosa P11-1 in larvae of M. sexta was examined (3). This second increase in lysozyme activity may

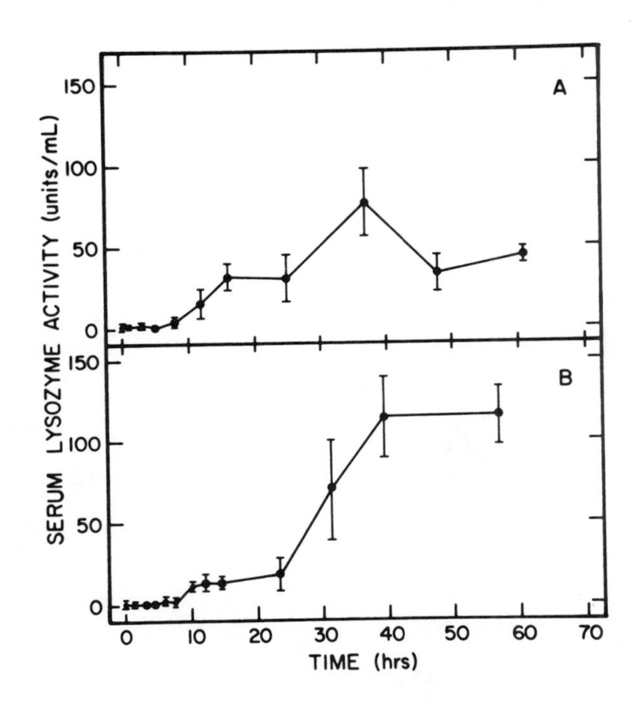

FIGURE 2. Serum lysozyme activity in M. sexta following injection of viable cells of P. aeruginosa 9027 (A) and P11-1 (B). Data represent mean ±S.D. (n=3).

be indicative of an increasing host response to the multiplying bacteria in the hemocoel, a dose-response. To test this possibility, naive larvae were injected with increasing concentrations of formalin-killed P. aeruginosa P11-1 and serum lysozyme activity measured at times after treatment. Injection of killed bacteria in doses greater than 10^3 cells per larva elicited an increase in serum lysozyme activity (Fig. 3). After a lag period of ca. 6 hr, activity increased for 24 to 48 hr, depending on dose. With increasing doses, activity rose at a greater rate and for a longer time. Doses of 10^3 cells or less did not elicit significant increase in lysozyme activity. Similar dose response curves to those shown in Fig. 3 were obtained with P. aeruginosa 9027 and E. coli D-31 (data not shown).

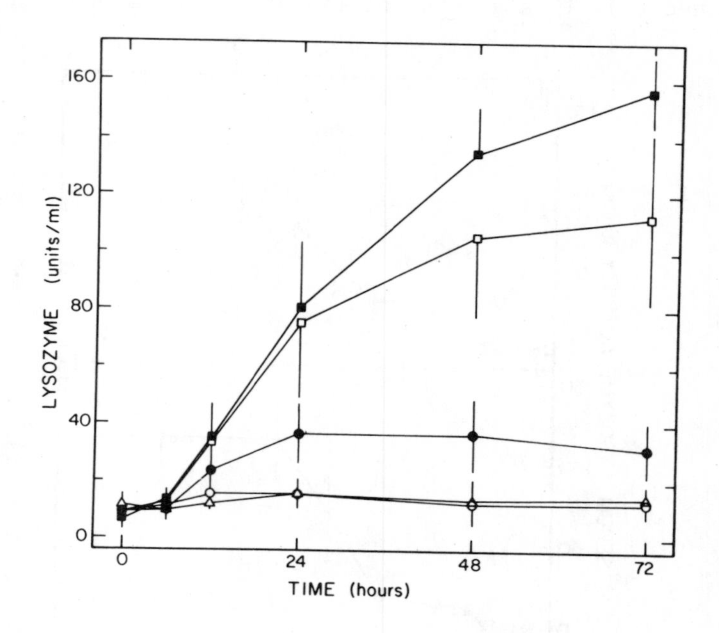

FIGURE 3. Serum lysozyme activity in M. sexta following injection of formalin-killed P. aeruginosa P11-1. Data represent the mean ± S.D. (n=10). Insects were injected with 10^3 (○), 10^5 (●), 10^7 (□), 10^9 (■) bacteria or with sterile saline (△).

Serum Protein Concentration

The serum protein concentration of naive larvae increased ca. six-fold from the second to fifth days of the fifth larval stadium. No significant differences (ANOVA) were detected when the serum protein concentrations of naive, saline-treated, and bacteria-treated larvae at comparable times were compared.

Effect of Inhibitors of RNA and Protein Synthesis

When administered 1 hr prior to treatment with formalin-killed bacteria, doses of 2.5 μg actinomycin D (Act D) or greater blocked the elevation of serum lysozyme activity (data not shown). Act D inhibition of the elevation of serum lysozyme activity in bacteria-treated insects suggested that RNA synthesis was required prior to increase in enzyme activity. To determine how soon after bacterial challenge this RNA synthesis occurred, groups of larvae were treated with Act D at varying times before and after injection of 10^8 killed cells of strain 9027 and serum lysozyme activity was measured at 24 hr PI (Table 1). Treatment of larvae with Act D at 1 hr before or after injection of bacteria blocked any increase in serum lysozyme and lowered enzyme activity to below that in saline-treated controls. Act D-treatment at 2 or 4 hr PI partially inhibited the increase in enzyme activity following bacterial injection while treatment at 6 hr PI had no effect on the level of lysozyme at 24 hr PI.

Pretreatment of larvae with inhibitors of protein synthesis (puromycin and cycloheximide, 100 μg/insect) had no effect on the elevation of serum lysozyme activity following injection of bacteria.

DISCUSSION

Assay of serum lysozyme activity in larvae treated with *P. aeruginosa* and *E. coli* demonstrated a marked increase in enzyme activity. However, injection of larvae with saline produced only a small transient increase in lysozyme activity. Thus, the bacteria-elicited increase is not due to non-specific wounding effects.

Schreiber (5) has also reported that hemolymph lysozyme activity increased in *M. sexta* larvae injected with killed cells of *P. aeruginosa* P11-1. Although a

TABLE 1
EFFECT OF TIME OF TREATMENT WITH ACTINOMYCIN D

Treatment at t=0 hr	Actinomycin D Treatment[1]	Serum Lysozyme Act[2] (units/mL)
saline	none	12 ± 3 a
bacteria	none	51 ± 21 b
bacteria	-1 hr	4 ± 3 c
bacteria	1 hr	4 ± 3 c
bacteria	2 hr	22 ± 6 d
bacteria	4 hr	45 ± 9 e
bacteria	6 hr	61 ± 9 b

[1]5 ug Act D in 5 ul saline.
[2]Mean ±S.D. (n=10). Values followed with the same letter are not significantly different (K-W test).

different assay was used, it appears that untreated insects in his study had an already elevated level of lysozyme. Further, untreated and saline-treated insects showed significant increases in enzyme activity and no initial lag was detected. The insects used in Schreiber's experiments were mass reared. In our laboratory, larvae are reared individually to prevent wounding due to cannabalism and subsequent infection. Prior to adopting stringent rearing conditions (3), we also observed elevated levels of lysozyme in untreated larvae and increased activity following wounding and saline injection. We conclude that Schreiber's insects may not have been naive.

Treatment of larvae with equal numbers of killed cells of P. aeruginosa strains 9027 and P11-1 elicited the same level of serum lysozyme activity. Assay of lysozyme activity in insects injected with viable cells of these strains demonstrated the appearance of enzyme during active infection. Further, similar lysozyme levels at varying times after injection were observed in insects receiving the same dose of killed and viable cells of P. aeruginosa 9027. This result is consistent with previous data which demonstrated qualitatively and quantitatively similar cellular responses of M. sexta to killed and viable cells of this bacterium (4). In contrast, two periods of increase in lysozyme activity were observed in larvae injected with a low dose of viable bacteria of the more virulent strain P11-1. The second increase corresponded to the onset of multiplication of these bacteria reported previously (3). This second increase in lysozyme activity suggested an increasing host response to the multiplying bacteria in the hemocoel. A dose response in the level of lysozyme following injection of increasing numbers of cells was verified in subsequent experiments with killed cells of three Gram negative bacteria representing two species.

The increase in serum lysozyme activity in response to injection of bacteria was not accompanied by change in total serum protein concentration. These results with bacteria-treated larvae differ from those of Dahlman (11) who reported significantly higher protein concentrations in the hemolymph of parasitized fifth instar larvae of M. sexta. Our result would suggest that synthesis of antibacterial protein neither contributes significant new mass of protein to hemolymph nor requires compensatory decrease in the synthesis of normal hemolymph proteins.

Experiments with the RNA synthesis inhibitor Act D have shown that the elevated levels of serum lysozyme elicited by bacteria require prior RNA synthesis which occurs within 4 hr after injection of bacteria. These data, together with the lag time prior to appearance of elevated serum lysozyme, are consistent with the synthesis of new enzyme protein rather than release of stored enzyme. Our inability to block elevation of serum lysozyme with inhibitors of protein synthesis may indicate that puromycin and cycloheximide are not readily taken up by the tissue that synthesizes lysozyme or that they are rapidly metabolized by fifth instar larvae. The observation that Act D lowered serum lysozyme activity to below the level in naive insects may indicate that the basal level of enzyme is maintained by continuous turnover.

These data on the synthesis of lysozyme may be viewed together with initial hemocyte-mediated defenses (2,4) as elements of a model of the antibacterial responses of M. sexta larvae. It is now possible to identify varying contributions of different cellular and humoral defensive responses during the process of bacterial clearance. However, these responses may not be discrete non-interacting systems operating during windows in time. The fact that RNA synthesis required for the production of defensive proteins occurs within 4 hr PI provides a possible link between initial hemocytic reactions and the subsequent production of defensive proteins.

ACKNOWLEDGEMENTS

The authors wish to thank Miss M. Wilson and Mrs. L. Pace for their excellent technical assistance and Mr. D. Ross for assistance in the preparation of figures.

REFERENCES

1. Dunn PE (1986). Biochemical aspects of insect immunology. Ann Rev Entomol 31:321.
2. Horohov DH, Dunn, PE (1982). Changes in the circulating hemocyte population of Manduca sexta larvae following injection of bacteria. J Invertebr Pathol 40:327.
3. Dunn PE, Drake, DR (1983). Fate of bacteria injected into naive and immunized larvae of the tobacco hornworm Manduca sexta. J Invertebr Pathol 41:77.
4. Horohov DH, Dunn, PE (1983). Phagocytosis and nodule formation by hemocytes of Manduca sexta larvae following injection of Pseudomonas aeruginosa. J Invertebr Pathol 41:203.
5. Schreiber FE (1977). "Induced antibacterial activity against Pseudomonas aeruginosa (Schroeter) Migula in the larvae of the tobacco hornworm, Manduca sexta (L.). Ph.D. Thesis, The Ohio State Univ., Columbus.
6. Hughes JA, Hurlbert RE, Rupp RA, Spence KD (1983). Bacteria-induced haemolymph proteins of Manduca sexta pupae and larvae. J Insect Physiol 29:625.
7. Spies AG, Karlinsey JE, Spence KD (1986). Antibacterial hemolymph proteins of Manduca sexta. Comp Biochem Physiol 83B:125.
8. Campbell DH, Garvey JS, Cremer NE, Sussdorf DH (1970). "Methods in Immunology," 2nd Ed. London: W. A. Benjamin, p 73.
9. Nie NH, Hull CH, Jenkins JG, Steinbrenner K, Bent DH (1975). "Statistical Package for the Social Sciences," 2nd Ed. New York: McGraw-Hill, p 269.
10. Conover WJ (1980). "Practical Nonparametric Statistics." New York: John Wiley and Sons, p 229.
11. Dahlman DL (1969). Haemolymph specific gravity, soluble total protein, and total solids of plant-reared, normal, and parasitized, diet-reared tobacco hornworm larvae. J Insect Physiol 15:2075.

Molecular Entomology, pages 391–401

A CRITICAL EVALUATION OF THE FACTORS ESSENTIAL FOR CORRECT EXPRESSION OF THE GENES CODING FOR DROSOPHILA YOLK PROTEINS

Mary Bownes, Alan D. Shirras and Robert D.C. Saunders

Department of Molecular Biology, University of Edinburgh, King's Buildings, Mayfield Road, Edinburgh EH9 3JR

ABSTRACT The 3 major yolk polypeptides (YP1, YP2 and YP3) in the yolk of *Drosophila* eggs are each encoded by a single copy gene on the X chromosome. The availability of the proteins for selective uptake into oocytes depends upon their correct tissue and sex-specific transcription, translation and transport into the haemolymph. We evaluate the roles of sex-determining genes, hormones, and DNA sequence organisation in these processes. New data are presented on the sequence of YP3 and of a female sterile mutant which has defective YP1 secretion.

INTRODUCTION

The yolk proteins of *Drosophila* provide a valuable system for studying eukaryotic gene regulation. The 3 major polypeptides (YP1, 2 and 3) are synthesised only at the adult stage of the life cycle in the fat bodies and ovarian follicle cells of females (Bownes and Hames 1978; Brennan *et al*., 1982). We are investigating regulatory factors and sequence requirements of the genes which lead to the tissue and sex-limited expression of these genes. Since the yolk proteins are synthesised externally to their site of accumulation, "expression" of the genes is used in the wide sense of availability of the final gene product for storage in the oocyte, thus covering transcription of the genes, translation into proteins and secretion of the proteins into the haemolymph.

[1]This work was supported by the Medical Research Council. RDCS is supported by a Science and Engineering Research Council Grant.

The 3 polypeptides have apparent molecular weights of approximately 47, 46, and 45 Kdaltons (Bownes and Hames, 1977). They are initially synthesised as slightly different precursors, YP1 and YP2 at 47 Kd and YP3 at 46 Kd. An amino terminal leader sequence is then cleaved from each reducing their size to 46, 46 and 45 Kd and YP1 is modified further to increase its size to 47 Kd (Brennan *et al.*, 1980). All these changes occur prior to secretion from the fat body or follicle cells, presumably within the rough endoplasmicreticulum, and the Golgi body. Each polypeptide is encoded by a single copy gene located on the X-chromosome (Barnett *et al.*, 1980). YP1 and YP2 are divergently transcribed and separated by only 1.2Kb of DNA (Hung *et al.*,1982). YP3 is located some distance away. The sequences of YP1 and YP2 and the intergenic spacer are published (Hung and Wensink, 1983; Hovemann *et al.*, 1981). Using P-element mediated transformation (Spradling and Rubin 1982), some of the DNA sequences 5' to YP1 and YP2 which are essential for the correct tissue and sex-specific expression of these genes, have been defined (Garabedian *et al.*, 1985). Besides these DNA sequences, a number of other factors must be present if the YP-genes are to be expressed correctly.

The sex-determination genes *ix* (*intersex*), *tra* (*transformer*), *tra-2* (*transformer-2*) and *dsx* (*doublesex*) must be in their wild-type configuration for the sex-specific expression of the YP genes. Mutations in these sex-genes can lead to inactivation of the YP genes in flies with two X-chromosomes which would normally develop into females and activation of the YP-genes in flies with one X-chromosome and one Y-chromosome which would normally develop into males (Bownes and Nöthiger, 1981).

Two insect hormones, 20-hydroxyecdysone and juvenile hormone, are also known to affect the expression of the YP-genes (Postlethwait and Shirk, 1981). The genes are transcribed and the proteins translated in males in reponse to high doses of 20-hydroxyecdysone. In females with low YP-gene expression, achieved either by ligating the abdomen (Jowett and Postlethwait, 1980) or feeding on a sugar diet (Bownes and Blair, 1986), the transcription of the YP-genes or stability of YP mRNA is increased in the fat body by juvenile hormone and 20-hydroxyecdysone. Juvenile hormone also leads to the appearance of the vitellogenic stages of oocyte development. Thus both these hormones can affect the expression of yolk protein genes.

In this paper we critically analyse the information available along with new sequence data on YP3 and a female sterile mutant affecting YP1 secretion to try to establish which factors are required for the correct expression of YP genes in wild-type females.

RESULTS AND DISCUSSION

Evidence for the Involvement of the Sex-determination Genes in Sex-limited Expression of the Yolk-protein Genes.

A number of autosomal genes have been identified which interfere with the sexual pathway, the resulting mutant adults being either intersexual flies or pseudomales. It has been found that all these genes must be in the wild-type state for the YP-genes to be correctly expressed (Postlethwait *et al.*,1980; Bownes and Nöthiger, 1981; Ota *et al.*,1981). All intersexual flies, whether they have one or two X-chromosomes, express the YPs, and all pseudomales have their YP-genes off, thus *ix*, *dsx*, *tra* and *tra-2* are thought to interact and regulate YP-gene expression along with other characteristics of the sexual phenotype.

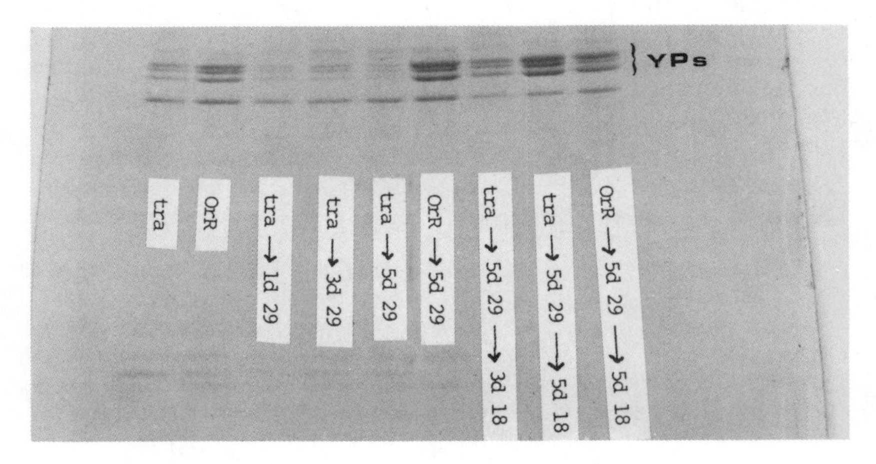

FIGURE 1. Wild-type (OrR) and *tra-2*^{ts} (tra) flies were collected and maintained at 18°C for 3 days. Subsequently they were split into groups and reared at the combinations of 18°C and 29°C. The haemolymph polypeptides were separated by SDS gel electrophoresis. d=days at that temperature in °C. ⟶ = temperature shift.

Using a temperature-sensitive *tra-2* mutation, Belote *et al* (1985) showed that the *tra-2* gene product was needed to initiate YP synthesis in females. Their experiments also suggested that the *tra-2* gene product was needed to maintain YP synthesis since the levels of YP production fell off much more rapidly at the restrictive temperature of 29°C in *tra-2*^{ts} flies than in wild-type flies. We have confirmed this interpretation by shifting flies reared at the permissive temperature (18°C) to 29°C and, after YP synthesis is much reduced, returning them to 18°C. The synthesis of YPs was resumed in these flies; thus the *tra-2* gene product is required to maintain YP synthesis (Fig.1).

Studies on sex-determination suggest that *tra* affects sexual phenotype via the *doublesex* (*dsx*) gene (Baker and Ridge, 1980; Belote *et al*., 1985; Nöthiger and Steinmann-Zwickey, 1985). This is also true for the yolk-protein genes since flies mutant for *tra* and *dsx* synthesise YPs even though those mutant for *tra* alone do not (Bownes and Nöthiger, 1981).

That the products of other genes are required to maintain YP synthesis has also been shown by using an inhibitor of protein synthesis. Cycloheximide treatment causes YP synthesis to cease, and then resume as the flies recover from the inhibition (Bownes *et al*., 1985). It is not known if it is the translation of the sex-determination gene transcripts that is inhibited by cycloheximide, but this seems to be a reasonable interpretation. The pathway of control of the YP genes is shown in Fig.2.

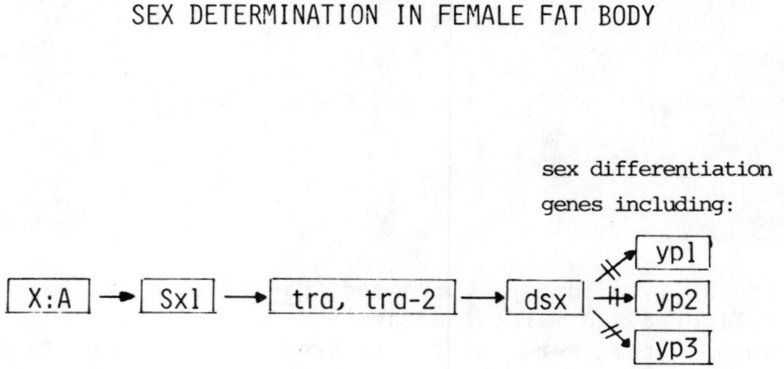

SEX DETERMINATION IN FEMALE FAT BODY

sex differentiation
genes including:
X:A → Sxl → tra, tra-2 → dsx ⊣ yp1, yp2, yp3

Figure 2. Simplified version of the genetic elements regulating sex-determination in the female fat body.

Fig.2 contd: An X:A ratio of 1:0 turns on Sxl (*sex-lethal*). This activates *tra* and *tra-2* (*transformer* and *transformer 2*) which in turn activate *dsx* (*doublesex*). Sexual differentiation in the female fat body includes activation of the three YP (yolk protein) genes. The unbroken arrows indicate established pathways (see Nöthiger and Steinmann-Zwickey, 1985),the arrow marked -//->shows that the *YPs* are under the control of *dsx*,but it is not clear if this is a direct interaction of the *dsx* gene product with *YP* regulatory DNA, or if there are other intermediate genes involved.

Are 20-hydroxyecdysone and juvenile Hormone required for normal Sex and Tissue-specific Expression of the YP-genes?

There is a large accumulation of evidence showing that both juvenile hormone and 20-hydroxyecdysone stimulate transcription of the YP-genes (Postlethwait and Handler, 1979; Jowett and Postlethwait, 1980; Postlethwait *et al.*, 1980). Juvenile hormone affects both ovarian and fat body synthesis while ecdysone affects only the fat body. However this should not be confused with proving that the two hormones are an essential part of the normal system for initiating and maintaining transcription of the YP-genes.

Juvenile hormone appears to be required for uptake of YPs from haemolymph (Postlethwait and Weiser, 1973; Jowett and Postlethwait, 1980). However, juvenile hormone is not needed for fat body synthesis of YPs since the mutant *ap*^{4} which has a defective *corpora allata* (Handler and Postlethwait, 1977) and thus reduced juvenile hormone, has active YP-genes in the fat body (Redfern and Bownes, 1982). Ovarian YP-synthesis can also be maintained in the absence of juvenile hormone. When late pupal ovaries are transplanted into 3rd instar larval hosts, which re-enter metamorphosis, they develop mature oocytes in a pupal environment (Bownes, 1982). We have now shown that there is no juvenile hormone present at this stage of the life cycle (M. Bownes and H. Rembold, unpublished). Thus juvenile hormone does not seem to be essential for maintenance of YP-synthesis.

20-hydroxyecdysone stimulates transcription of the YP-genes in male and female fat bodies. Hung and Wensink (1983) found a consensus sequence in the 5' flanking DNA of YP1 and YP2 which resembled the progesterone/receptor complex binding site of chick ovalbumin and gave strong support to the idea that 20-hydroxyecdysone, also a steroid hormone, could activate transcription of these genes. We have now sequenced YP3 and its flanking DNA and find no such consensus sequence in 730 bp 5' to the gene (Fig.3).

```
     -226
YP1  ATCAGTGGGGTCAGCTATAGGTAGGCCCCGTGTCTATTTTGTATGTATACAATTTATTCC
YP2  CTGTTACCCGATGGATACTTAATAACCAAAAAAAAAAAAAAAAAAAGGGGAACGAAATTT
YP3  CCGGCTATTTGCTAATCGAATATTTCCCTTGACTTGCACCTCTTATACACCGGCGACATG

     -166
YP1  GCTATCGATAGCATATACACTCGATCCGATTCCCAGGCACCCGAAAACCCTTACTCAGCA
YP2  TAGAGCTCATCAGAAGTGTTGCAAAATCACTGACGCTGCGAATCAATGCATTTGTGCTGA
YP3  ATCAGCAGGAAACGAAAGGGGTGGGAAAAAACTGGAAGCCTAGACAGCCGACAACGACG

     -106
YP1  CAAGTGACCGATTAAGGCCTGAGCCAGCGAAAAGCAAGTCGGAAAATGGGAAATCGCTCA
YP2  TCCAGAAATCGGTTGCTGAAATCAATTGCTTCCGGCGCGTGCATTATAGTGTGTATAGAC
YP3  ACAACGACGACGACGACGACGACTTCCTGTGGTCAGCAGAAAATCGCTGGCAGTGCGCTA

     -46                                           +1
YP1  GCGTAAATTGTGGTATATAAACCACCATCGTTGGATTTGGAAGGCCAGTTCAACTCACTC
YP2  TACCGATTCCAAGGGGTATAAAATGCATTGAGTCGCAGCAGTGGGCATCCAGTACAATTT
YP3  TCGGGAATCGGAGCTATATAAGCCAGAGATGGGGCTGAAGGAAGCCAATCGAAACGTCGT

     +15
YP1  AGTGTTGAAGTCGCATCCGCAGGACCAAATCCCAAATCCGAACCATGAACCCCATGAGAG
YP2  GGTACGGTGTCTGAAAAAGTCGAACTTGGAAGCCACAATGAATCCTCTGCGCACCCTTTG
YP3  TTAGCGTTTGGCCCTGATCTGATTCAATTCCGGATTTGCACCAAAATGATGAGTCTAAGG
```

Figure 3. Nucleotide sequences of the YP1, YP2 and YP3 genes from 226 bases before the mRNA start sites to 75 bases after the mRNA start, aligned at the start site (+1). Probable initiation codons are boxed. The TATA boxes at around -30 are underlined. The homologous sequences found between -113 and -101 (YP1) and -161 and -152 (YP2) which

Fig.3 contd: have been shown to be similar to a consensus progesterone receptor-binding sequence (Hung and Wensink, 1983) are similarly indicated. There is no corresponding sequence within 730 bases upstream of the YP3 mRNA start site. The YP1 and YP2 sequences are taken from Hung and Wensink, 1981 and Hung and Wensink, 1983.

Since all three genes are transcribed in males in response to 20-hydroxyecdysone, it seems likely that this consensus sequence is not the hormone-binding site. Furthermore, P-element mediated transformation studies with YP1 have shown that a 125bp YP1 flanking sequence, fused to a lacZ gene, is sufficient for normal fat body expression (Wensink et al., in press), yet this sequence does not appear to contain the putative hormone/receptor recognition sequence. We conclude, therefore, that this sequence is not essential for normal expression of YP1 or YP3.

The mutant ecd^{1} has reduced ecdysone titres at 29° (Garen et al., 1977), yet fat body YP synthesis is maintained (Redfern and Bownes, 1983). Thus it seems unlikely that ecdysone is needed to maintain fat body synthesis of the YP genes. It is not possible to determine if ecdysone is essential for initiating YP synthesis as it is required for fat body metamorphosis.

All three results taken together suggest that normal sex- and tissue-specific expression of the YP-genes is not directly dependent upon ecdysone or juvenile hormone, but that these hormones can modulate the expression of the genes under certain circumstances.

Protein Structure other than the Leader Sequence is required for YP1 Secretion.

The mutant fs(1)1163 was first isolated in a large screen for first chromosomal female sterile mutants (Gans et al., 1975), in which a v^{24} strain was mutagenised. It was subsequently identified as a YP1 variant (Bownes and Hames, 1978; Bownes and Hodson, 1980). Flies heterozygous for fs(1)1163 are sterile at 29°C, though not at 18°C. The fs(1)1163 homozygotes are sterile at both temperatures, and appear to be defective for YP1 secretion, having virtually no YP1 in the haemolymph, and a high YP1 concentration in the fat bodies. Genetic studies map the fs(1)1163 lesion to the vicinity of the YP1 locus (Bownes and Hodson, 1980), suggesting the female sterile phenotype may result from a mutation within the YP1 locus itself.

The YP1 genes from fs(1)1163 and v^{24} homozygotes have been cloned and $YP1^{1163}$ completely sequenced by the dideoxynucleotide chain termination method. Two nucleotide changes have been identified. One variation is within the intron, and is at the site of previously observed sequence diversity. The only mutation leading to an amino acid change is at amino acid residue 92, where Ile^{92} in YP1 is replaced by Asn^{92} in $YP1^{1163}$ (Table 1). This region has been sequenced in YP^{v24}, and residue 92 shown to be Ile.

Ser	Gly	Ile	Gln	Val	(1,2)
AGC	GGC	ATC	CAG	GTC	
Ser	Gly	Asn	Gln	Val	(3)
AGC	GGC	AAC	CAG	GTC	

Table 1. Sequence differences between YP^{1163}, and published Canton S YP1 sequences. 1 and 2 are Canton S sequences (Hovemann et al., 1981; Hung et al., 1981). 3 is the $YP1^{1163}$ sequence.

The possible effect of an Ile^{92} to Asn^{92} alteration was investigated using the PEPPLOT program (Gribster et al., 1986). The results suggest that a peak of hydrophobicity (Kyte and Doolittle, 1982) at position 92 is lost in $YP1^{1163}$, with a consequent effect on the Beta and Alpha hydrophobic moment parameters (Eisenberg et al., 1984).

Ile^{92} is conserved in comparisons between YP1 and the closely related YP2 sequences. The sequence of the $YP1^{1163}$ leader peptide is identical to that in YP1, in agreement with in vitro translation data (Minoo, 1982) which suggests that $YP1^{1163}$ is not defective in translocation across the rough endoplasmic reticulum membrane. The Ile^{92} to Asn^{92} change does not create an N-linked glycosylation consensus sequence, and YP1 has been reported to contain no detectable carbohydrate (Mintzas and Kambysellis, 1982). It seems likely that the mutation causes a sufficiently altered secondary or tertiary protein structure to prevent correct processing or secretion of $YP1^{1163}$.

ACKNOWLEDGMENTS

Thanks to Betty McCready for typing the manuscript, and Ann Scott for running the protein gel. MB is grateful to the Royal Society and the Wellcome Foundation for their help in attending this meeting.

REFERENCES

1. Barnett T, Pachl C, Gergen JP, Wensink PC (1980). The isolation and characterisation of Drosophila yolk protein genes. Cell 21:729-738.
2. Baker BS, Ridge KA (1980). Sex and the single cell. I. On the action of major loci affecting sex-determination in Drosophila melanogaster. Genetics 94:383-423.
3. Belote JM, Handler AM, Wolfner MF, Livak KJ, Baker BS (1985). Sex-specific regulation of yolk protein gene expression in Drosophila. Cell 40:339-348.
4. Bownes M, Hames BD, (1977). Accumulation and degradation of three major yolk proteins in Drosophila melanogaster. J Exp Zool 200:149-156.
5. Bownes M, Hames BD, (1978). Analysis of the yolk proteins in Drosophila melanogaster. FEBS Lett 96: 327-330.
6. Bownes M, Hames BD, (1978). Genetic analysis of vitellogenesis in Drosophila melanogaster: the identification of a temperature-sensitive mutation affecting one of the yolk proteins. J E E M 47:111-120.
7. Bownes M, Hodson BA, (1980). Mutant fs(1)1163 of Drosophila melanogaster alters yolk protein secretion from the fat body. Mol Gen Genet 180:411-418.
8. Bownes M, Nöthiger R (1981). Sex-determining genes and vitellogenin synthesis in Drosophila melanogaster. Mol Gen Genet 182:222-228.
9. Bownes M (1982). Ovarian yolk-protein synthesis in Drosophila melanogaster. J Insect Physiol 28:953-960.
10. Bownes M, Smith T, Blair M (1985). The regulation of expression of the genes coding for the yolk-proteins in Drosophila. Int J Invertebr Reprod Dev. In press.
11. Bownes M, Blair M, (1986). The effects of a sugar diet and hormones on the expression of the Drosophila yolk-protein genes. J Insect Physiol in press.

12. Brennen MD, Warren M, Mahowald AP (1980). Signal peptides and signal peptidase in *Drosophila melanogaster*. J Cell Biol 87:516-520.
13. Brennen MD, Weiner AJ, Goralski TJ, Mahowald AP (1982). The follicle cells are a major site of vitellogenin synthesis in *Drosophila melanogaster*. Dev Biol 89:225-236.
14. Eisenberg D, Weiss RM, Terwilliger TC (1984). The hydrophobic moment detects periodicity in protein hydrophobicity. Proc Natl Acad Sci USA 81:140-141.
15. Gans M, Audit C, Masson RM (1975). Isolation and characterisation of sex-linked female-sterile mutants in *Drosophila melanogaster*. Genetics 81:683-704.
16. Garabedian M, Hung MC, Wensink PC (1985). Independent control elements that determine yolk protein gene expression in alternative *Drosophila* tissues. Proc Natl Acad Sci USA 82: 1396-1400.
17. Garen A, Kauver L, Lepesant J-A (1977). Roles of ecdysone in *Drosophila* development. Proc Natl Acad Sci USA 74:5099-5103.
18. Gribster M, Burgess RR, Devereux J (1986). PEPPLOT, a protein secondary structure analysis program for the UWGCG sequence analysis software package. Nucl Acids Res 14:327-336.
19. Handler AM, Postlethwait JH (1977). Endocrine control of vitellogenesis in *Drosophila melanogaster*. Effects on the brain and *corpus allatum*. J Exp Zool 202: 3339-3402.
20. Hovemann B, Galler R, Waldorf U, Kupper H, Bantz EKF (1981). Vitellogenin in *Drosophila melanogaster*. Sequence of the yolk protein I gene and its flanking regions. Nucl Acids Res 9:4721-4736.
21. Hung MC, Wensink PC (1981) The sequence of the *Drosophila melanogaster* gene for yolk protein I. Nucl Acids Res 9:6407-6419.
22. Hung MC, Barnett T, Woolford C, Wensink P (1982). Transcript maps of *Drosophila* yolk protein genes. J Mol Biol 154:581-602.
23. Hung MC, Wensink PC (1983). Sequence and structure conservation in yolk proteins and their genes. J Mol Biol 164:481-492.
24. Jowett T, Postlethwait JH (1980). The regulation of yolk polypeptide synthesis in *Drosophila* ovaries and fat body by 20-hydroxyecdysone and a juvenile hormone analogue. Dev Biol 80:225-234.

25. Kyte J, Doolittle RF (1982). A simple method for displaying the hydropathic character of a protein. J Mol Biol 157:105-132.
26. Minoo P, (1982). Molecular analysis of a quantitative yolk polypeptide variant in Drosophila melanogaster. Ph.D. Thesis, University of Eugene, Oregon.
27. Mintzas AC, Kambysellis MP (1982). The yolk proteins of Drosophila melanogaster: Isolation and characterisation. Insect Biochem 12:25-33.
28. Nöthiger R, Steinmann-Zwicky (1985). Sex-determination in Drosophila. TIGS 1:209-215.
29. Ota T, Fukunaga A, Kawabe M, Oishi K (1981). Interaction between sex transformation mutants of Drosophila melanogaster 1. Haemolymph vitellogenins and gonad morphology. Genetics 99:429-441.
30. Postlethwait JH, Weiser K (1973). Vitellogenesis induced by juvenile hormone in the female sterile mutant apterous-four in Drosophila melanogaster. Nature New Biol 244:284-285.
31. Postlethwait JH, Handler AM (1979). The roles of juvenile hormone and 20-hydroxyecdysone during vitellgenesis in isolated abdomens of Drosophila melanogaster. J Insect Physiol 25:455-460.
32. Postlethwait JH, Bownes M, Jowett T (1980). Sexual phenotype and vitellogenin synthesis in Drosophila melanogaster. Dev Biol 79:379-387.
33. Postlethwait JH, Shirk PD (1981). Genetic and endocrine regulation of vitellogenesis in Drosophila. Am Zool 21:687-700.
34. Redfern CPF, Bownes M (1983). Pleiotiopic effects of the 'ecdysoneless-1' mutation of Drosophila melanogaster. Mol Gen Genet 189:432-440.
35. Redfern CPF, Bownes M (1982). Yolk polypeptide synthesis and uptake in ovaries of the $apterous^4$ mutant of Drosophila melanogaster. Molec Gen Genet 185:181-183.
36. Spradling A, Rubin GM (1982). Transposition of cloned P elements in Drosophila germ line chromosomes. Science 218:341-347.
37. Wensink PC, Shepherd B, Garabedian M, Hung MC (1986). Developmental control of Drosophila yolk protein 1 gene by cis-acting DNA elements. In press.

Molecular Entomology, pages 403–413

ANALYSIS OF MOSQUITO YOLK PROTEIN BY MONOCLONAL ANTIBODIES[1,2]

Alexander S. Raikhel and Arden O. Lea

Department of Entomology, University of Georgia, Athens, Georgia 30602, USA

ABSTRACT We have produced a library of 45 monoclonal antibodies (mAB) against yolk protein of the mosquito _Aedes aegypti_, which serve as molecular probes for studying vitellogenesis. Three groups of mABs were revealed by immunoblot analysis recognizing either (1) a 200 kDa, (2) a 68 kDa, or (3) both of these polypeptides. All mABs recognized these polypeptides only in extracts from vitellogenic fat bodies and ovaries. Immunoprecipitation experiments with mABs demonstrated that 200-kDa and 68-kDa polypeptides are indeed subunits of mosquito vitellogenin-vitellin. Video-enhanced immunofluorescence showed that yolk polypeptides, recognized by mABs, originated in the fat body and accumulated in the oocytes. The immunolocalization also indicated that our panel of mABs directed against different steps in the biosynthetic processing of yolk protein in trophocytes of the fat body, as well as during its accumulation in oocytes. Ultrastructural localization provided evidence of simultaneous processing, secretion, and internalization of yolk protein subunits.

[1]This work was supported by NIH Grants AI-20371 and AI-17297.

[2]The monoclonal facility of the Botany Department of the University of Georgia is supported in part by NSF grant PCM 8315882.

INTRODUCTION

In insects, vitellogenesis requires the correlated activity of trophocytes of the fat body, which synthesize large amounts of yolk protein precursor(s), and of oocytes, which specifically accumulate the yolk protein(s). These activities are regulated by a complex of developmental and hormonal factors. To understand insect vitellogenesis fully requires the solution to a wide range of problems, from the hormonal control of vitellogenin gene expression to receptor-mediated endocytosis of yolk proteins. Although, the factors involved in regulation of vitellogenesis in the mosquito are not completely understood (1-5), this important vector has become one of the model systems for studying insect vitellogenesis (3-13).

The diverse applications of mABs have revolutionized many areas of immunology, cellular and molecular biology (14; 15), and we believe they will provide powerful tools for studying structural and functional properties of yolk proteins.

We have described the production of a library of mABs against the yolk protein of *A. aegypti* (13), which we are using for the analysis of yolk protein.

MATERIAL AND METHODS

Monoclonal Antibody Production. Preparation of antigen, immunization and hybridoma production has been described in detail elsewhere (13). After the initial screening, 45 hybridoma cell lines were selected and cloned; mABs were characterized by a combination of enzyme-linked immunosorbent assay (ELISA), immunoblot, and immunocytochemistry.

Gel electrophoresis. Polypeptides were separated by a 5-15% gradient dodecyl sulphate polyacrylamide gel electrophoresis (SDS PAGE) (16).

Immunoblotting. Immunoblotting was performed according to Towbin *et al*. (17). Monoclonal ABs used for immunoblotting were: a) hybridoma culture medium containing the mAB, b) mABs, concentrated 20-fold by ammonium sulphate precipitation and diluted 5 to 10-fold, or c) affinity-purified and then diluted mABs, at concentrations of 1 to 5 ug/ml.

Immunoprecipitation. Extracts of yolk protein from vitellogenic ovaries and homogenates of vitellogenic fat bodies were pretreated with Pansorbin (Calbiochem) and incubated with mABs. Extracts were incubated with rabbit antibodies to mouse immunoglobulins coupled with Pansorbin, washed, and sedimented by centrifugation. The immunoprecipitates were then subjected to SDS PAGE (18).

Enzyme-linked immunosorbent assay. An indirect ELISA, performed essentially as described by Voller *et al.* (19), was used throughout to screen the mABs.

Immunocytochemistry with monoclonal antibodies. To localize the proteins recognized by mABs at the light microscopical level, we used video-enhanced immunofluorescence. Tissue preparation, cryosectioning, immunocytochemical methodology, and the use of video-enhanced optics were previously described (7; 12). By using unlabelled rabbit antibodies to mouse IgG as a bridge, the specific immunopositive signals were significantly increased.

For electron microscopical localization, we utilized a postembedding technique with protein A-colloidal gold as previously described (9; 11). The immunoreaction was enhanced by a bridge similar to that in the immunofluorescence technique.

RESULTS

Analysis of mosquito yolk protein by monoclonal antibodies. All mABs were screened against extracts from fat bodies and ovaries. None of the mAB reacted with proteins from non-vitellogenic ovaries or non-vitellogenic fat bodies (not shown).

Three groups of mABs were identified, based on an immunoblot analysis utilizing extracts of the fat bodies and ovaries 24 h after initiation of vitellogenin synthesis by a blood meal (Figs. 1-4). The first group of mABs recognized polypeptides of about 200 kDa, a second group recognized a polypeptide of about 68 kDa, and a third group of mABs immunostained both the 200-kDa and 68-kDa polypeptides. Several other mAB, selected because of their strong reaction in the ELISA, did not recognize denatured polypeptides on immunoblots.

Next, we investigated whether polypeptides recognized by mABs are subunits of the same molecule, or whether they represent different proteins synthesized. Crude extracts of the fat body and of the ovaries 24 h after a blood meal were immunoprecipitated by mABs from either the first or the second group and subjected to SDS PAGE. The electrophoretic pattern of the recovered proteins was similar regardless of the particular mAB used: both 200-kDa and 68 kDa polypeptides were present (Fig. 5.). Solubilization of immunoprecipitants in the absence of reduction (without DL-dithiothreitol or 2-mercaptoethanol in the sample buffer) still yielded 200-kDa and 68-kDa polypeptides after SDS PAGE (not shown).

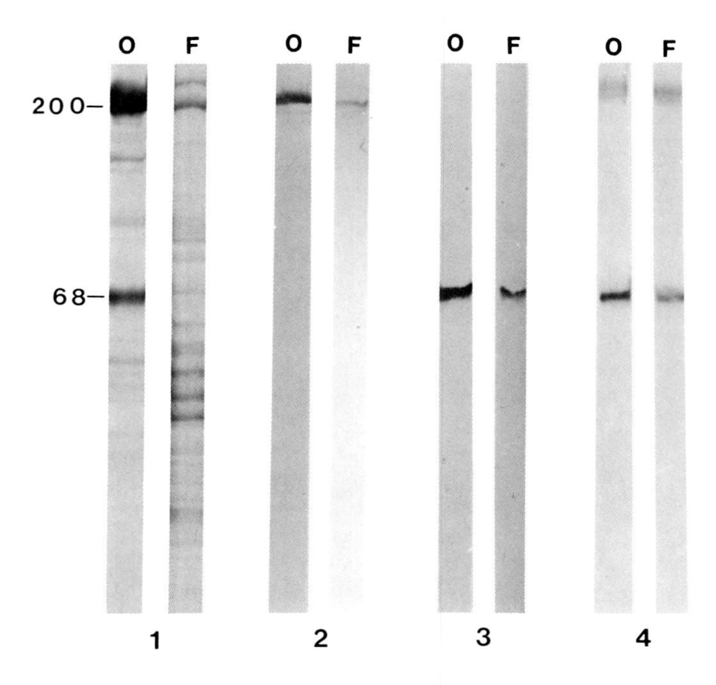

FIGURES 1-4. Immunoblot analysis of yolk polypeptides with mABs. Fig. 1 - electrophoretic patterns of crude extracts, used for immunoblot, of ovaries (O) and fat bodies (F) 24 h after a blood meal. Numbers indicate molecular weights X 10^3. Fig. 2 - mAB A6D7 recognizing a 200-kDa yolk polypeptide. Fig. 3 - mAB 2G1 recognizing a 68-kDa polypeptide. Fig. 4 - a mAB of the third group immunostained both 200-kDa and 68-kDa polypeptides. Note that the mABs recognize polypeptides in the ovary and fat body with different relative intensities.

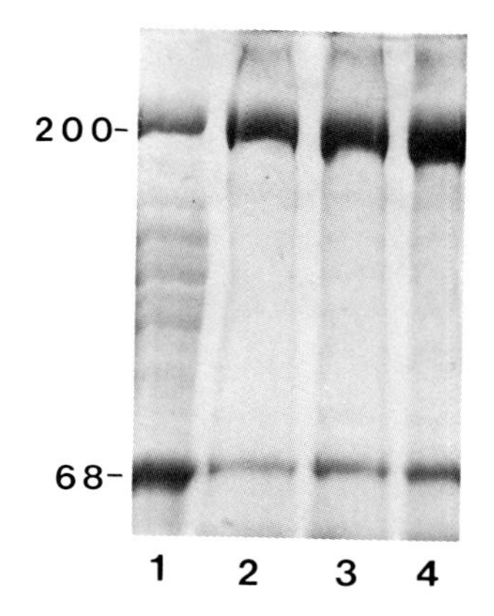

FIGURE 5. Identification of yolk polypeptides by immunoprecipitation with mABs. The first lane is the SDS PAGE pattern of a crude extract from ovaries, 24 h after a blood meal, used for immunoprecipitation. Lanes 2-4 are SDS PAGE patterns of yolk polypeptides immunoprecipitated from the ovarian extract: (lane 2) by mAB A1D12 recognizing a 200-kDa polypeptide, (lane 3) by mABs 2G1 and (lane 4) by mAB 2D4 both recognizing a 68-kDa polypeptide.

Immunofluorescence analysis with monoclonal antibodies. We tested all 45 mABs and found that they only reacted positively with vitellogenic tissues, in both fat bodies and ovaries, while non-vitellogenic tissues were free of any background signals (Figs. 6 and 7). Thus, immunofluorescence confirmed the immunoblot analysis, indicating that our mABs were directed against stage-specific polypeptides that originated in the fat body and accumulated in the oocytes.

In trophocytes of the fat body 24 h after a blood meal, immunostaining was distributed throughout the cytoplasm. Small granules, likely secretory granules, were visible. This immunolocalization pattern was characteristic of all mABs of the first group, i.e., those directed against the 200-kDa polypeptides, most of the mABs from the second group, i.e., those directed against the 68-kDa polypeptide, and most of the mABs of the third group (Figs. 8 and 9). The intensity of immunofluorescence produced by different mABs varied from very strong to weak.

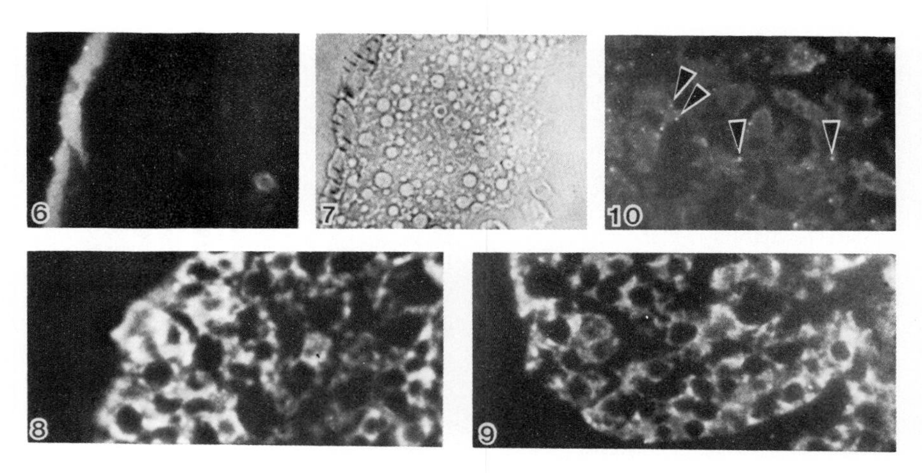

FIGURES 6-10. Video-enhanced immunofluorescence analysis of yolk polypeptides in trophocytes by mABs. Fig. 6 - non-vitellogenic trophocytes from a 3-day-old female. The cuticle in Fig. 6 exhibits an autofluorescence. Fig. 7 - phase-contrast image of the same tissue. Fig. 8 and 9 - immunostaining pattern of trophocytes, 24 h after a blood meal, visualized with mAB A6D7, which recognizes only the 200-kDa polypeptide (Fig. 8) and with mAb 13D3, which recognizes only the 68-kDa polypeptide (Fig. 9). Fig. 10 - preferential staining of secretory granules (arrowheads) by mAB 10A5, which recognizes both 200-kDa and 68-kDa polypeptides.

A few mABs gave a different distribution of immunostaining, which indicated that these mABs possibly recognize yolk polypeptides only at certain stages of their processing in trophocytes (Fig. 10).

In vitellogenic oocytes, some mABs stained only small granules (Fig. 11), while others stained larger granules in the cortex of the oocyte. Several mABs recognized yolk polypeptides in mature yolk bodies, while some stained only their outer rims and other stained entire yolk bodies (Figs. 12 and 13).

Ultrastructural immunolocalization of yolk polypeptides. The specificity of immunolocalization of yolk polypeptides observed with our mABs using immunofluorescence was confirmed at the ultrastructural level using the immunogold technique. In vitellogenic trophocytes, most of the mABs tested showed a low level of labelling over rough endoplasmic reticulum and a more intense labelling of Golgi com-

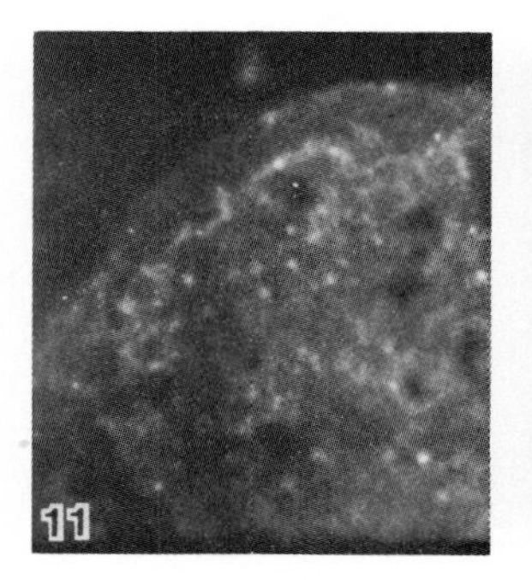

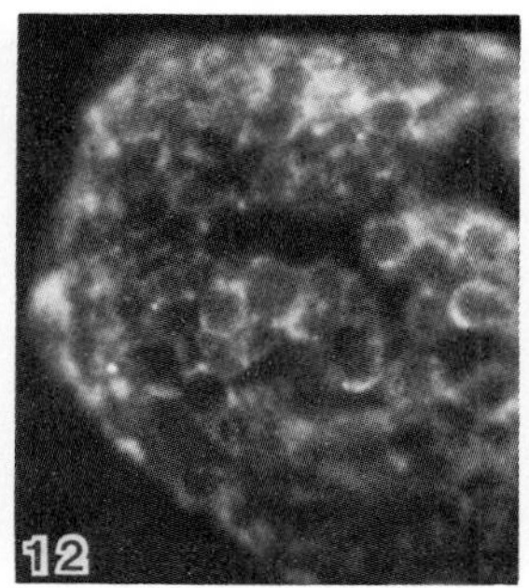

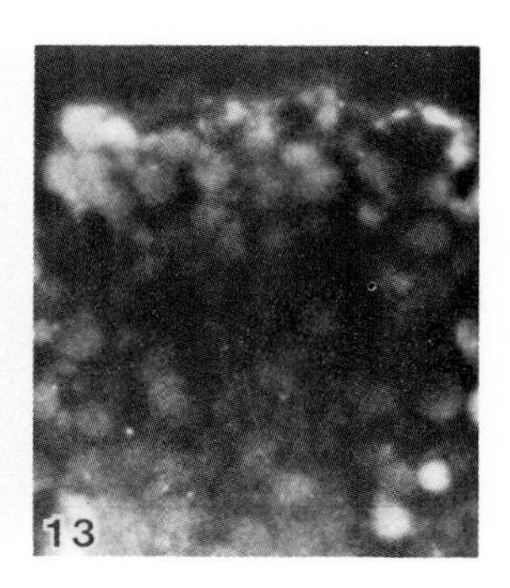

FIGURES. 11-13. Immunofluorescent analysis of yolk polypeptides visualized by different mABs in vitellogenic oocytes, 24 h after a blood meal. Fig. 11 - immunostaining by a mAB that likely recognizes yolk polypeptides early in their internalization, i. e. in coated vesicles and endosomes. Figs. 12 and 13 - immunostaining patterns in a mature yolk bodies: immunopositive material, presumably non-crystalized yolk polypeptides, at the rim of mature yolk body (Fig. 12) or homogeneous staining of crystalline yolk (Fig. 13).

plexes and secretory granules (Fig. 14). Both 200-kDa and 68-kDa polypeptides were localized in these organelles of the same cell as demonstrated by labeling of adjacent sections. As with immunofluorescence, some mABs recognized yolk polypeptides only in mature secretory granules (Fig. 15).

In vitellogenic oocytes, mABs recognized both 200-kDa and 68-kDa polypeptides in compartments of the vitellogenin accumulative pathway: coated vesicles, endosomes and yolk bodies (Fig. 16). In the mature yolk bodies, some of the mABs preferentially labelled crystalline yolk (Fig. 17).

DISCUSSION

Methodology. The availability of a library of mABs against *Aedes aegypti* yolk polypeptides allows us to analyze the synthesis and processing of these molecules by the fat body as well as their specific accumulation in the oocytes. Together with numerous advantages of mABs over conventional polyclonal antibodies, there are some methodological prob-

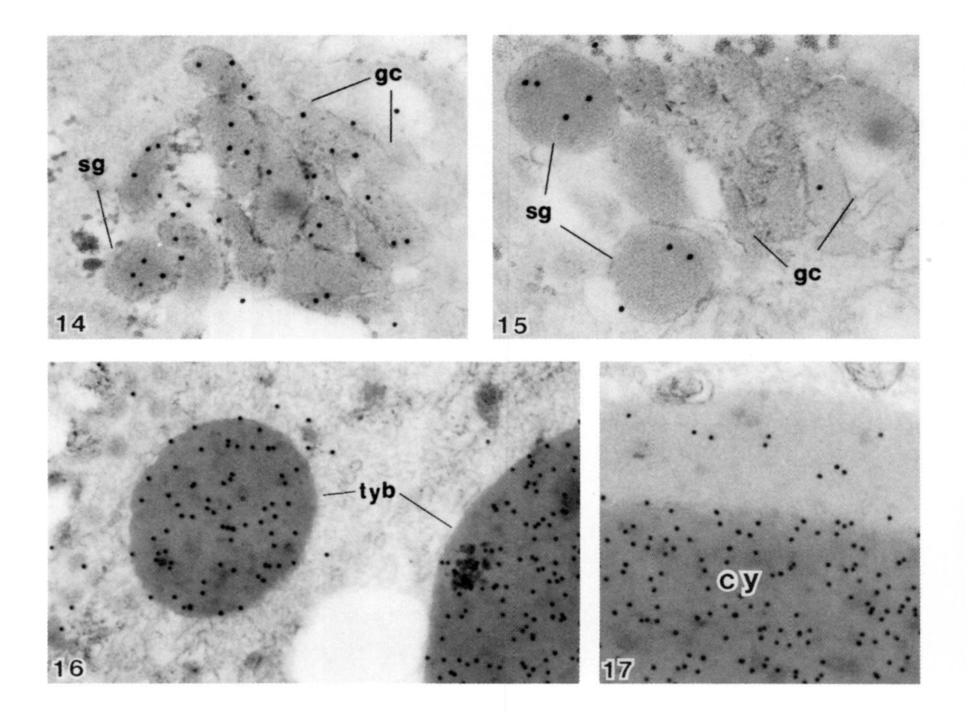

FIGURES 14-17. Immunogold localization of yolk polypeptides with mABs. Fig. 14 - typical immunolabelling of the Golgi complex and secretory granules (sg) in vitellogenic trophocytes. Note that all compartments of the Golgi complex (gc) are labelled. Fig. 15 - preferential staining of mature secretory granules (sg) by mAB 10A5. Note that compartments of the Golgi complex are not labelled. Figs. 14 and 15 - X 60,000. Figs. 16 and 17 - immunolabelling of the cortex of a vitellogenic oocyte. Fig. 16 - intense labelling of transitional yolk bodies (tyb). Fig. 17 - a section of a mature yolk body. Note that the label is concentrated over the crystalline yolk (cy). Figs. 16 and 17 - X 35,000.

lems in their applications; for example, some mABs recognize an antigen in an immunoblot but not in an ELISA, or by immunocytochemistry, or vice versa (14; 15). Therefore, we have characterized our mABs by using a combination of ELISA, immunoblot, and immunocytochemistry and have identified the mABs that are the most suitable for use in specific types of assays (13). In this work, the immunoblot and immunoprecipitation analyses with our mABs have allowed us to begin to investigate the molecular composition of protein yolk, and the immunocytochemical survey has localized the antigens at specific sites in tissues. In addition, we have utilized the library of mABs to develop an ELISA to quantify yolk protein (Raikhel and Lea, unpublished).

The composition of the mosquito protein yolk. It has been debated whether mosquito protein yolk is composed of large subunits only (20), or whether it consists of both large and small subunits (6; 21; 22). Our analysis utilizing immunoblot and immunoprecipitation with mABs has clearly demonstrated that both vitellogenin from the fat body and vitellin from the oocytes consists of 200-kDa and 68-kDa polypeptides. These subunits are not associated by disulfate bonds, since solubilization in the absence of reduction still yields these polypeptides after SDS PAGE.

The intracellular localization of yolk polypeptides. Our immunolocalization study, utilizing a panel of mABs in combination with video-enhanced immunofluorescence and ultrastructural immunogold technique, demonstrates the simultaneous processing and secretion of 200-kDa and 68-kDa polypeptides in the fat body and their internalization by oocytes. These data support our biochemical analysis which indicates that these polypeptides are indeed subunits of the yolk protein molecule in the mosquito.

Furthermore, the immunocytochemical survey indicates that our panel of mABs recognizes different stages of yolk polypeptides both in trophocytes, during their biosynthetic processing, and in oocytes during their accumulation and deposition in mature yolk bodies. This provides us with a powerful tool with which to develop further insights into the molecular mechanisms of egg maturation in insects.

ACKNOWLEDGEMENTS

We thank Dr. L. Pratt for consultations and Mrs. M. Sledge for excellent technical assistance in the production of mABs; Dr. B. Palevitz for the use of his video-intensification system; Mr. C. Maier and Mr. D. Fendley for technical assistance.

REFERENCES

1. Fuchs MS, Kang SH (1981). Ecdysone and mosquito vitellogenesis: a critical appraisal. Insect Biochem 11: 627.
2. Lea AO (1982). Artifactual stimulation of vitellogenesis in Aedes aegypti by 20-hydroxyecdysone. J Insect Physiol 28:173.
3. Hagedorn HH (1983). The role of ecdysteroids in the adult insect. In Downer RGH, Laufer H (eds): "Endocrinology of Insects", New York: Alan R. Liss, p 271.
4. Hagedorn HH (1985). The role of ecdysteroids in reproduction. In Kerkut GA, Gilbert LI (eds): "Comprehensive Insect Physiology, Biochemistry and Pharmacology", Pergamon Press, vol 8, p 205.
5. Borovsky D, Thomas BR, Carlson DA, Whisenton LR, Fuchs MS (1985). Juvenile hormone and 20-hydroxyecdysone as primary and secondary stimuli of vitellogenesis in Aedes aegypti. Arch Insect Biochem Physiol 2:75.
6. Racioppi JV, Hagedorn HH, Calvo JM (1984). Physiological mechanisms controlling the reproductive cycle of the mosquito Aedes aegypti. In Engels W (ed): "Advances in Invertebrate Reproduction", Elsevier Science Publishers, vol 3, p 259.
7. Raikhel AS, Lea AO (1983). Previtellogenic development and vitellogenin synthesis in the fat body of a mosquito: an ultrastructural and immunocytochemical study. Tissue Cell 15:281.
8. Raikhel AS, Lea AO (1985). Hormone-mediated formation of the endocytic complex in mosquito oocytes. Gen Comp Endocr 57:422.
9. Raikhel AS (1984). The accumulative pathway of vitellogenin in the mosquito oocyte: a high-resolution immuno- and cytochemical study. J Ultrastr Res 87:285.
10. Raikhel AS (1984). Accumulations of membrane-free clathrin-like lattices in the mosquito oocyte. Eu J Cell Biol 35:279.
11. Raikhel AS (1986). Lysosomes in the cessation of vitellogenin secretion by the mosquito fat body; selective degradation of Golgi complexes and secretory cells. Tissue Cell 18:125.
12. Raikhel AS (1986). Role of lysosomes in regulating of vitellogenin secretion in the mosquito fat body. J Insect Physiol (in press).
13. Raikhel AS, Pratt LH, Lea AO (1986). Monoclonal antibodies as probes for processing of yolk protein in the

mosquito; production and characterization. J Insect Physiol (in press).
14. Yelton DE, Schariff MD (1981). Monoclonal antibodies: a powerful new tool in biology and medicine. Ann Rev Biochem 50:657.
15. Milstein C (1982). Monoclonal antibodies from hybrid myelomas: theoretical aspects and some general comments. In McMichael A, Fabre JW (eds): "Monoclonal Antibodies in Clinical Medicine", Academic Press, p 3.
16. Laemmli UK (1970). Cleavage of structural proteins during the assembly of the head of bacteriophage T4. Nature 227:680.
17. Towbin H, Staehelin T, Gordon J (1979). Electrophoretic transfer of proteins from polyacrylamide gels to nitrocellulose sheets: procedure and some applications. Proc Natl Acad Sci USA 76:4350.
18. Hubbard AL, Bartles JR, Braiterman LT (1985). Identification of rat hepatocyte plasma membrane proteins using monoclonal antibodies. J Cell Biol 100:1115.
19. Voller A, Bidwell D, Bartlett A (1980). Enzyme-linked immunosorbent assay. In Rose NR, Frieman H (eds): "Manual of Clinical Immunology", Washington D.C.: Am Soc Microbiol, p 359.
20. Harnish DG, White BN (1982). Insect vitellins: identification, purification, and characterization from eight orders. J Exp Zool 220:1.
21. Atlas SJ, Roth TF, Falcone AJ (1978). Purification and partial characterization of _Culex pipiens fatigans_ yolk protein. Insect Biochem 8:111.
22. Ma M, Newton PB, He G, Kelly TJ, Hsu HT, Masler EP, Borkovec AB (1984). Development of monoclonal antibodies for monitoring _Aedes atropalpus_ vitellogenesis. J Insect Physiol 30:529.

Molecular Entomology, pages 415–424
Published 1987 by Alan R. Liss, Inc.

20-HYDROXYECDYSONE SUPPRESSES YOLK PRODUCTION IN THE INDIANMEAL MOTH

Paul D. Shirk and Victor J. Brookes[1]

Insect Attractants, Behavior, and Basic Biology Research Laboratory, Agric. Res. Serv., USDA Gainesville, Florida 32604

ABSTRACT As in other moths that have short-lived adult stages, egg maturation takes place during pharate adult development in the Indianmeal moth, *Plodia interpunctella* (Hübner). Treatment of pharate adult females with 20-hydroxyecdysone blocked vitellogenesis. Inclusion of varying concentrations of 20-hydroxyecdysone in the culture medium of ovarioles showed that maximal synthesis of yolk polypeptide-2 occurred at 10^{-8} M. Concentrations of 20-hydroxyecdysone higher than 10^{-8} M suppressed yolk polypeptide-2 synthesis. Total RNA was isolated from ovaries of pharate females that had been treated with 20-hydroxyecdysone, and the RNA was translated in a cell-free reticulocyte lysate. Cleaveland peptide mapping of the pre-yolk polypeptide-2 showed the primary translate to be similar structurally to the mature secreted form of yolk polypeptide-2. Quantitation of the relative amounts of translation products showed that 20-hydroxyecdysone suppressed the accumulation of yolk polypeptide-2 mRNAs in the ovaries to nearly 35% of normal. These data suggest that 20-hydroxyecdysone exerts a negative control over vitellogenesis and on the rate of yolk polypeptide synthesis.

[1]Department of Entomology, Oregon State University, Corvallis, OR 97331

INTRODUCTION

Adults of the Indianmeal moth, Plodia interpunctella (Hübner), are non-feeding and consequently live for only a few days. To achieve sexual competency rapidly as an adult, the majority of vitellogenesis and egg maturation is completed during pharate adult development, and the females emerge with oocytes that are ready for fertilization by the time of mating. The mature eggs of P. interpunctella were found to contain two major yolk proteins that were each comprised of two unique subunits (1,2). The four major yolk polypeptide subunits (YPs) were designated YP1 (153 kDa), YP2 (69 kDa), YP3 (43 kDa), and YP4 (33 kDa) (1). Vitellogenin was comprised of two subunits, YP1 and YP3, which were synthesized by the fat body. Purified vitellin had an apparent molecular mass of 462 kDa. The yolk also contained a major protein with an apparent molecular mass of between 133 kDa and 260 kDa that was synthesized within the ovarioles and consisted of YP2 and YP4 as subunits. The vitellin was found to contribute about 40% of the protein to the yolk as did the YP2/YP4 protein.

Endocrine control of vitellogenesis in most insects has been shown to rely on the stimulation of yolk production by ecdysteroids, juvenile hormone, or a combination of these and additional hormones during the adult stage (3,4). However, inclusion of egg maturation within the format of metamorphosis rather than after the emergence of the adult should place the physiological control(s) of vitellogenesis in the Indianmeal moth under the influence of the regulatory mechanism(s) that coordinate the generation of the various adult organ systems. The rate of metamorphosis (5) and the events of eclosion (6,7) appear to be regulated by declining levels of ecdysteroids during pharate adult development in the moth, Manduca sexta (L.). Inhibition of metamorphic processes by ecdysteroids have been observed both in vitro and in vivo in several other insects as well. Cultured imaginal wing discs of P. interpunctella began the synthesis of cuticle that contained chitin only after the discs had been exposed to a 24-h pulse of 20-hydroxyecdysone (20HE) (8). However, the cultured discs would not synthesize chitin if maintained continuously in medium that contained 20HE. Similarly, continuous exposure of organ cultures of pupal cuticle (9) larval crochets (10) to

20HE prevented tanning of the epidermal structures. In addition, imaginal discs from Drosophila cultured in the presence of 20HE produced epicuticle and began deposition of procuticle, an indication of advancing metamorphosis, only when the levels of 20HE were reduced (11).

As seen in *M. sexta*, injection of 2-day-old pharate adult female *Bombyx mori* (L.) with a large single dose of 20HE (100 μg) delayed adult eclosion by as much as 8 days (12). The hormonal treatment had the additional effect of initially delaying the normal increase in ovarian weight, but when the treated females emerged as adults they deposited heavier and larger than normal eggs. A similar phenomenon appears to be operating on the control of vitellogenesis in *P. interpunctella* since 20HE was found to be exerting an inhibitory control on the processes of egg maturation in pharate adults (13). Egg production was inhibited in a dose responsive manner by injecting previtellogenic females twice daily throughout the final days of adult development. The females treated with 250 ng 20HE per dose had no oocytes that were at least 50% vitellogenic, whereas, females injected with saline contained an average of nine 50% vitellogenic oocytes per ovariole. Further, the total amount of protein or the total amount of radiolabeled proteins accumulating in the ovarioles was decreased to approximately 20% of normal when the moths were injected with a dose of 250 ng 20HE per treatment. The suppression of vitellogenesis did not appear to be the result of a general steroid effect since injection of 22-isoecdysone, an inactive ecdysteroid analogue (14), had no effect on the accumulation of protein in the oocytes. Although data showed that 20HE had a general inhibitory effect on vitellogenesis, the specific action of 20HE on the control of egg production was not determined. The data presented here demonstrate that 20HE had specific inhibitory actions on the synthesis of the YPs in the ovarioles.

RESULTS

20HE Controls the Synthesis of YP2 in Cultured Ovarioles

To examine the specific response of an organ to 20HE, ovarioles from early vitellogenic females were cultured *in vitro* in the presence of varying concentrations of 20HE, and the rate of YP2 synthesis was measured. An autoradiogram showed that after 24 h of culture the rate of YP2 synthesis was greatest in 10^{-8} M 20HE (Fig. 1) and represented nearly 19% of the total proteins secreted into the medium (Table 1). The rate of YP2 synthesis was found to be similar to the rate observed for ovarioles that were incubated for only 2 h after dissection from the females (data not shown). If the concentration of 20HE were less than 10^{-8} M, the synthesis of YP2 represented only 4-7.5% of the total secreted protein. That the synthesis of YP2 occurred in ovarioles cultured without 20HE showed that the tissues had achieved a developmental state where the expression of YP2 was open. However, the tissues required the presence of 20HE to support normal levels of YP2 synthesis. When the concentration of 20HE was raised above 10^{-8} M, a progressively decreasing rate of YP2 synthesis was observed. These data suggested that the rate of YP synthesis was dependent on the presence of 20HE and was related to the concentration of the hormone. Preliminary estimations of the concentration of 20HE in *P. interpunctella* support this hypothesis since the levels range from 10^{-6} to 10^{-5} M at the time of pupation to 10^{-8} at adult eclosion (Brookes, unpublished).

20HE Suppresses the Accumulation of Translatable YP2 Transcripts in Ovaries

To determine what effect 20HE had on the synthesis of YP2, total RNA from ovaries of females that had been treated with 20HE was translated in a cell-free reticulocyte lysate. Poly(A) RNA isolated from ovaries contained a transcript that translated a product that was immunoprecipitable with antiserum to yolk proteins, had a molecular mass of 70 kDa (data not shown), and was considered to be the primary translate for YP2 (preYP2). PreYP2 and YP2 secreted from ovarioles (previously shown to be identical with YP2 accumulating in the oocytes (1)) were subjected to peptide mapping to determine their structural relatedness. The autoradiogram showed that the digestion patterns for the two polypeptides were nearly identical when treated with either V8 protease or α-chymotrypsin (Fig. 2). The digestion fragments for α-chymotrypsin showed the presence of a single fragment that was of slightly smaller size in the digests of YP2 than in the preYP2 digests. This suggests there may be a posttranslational cleavage of a signal peptide from the preYP2. A size shifting of some of the fragments of higher molecular mass

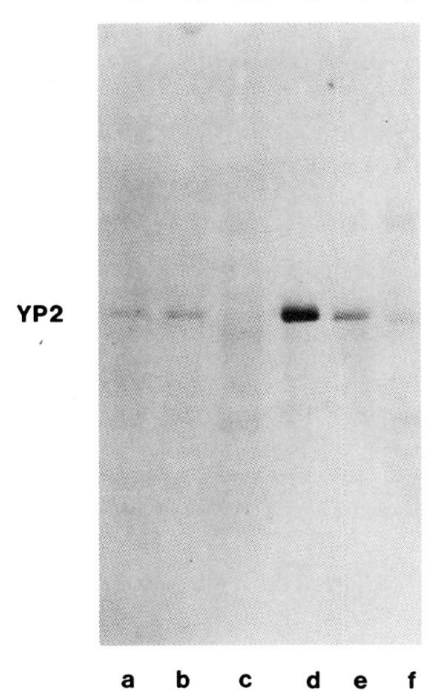

FIGURE 1. Dependence of the rate of YP2 synthesis by cultured ovarioles on the concentration of 20HE. Ovarioles from day-5 pharate adult females were placed in Grace's insect medium with varying concentrations of 20HE. After 22 h, the media were removed and replaced with media that contained 0.5 μCl/μl ^{35}S-methionine for an additional 2-h incubation. Equal amounts of TCA precipitable radiolabeled proteins were loaded on each lane of an 8-15% gradient SDS-PAGE, and the dried gel was autoradiographed. Lanes: a) no 20HE; b) 10^{-12} M 20HE; c) 10^{-10} M 20HE; d) 10^{-8} M 20HE; e) 10^{-6} M 20HE; f) 10^{-4} M 20HE.

also was seen (c.f. Fig. 2E & F), which may be accounted for by further posttranslational modifications. The autoradiogram of the translation products showed that treatment with 20HE decreased the level of YP2 transcripts specifically (Fig. 3). The amount of preYP2 produced by ovarian RNA from females treated with 250 ng 20HE

TABLE 1
20-HYDROXYECDYSONE SUPPRESSES SYNTHESIS OF YP2 BY OVARIOLES CULTURED IN VITRO AND THE ACCUMULATION OF TRANSLATABLE YP2 TRANSCRIPT IN OVARIES.[a]

20HE conc. (molar)	Synthesis of YP2 as % total secreted protein (SD)	20HE Treat-ment	Accumulation of YP2 transcript shown as % total translation products (SD)
0	7.3 (2.3)	Control	7.1 (1.2)
10^{-12}	5.0 (1.6)	Saline	5.5 (0.4)
10^{-10}	4.3 (1.6)	10 ng	5.0 (1.0)
10^{-8}	18.9 (4.9)	50 ng	3.4 (0.2)
10^{-6}	10.8 (4.8)	250 ng	2.5 (1.6)
10^{-4}	8.7 (2.1)		

[a]The amount of radiolabeled proteins and translation products was quantified by integrating the band areas on an autoradiogram with an LKB 2222 laser microdensitometer (LKB Instruments).

represented only 2.5% of the total radiolabeled translation products whereas in normal nontreated females preYP2 was 7.1% of the total translation products (Table 1). The 20HE treatment resulted in a decrease of YP2 mRNA to 35% of the normal females. The amount of transcript for YP4 as well as several polypeptides appears to be decreased by 20HE treatment, but the changes were not quantified here. However, the 20HE treatment generally did not affect the relative percent contribution of all proteins since the percent of total translation products for a 43 kDa protein and a 96 kDa protein were found not to change with the increasing hormonal treatment.

DISCUSSION

Vitellogenesis in P. interpunctella appears to be regulated by mechanisms similar to those controlling the maturation of other organ systems in pharate adults. As observed in M. sexta (5), 20HE or some related ecdysteroid exerts an inhibitory effect on the progress of metamorphosis, and only when the levels of ecdysteroids

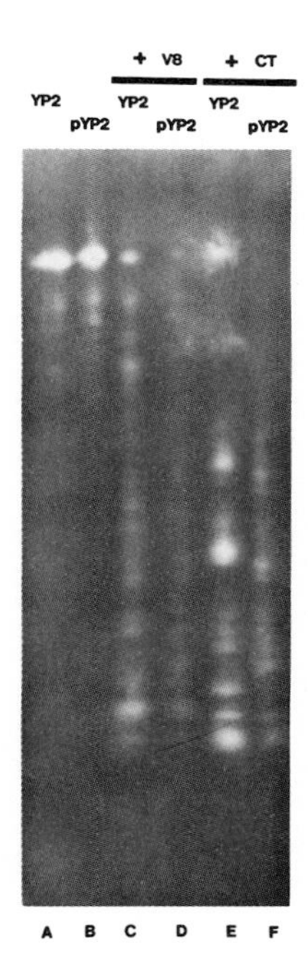

FIGURE 2. Peptide maps of the primary translation product for YP2 (preYP2) and YP2 secreted into culture medium. The immunoprecipitated radiolabeled polypeptides were treated essentially as described by Cleaveland et al. (15) with either Staphylococcus aureus V8 protease (V8) or α-chymotrypsin (CT). The digestion fragments were resolved on an 8-15% gradient SDS-PAGE, and the dried gel was autoradiographed. The presence of fragments of smaller molecular mass than YP2 or preYP2 in lanes A and B were the result of protease contamination. Lanes: A) YP2; B) preYP2; C) YP2 plus V8 protease; D) preYP2 plus V8 protease; E) YP2 plus α-chymotrypsin; F) preYP2 plus α-chymotrypsin.

decrease does organ maturation take place. The effects of 20HE on the synthesis of YP2 were shown to be specific

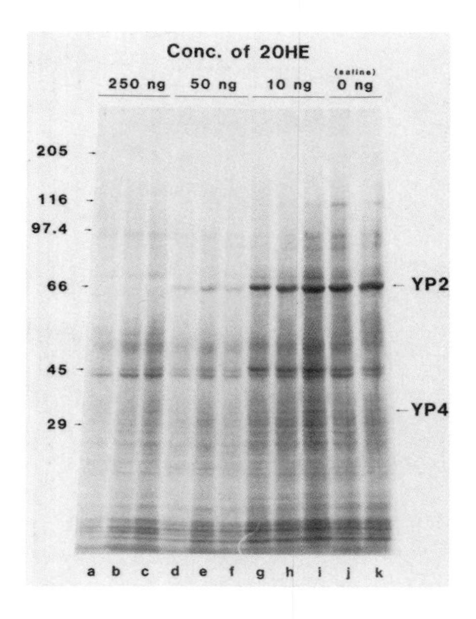

FIGURE 3. Translation products of total RNA isolated from ovaries of pharate adult females treated with 20HE. Previtellogenic females were injected with 20HE twice daily for 4 days. RNA was extracted from the ovaries with LiCl/urea (16,17) and equal A_{260} units were translated. Equal quantities of TCA precipitable radiolabeled proteins were loaded on each lane of an 8-15% gradient SDS-PAGE, and the dried gel was autoradiographed. Females were treated with 20HE at a dose of 250 ng, lanes a-c; 50 ng, lanes d-f; 10 ng, lanes g-i; and lanes j and k saline injected.

to the control of the accumulation of YP mRNA in the ovarioles. Unlike the observations on 20HE, regulation of chitin synthesis in cultured wing discs or epidermis (8-11), the synthesis of YP2 was shown to be dependent on continuous exposure of the ovarioles to 20HE and at specific concentrations. Inhibition of gene activity has been observed for other steroid hormone regulated proteins. Glucocorticoids were found to inhibit transcription of the pro-opiomelanocortin gene as measured by nuclear run-on transcription assays of nuclei isolated from hormonally treated primary cultures of rat anterior pituitary cells

(18). Similarly, the transcription of the 74 kDa serum albumin gene in Xenopus laevis was shown to be inhibited by estradiol treatment of male hepatocytes (19). However, the estradiol was found also to cause a three-fold destabilization of the albumin RNA. The combined effect of the inhibition of transcription and RNA destabilization accounted for the changes in the steady-state levels of the albumin mRNA. The nature of the effect of 20HE on the accumulation of YP mRNAs is not known. We are developing cloned copies of the genes for the YPs in order to determine the molecular action of the hormone on these genes.

ACKNOWLEDGMENT

This work was supported in part by a grant from the Department of Defense (No. DAAG 29-81-K-0124) to VJB. Mention of a proprietary product does not constitute an endorsement by USDA.

REFERENCES

1. Shirk PD, Bean D, Millemann M, Brookes VJ (1984). Identification, synthesis, and characterization of the yolk polypeptides of Plodia interpunctella. J Exp Zool 232: 87.
2. Bean D, Shirk PD, Brookes VJ (1986). Charactericätion of yolk proteins from the eggs of the Indianmeal moth, Plodia interpunctella. In press.
3. Hagedorn H H (1985). The role of ecdysteroids in reproduction. In Kerkut GA, Gilbert LI (eds.): "Comprehensive Insect Physiology, Biochemistry and Pharmacology. Vol. 8. Endocrinology II", New York, Pergamon Press, p. 205.
4. Koeppe JK, Fuch M, Chen TT, Hunt L-M, Kovalick GE, Briers T (1985). The role of juvenile hormone in reproduction. In IBID, p. 165.
5. Schwartz LM, Truman JW (1983). Hormonal control of rates of metamorphic development in the tobacco hornworm Manduca sexta. Devel Biol 90: 103.
6. Truman JW (1981). Interaction between ecdysteroids, eclosion hormone, and bursicon titres in Manduca sexta. Am Zool 21: 655.

7. Truman JW, Roundtree DB, Reiss SR, and Schwartz LM (1983). Ecdysteroids regulate the release of eclosion hormone in moths. J Insect Physiol 29: 895.
8. Oberlander H, Ferkovich SM, Van Essen F, Leach CE (1978). Chitin biosynthesis in imaginal discs cultured in vitro. Wilhelm Roux's Archiv 185: 95.
9. Mitsui T, Riddiford LM (1978). Hormonal requirements for the larval-pupal transformation of the epidermis of Manduca sexta in vitro. Devel Biol 62: 193.
10. Fain MJ, Riddiford LM (1977). Requirements for molting of the crochet epidermis of the tobacco hornworm larva in vivo and in vitro. Wilhelm Roux's Archiv 181: 285.
11. Fristrom JW, Yund MA (1980). A comparative analysis of ecdyseroid action in larval and imaginal tissues of Drosophila melanogaster. In Hoffman JA (ed.): "Progress in Ecdysteroid Research," NY: Elsevier, p. 349.
12. Hamada N, Koga K, Hayashi K (1983). Effects of a large dose of ecdysteroid in development of ovary in Bombyx pupae. J Fac Agr Kyushu Univ 28: 23.
13. Shirk PD, Bean D, Brookes VJ (1986). Suppression of vitellogenesis by 20-hydroxyecdysone in the Indianmeal moth, Plodia interpunctella. In press.
14. Oberlander H (1972). α-Ecdysone induced DNA synthesis in cultured wing disks of Galleria mellonella: Inhibition by 20-hydroxyecdysone and 22-isoecdysone. J Insect Physiol 18: 223.
15. Cleaveland DW, Fischer SG, Kischner MW, Laemmli UK (1977). Peptide mapping by limited proteolysis in sodium dodecylsulfate and analysis by gel electrophoresis. J Biol Chem 252: 1102.
16. Auffray C, Rougeon F (1980). Purification of mouse immunoglobin heavy-chain messenger RNAs from total myeloma tumor RNA. Eur J Biochem 107: 303.
17. LeMeur M, Glanville N, Mandel JL, Gerlinger P, Palmiter R, Chambon P (1980). The ovalbumin gene family: Hormonal control of X and Y gene transcription and mRNA accumulation. Cell 23: 561.
18. Ganger J-P, Drouin J (1985). Opposite regulation of pro-opiomelanocortin gene transcription by glucocorticoids and CRH. Molec Cell Endocrinol 40: 25.
19. Wolfe AP, Glover JF, Martin SC, Tenniswood MPR, Williams JL, Tata JR (1985). Deinduction of transcription of Xenopus 74-kDa albumin genes and destabilization of mRNA by estrogen in vivo and in hepatocyte cultures. Eur J Biochem 146: 489.

Molecular Entomology, pages 425–432

MICROVITELLOGENIN, A 31,000 DALTON FEMALE SPECIFIC PROTEIN[1]

John K. Kawooya, Ellie O. Osir[2] and John H. Law

Department of Biochemistry, University of Arizona
Tucson, Arizona 85721

Microvitellogenin is a 31,000 dalton female specific protein found in the hemolymph and eggs of the tobacco hornworm, *Manduca sexta*. The protein has been purified to homogeneity, and characterized extensively. Microvitellogenin is synthesized by the fat body, and is secreted into the hemolymph where it first appears 17 days before eclosion. The protein is then sequestered from the hemolymph into the eggs where it accumulates without undergoing any detectable modifications. Microvitellogenin bears no physical, chemical or immunological identity to the major egg yolk protein, vitellogenin. However, its sex specificity and its accumulation in the egg justifies the name microvitellogenin, first given to an analogous protein of the giant silkmoth, *Hyalophora cecropia*.

INTRODUCTION

An insect egg is the product of an assembly of various components synthesized by the ovary and those produced by other organs located at remote sites within the insect hemocoele. The assembly of these components in the ovary is a prerequisite for the formation of a mature insect egg. Proteins constitute a large percentage of the

[1]This work was supported by GM 29238 and PCM 8302670.

[2]Present address: Department of Zoology, University of Washington, Seattle, WA 98195.

insect egg components, and have therefore become the subject of extensive studies. The most well studied of these proteins is vitellogenin, the major egg yolk protein (1, 2, 3 and 4). In most insects, vitellogenin is synthesized by the cells of the fat body, secreted into the hemolymph, and sequestered into the ovary by an endocytotic process (5, 6). In Diptera, however, some vitellogenin is synthesized by ovary itself (7, 8). Vitellogenin forms more than half of the buffer soluble insect egg proteins (1, 2, 3). In addition to vitellogenin, several other non-sex specific hemolymph proteins that include insecticyanin, a blue biliprotein (9), arylphorin (10) and lipophorin, the major insect lipoprotein (11) are also found in the eggs of _M. sexta_. Recently, a second major female specific protein, microvitellogenin (mVg), has been discovered in the hemolymph and egg of _M. sexta_.

MICROVITELLOGENIN

During the course of isolating the insect lipophorin apoprotein, apolipophorin III (apoLp-III) from the hemolymlph of adult _M. sexta_, we observed a 31,000 dalton protein that persistently contaminated the apoLp-III preparations from the female animals, but was absent in the males (12, 13). A similar protein was also observed in the eggs of these animals. At about the same period, an analogous protein was reported in the giant silkmoth, _Hyalophora cecropia_. This protein, originally referred to as reluctin (6), was later named microvitellogenin (4). We adopted the name microvitellogenin for the 31,000 dalton protein from _M. sexta_ after observing that it was immunologically similar to that of _H. cecropia_.

Purification of Microvitellogenin.

Microvitellogenin was isolated from the hemolymph and from the egg of _M. sexta_ by several chromatographic steps (12, 15). These procedures yielded a homogeneously pure protein as ascertained by SDS-polyacrylamide gel electrophoresis (Fig. 1), isoelectric focusing and by N-terminal amino acid sequencing (15).

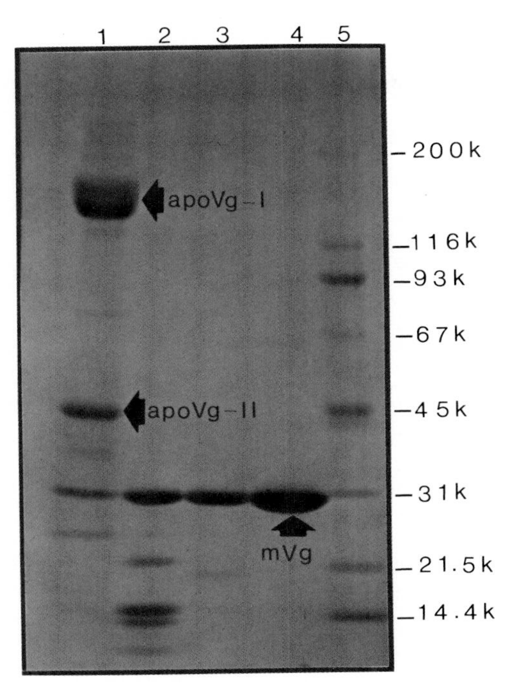

FIGURE 1. Steps for purification of microvitellogenin from the egg. 1) Buffer soluble egg proteins. 2) After gel permeation chromatography on Sephadex G-75. 3) After cation exchange chromatography on SP-Sephadex C-25. 4) After adsorption chromatography on hydroxylapatite, Biogel-HT. 5) Molecular weight standards. Abbreviations used are apoVg-I = large apoprotein of vitellogenin; apoVg-II = small apoprotein; mVg = microvitellogenin.

Physical and Chemical Properties.

The molecular weight of microvitellogenin ($M_r \simeq$ 31,000) was established by different measurements that included gel permeation chromatography on a calibrated Sephadex G-75 column, SDS-polyacrylamide gel electrophoresis and by analytical ultracentrifugation. On gel filtration, microvitellogenin co-eluted with apoLp-III $M_r \cong 17,000$, because apoLp-III has a prolate elipsoid shape which caused it to behave as a larger molecule than it really is (16). Microvitellogenin is soluble in both high and low ionic strength buffers. However, upon lyophilizing, the protein could only be solubilized in 10 mM Na-phosphate buffer containing 6M guanidinium HCl, pH 7.0. The isoelectric point of microvitellogenin was pI = 7.3. Both egg and hemolymph microvitellogenin have identical N-terminal amino acid sequence, amino acid composition, isoelectric point, immunological reaction and behavior on SDS-polyacrylamide gels, thus suggesting the protein not to undergo any detectable modification upon sequestration into the egg (15). Unlike vitellogenin, microvitellogenin is non-lipoglycosylated. The secondary structure of microvitellogenin was determined by circular dichroic measurements. The protein was found to contain 36 percent α-helix, 25 percent β-structure and 39 percent random coil, thus suggesting the protein to be globular. The CD-spectrum showed an unusually strong positive band between 220-240 nm (15) that may be attributed to the orientation of the aromatic amino acid residues along the peptide backbone (17). Although microvitellogenin and vitellogenin are both female specific proteins that are sequestered into the egg during oogenesis, the two proteins differ in the various properties shown in Table 1.

TABLE 1
PROPERTIES OF VITELLOGENIN AND MICROVITELLOGENIN FROM *M. SEXTA*

Properties	Vitellogenin[a]	Microvitellogenin
Molecular weight	$M_r \cong 500{,}000$	$M_r \cong 31{,}000$
Polypeptides	2	1
Component (%)		
Protein	84	100
Lipids	13	none
Carbohydrate	3	none
Bound phosphorous	0.6	none

[a]Data from (4).

Sex Specificity and Immunological Properties.

When antibodies directed against microvitellogenin were tested by immunodiffusion against the buffer soluble egg proteins, female hemolymph and male hemolymph, the antibodies reacted only with the egg and female hemolymph (Fig. 2). A similar experiment performed using immunoblotting method followed by autoradiography showed an identical result. In this experiment there was no reaction between microvitellogenin antibodies and vitellogenin or vitellin apoproteins (15). Another immunoblotting experiment showed microvitellogenin antibodies to react with a protein of similar electrophoretic mobility in whole hemolymph of _H. cecropia_ and _Tenebrio molitor_. These experiments showed microvitellogenin to be a female specific protein with no immunological relationship to the major egg yolk protein, vitellogenin and that microvitellogenin may be fairly wide spread among different insect species.

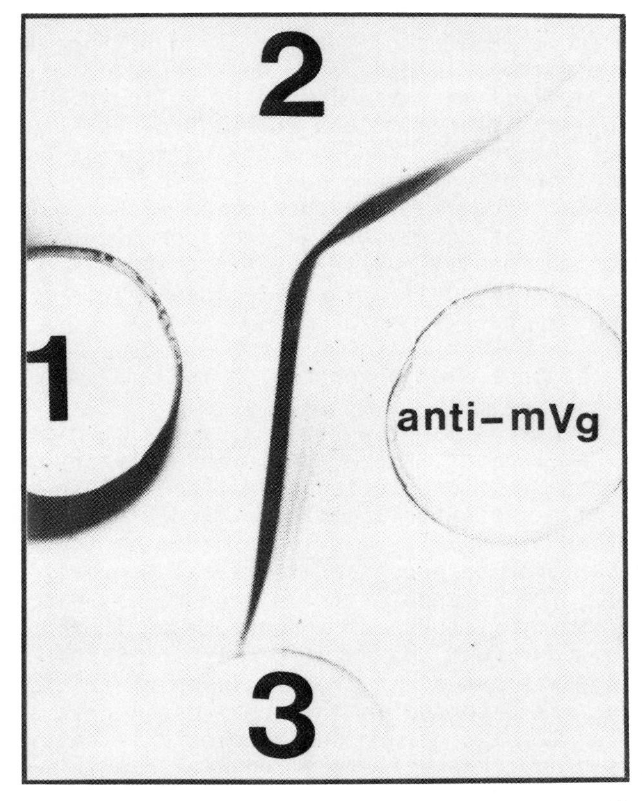

FIGURE 2. Double radial immunodiffusion. 1) Buffer soluble egg proteins. 2) Female hemolymph. 3) Male

hemolymph against antiserum to microvitellogenin.

Site of Synthesis.

The insect fat body has been established as the site of synthesis of most insect proteins including vitellogenin (1, 2, 19). In order to define the site of synthesis of microvitellogenin, we incubated the fat body (from 12 h old females) in lepidopteran saline (18) at 25°C for 6 h in the presence of ^{35}S-methionine. During this period ^{35}S-methionine was incorporated into a variety of proteins. Treatment of these proteins with microvitellogenin antibodies led to the immunoprecipitation of microvitellogenin containing ^{35}S-methionine. The immunoprecipitated microvitellogenin was subjected to SDS-PAGE, and then electrotransferred to nitrocellulose paper. The nitrocellulose paper was incubated with antibodies against microvitellogenin, followed by another incubation with Staphylococcus aureus protein A labeled with ^{125}I. The protein bands on the nitrocellulose paper were then visualized by auto- radiography (15). When a similar experiment was performed with isolated ovaries, there was no detectable incorporation of ^{35}S-methionine into microvitellogenin. Thus in M. sexta, microvitellogenin, like vitellogenin, is synthesized by the fat body.

Time of Appearance and Concentration in Hemolymph and Egg.

When hemolymph is drawn from M. sexta at different stages of development, subjected to SDS-polyacrylamide gel electrophoresis, immunoblotted on nitrocellulose membrane, incubated with microvitellogenin antibodies, followed by an incubation with ^{125}I S. aureus protein A, the membrane can then be processed to reveal microvitellogenin bands (15). The immunoblotting procedure was also used to determine the time course of appearance of microvitellogenin during M. sexta life cycle. Microvitellogenin was absent in the hemolymph of all larval stages, but was detected in the hemolymph 4 days after pupation (or 17 days before adult eclosion). A similar method was used to quantitate microvitellogenin in the hemolymph of adults and in the egg (15). The quantitative assays showed the concentration of microvitellogenin to be

5.83 ± 0.13 (n = 5) of total hemolymph protein and 66.51 ± 0.15 μg/mg (n = 5) of the total egg buffer soluble proteins. Thus there is a significant accumulation of microvitellogenin from the hemolymph into the egg. The mechanism of microvitellogenin uptake into the egg is currently being investigated in *H. cecropia* by Dr. W. Telfer and his co-workers (University of Pennsylvania, PA).

The biological function of microvitellogenin remains obscure. However, its accumulation in the egg would suggest that it is an important component of a mature insect egg. The basic data presented on microvitellogenin may provide a platform for defining the biological significance of this protein in insect reproduction.

REFERENCES

1. Engelmann F (1979). Insect vitellogenin--identification, biosynthesis and role in vitellogenesis. Adv Insect Physiol 14:49.
2. Hagedorn HH, Kunkel JG (1979). Vitellogenin and vitellin in insects. Annu Rev Entomol 24:475.
3. Kunkel JG, Nordin JH (1985). Yolk proteins. In Kerkut GA, Gilbert LI (eds): "Comprehensive Insect Physiology, Biochemistry and Pharmacology," London: Pergamon Press, p 83.
4. Osir EO, Wells MA, Law JH (1986). Chemical and immunological characterization of insect vitellogenin. Arch Insect Biochem Physiol, in press.
5. Pan ML (1971). The synthesis of vitellogenin in the cecropia silkworm. J Insect Physiol 17:677.
6. Telfer WH, Rubenstein E, Pan ML (1981). How the ovary makes yolk in Hyalophora. In Sehnal F, Zabza A, Menn JJ, Cymborowski B (eds): "Regulation of Insect Development and Behavior," Wroclaw: Wroclaw Technical University Press, p 637.
7. Heubner E, Tove SS, Davey KG (1975). Structural and functional dynamics of oogenesis in *Glossina austeni*: Vitellogenin with special reference to follicular epithelium tissue. Cell 7:535.
8. Jowett T, Postlethwait JH (1980). The regulation of yolk polypeptide synthesis in *Drosophila melanogaster* ovaries and fat body by 20-OH-ecdysone and juvenile hormone analog. Dev Biol 80:225.
9. Cherbas PK (1973). Biochemical studies of insecticyanin. Ph.D. thesis, Harvard University.

10. Osir EO, Law JH. Unpublished.
11. Kawooya JK, Osir EO, Law JH. Unpublished.
12. Kawooya JK, Law JH (1983). Purification and properties of microvitellogenin of Manduca sexta. Role of juvenile hormone in appearance and uptake. Biochem Biophys Res Commun 117:643.
13. Kawooya JK, Keim PS, Ryan RO, Shapiro JP, Samaraweera P, Law JH (1984). Insect apolipophorin III. Purification and properties. J Biol Chem 259:3680.
14. Telfer WH, Kulakosky PC (1984). Isolated hemolymph proteins as probes of endocytotic yolk formation in Hyalophora cecropia. In Engels W, Went D, Clark W (eds): "Invertebrate Reproduction," Amsterdam: Elsevier, p 81.
15. Kawooya JK, Osir EO, Law JH (1986). Physical and chemical properties of microvitellogenin, a protein from the egg of the tobacco hornworm moth, Manduca sexta. J Biol Chem, in press.
16. Kawooya JK, Meredith SC, Wells MA, Kézdy FJ, Law JH (1986). Physical and surface properties of insect apolipophorin III. J Biol Chem, in press.
17. Woody RW (1978). Aromatic side chain contribution to the far ultraviolet circular dichroism of peptides and proteins. Biopolymers 17:1451.
18. Jungreis AM, Jatlow P, Wyatt GR (1973). Inorganic composition of hemolymph of the cecropia silkmoth: Changes with diet and ontogeny. J Insect Physiol 19:225.

Molecular Entomology, pages 433–442

ACCESSORY GLAND DEVELOPMENT IN MEALWORM BEETLES[1]

George M. Happ

Department of Zoology, University of Vermont
Burlington, Vermont 05405

ABSTRACT 20-Hydroxyecdysone accelerates cell cycles in the bean-shaped accessory glands of male pupae of *Tenebrio molitor*. The hormone acts within an hour in the G_2 phase of the cell cycle *in vitro*. Its action coincides with a change in competence which becomes expressed days later. We report the isolation of a proline-rich adult-specific secretory protein that is indistinguishable from a structural protein of spermatophore.

INTRODUCTION

Insect reproduction usually involves both sexes, yet the vast majority of the research on insect reproductive physiology has emphasized chorion formation and vitellogenesis in females. Male physiology is equally important to most insect species, and disruption of male function has great potential for control of agricultural pests and disease vectors. Recently, increasing attention has been paid to male physiology, especially the endocrine regulation of maturation in testes (1) and accessory glands (2,3). Work in our laboratory concerns hormonal control of development in male accessory glands of a model insect, *Tenebrio molitor*.

Mealworm beetles have two pairs of accessory reproductive glands --- the smaller tubular accessory glands (TAGs) and the larger bean-shaped accessory glands (BAGs). Primary organogenesis of these glands takes place in the last larval instar and the two organ rudiments can

[1]This work was supported by grants AI-15662 and GM-26140 from the National Institutes of Health.

be readily located by dissection in the newly-ecdysed pupa. Over the ensuing pupal and adult stages, the TAGs and the BAGs grow in volume (100-fold and 10-fold, respectively) and in cell numbers (15-fold and 5-fold, respectively) (4-6). In the late pupa, after ecdysteroid levels have fallen, terminal differentiation begins (5,9).

20-HYDROXYECDYSONE ACCELERATES CELL CYCLES IN THE PUPA

Mitoses in the BAGs and TAGs persist through the major pupal ecdysteroid peak (6) while at the same time there is mitotic arrest in the sternal epidermis (8). This correlation suggests that ecdysteroids might stimulate cell cycling in the accessory glands while they inhibit mitoses in the epidermis. To investigate the latter possibility, BAGs were cultured _In vitro_. When glands were cultured in Landureau's medium, physiological concentrations of 20-hydroxyecdysone accelerated mitoses (7).

We decided to investigate the actions of 20-hydroxyecdysone on the cell cycle in the BAGs by a variant of the classic stathmokinetic design of Puck and Steffan (10). Glands were cultured in the presence of hydroxyurea to block the cells in early S-phase, near the G_1-S boundary. At various times, the experiment was terminated and a nuclear pellet prepared from the organs. After staining with propidium iodide, the nuclei were examined by flow cytometry. In these experimental conditions, 20-hydroxyecdysone accelerated flow of cells from G_2 into G_1. (Figure 1). Calculated doubling times were 61 hrs with hormone and 151 hrs without hormone.

In the late pupa and the adult, the cells of the BAGs are almost all in G_2. Similar G_2 arrests have been associated with increased levels of ecdysteroids in cells of _Drosophila_ (11) and epidermal cells of _Tenebrio_ (8). To determine whether G_2 arrest in the pupal BAGs requires hormone, we cultured 0 day pupal glands with and without hormone for 3 days and determined each cell fraction. Under these conditions, the majority of the cells did not arrest in G_2. In fact, addition of 20-hydroxyecdysone increased the proportion of cells in G_1 in these cultures. From doubling times and G_1:S:G_2 fractions, we calculated the length of each phase of the cell cycle. The calculations show that addition of hormone _in vitro_ shortened the length of G_2 from 94 hours to 24 hours. At what point in the cell cycle did ecdysteroid act? Since the effect persisted in the presence of hydroxyurea, 20-hydroxyecdysone acts later than early S (and therefore not in G_1). The lag between addition of hormone and the first detectable effect (intersection of the regression lines in Figure 1) was 5 hours. Our data suggest the presence of a control point 10-20 hours into G_2.

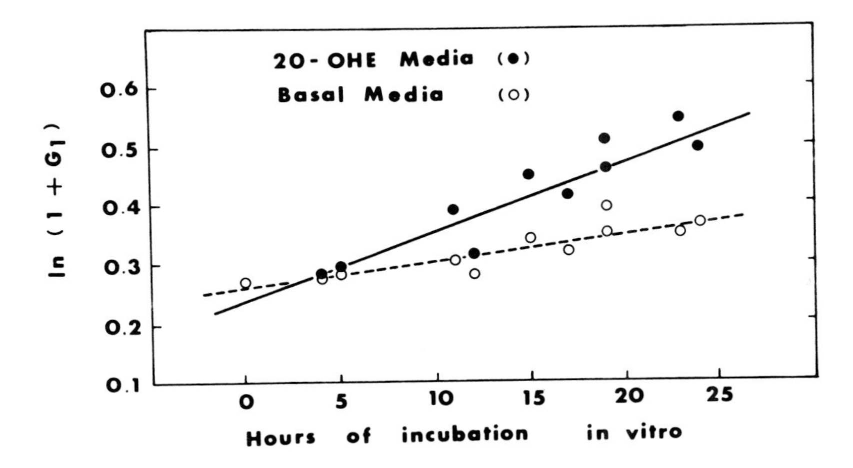

FIGURE 1. The rate at which cells from pupal BAGs enter the G_1 compartment after addition of hydroxyurea (5 mM) in basal media and 20-hydroxyecdysone-containing Landureau's S-20 media. Glands were dissected from 0 day pupae and cultured at 25° C before nuclear isolation, staining, and cytometry (6). Slopes of the regression lines are 0.0045 (r=0.862) for basal media and 0.0114 (r=0.943) for hormone-containing media.

How long must the hormone be present to produce a response? Exposures as short as 1 hour trigger a response that persists for a full day after returning the glands to basal media. In addition, 20-hydroxyecdysone did not accelerate cell cycles in the presence of α-amanitin (5×10^{-5} M) or low doses of Actinomycin D (5×10^{-9} M). Our data are consistent with a requirement for transcription before ecdysteroid accelerates the flow into G_1.

A CHANGE IN COMPETENCE COINCIDES WITH THE ECDYSTEROID PEAK

At the start of the pupal stage, each BAG consists of a coat of presumptive muscle cells surrounding an epithelial monolayer that will be secretory in the mature adult. The patterns of Coomassie-stained protein bands on SDS-gels and the patterns of ^{3}H-leucine incorporation on fluorographs differ markedly between young pupal glands and adult ones (5). Many of the adult-specific spots appear to be suitable as markers for the onset of terminal differentiation. Two particularly useful spots were those designated #24 (apparent molecular weight 24 kd, pI 6.8) and #58 (apparent molecular weight 84 kd, pI 5.2) (5).

We suspected that the pupal ecdysteroid peak which affected cell cycling might be a prerequisite for subsequent synthesis of adult-specific proteins. To test this possibility, we transplanted young pupal glands, before and after exposure to high ecdysteroid *in situ*, into adult female hosts which have low titers of 20-hydroxyecdysone. Leucine incorporation and fluorography of two-dimensional gels was used to detect the synthesis of the adult specific proteins in cultured glands.

Glands were dissected from 0, 1, and 2 day pupae (before the pupal ecdysteroid peak) or from 4 and 5 day pupae (at the time of high ecdysteroid titers), aspirated into a transfer needle, and introduced into the abdomens of 0 day female pupae or adults. When the hosts had matured to become 9 day adults, the implants were recovered by dissection and incubated *in vitro* with ^{3}H-leucine. 0 Day pupal implants placed in 0 day *pupal* hosts grew and incorporated leucine into many adult-specific spots, most notably #24 and #58. When 0, 2, or 3 day pupal glands were implanted into 0 day *adult* hosts, no incorporation of leucine into adult-specific proteins was detected nine days later. But when older pupal BAGs (4 or 5 days) were placed in 0 day adults, spots #24 and #58 were visible on the x-ray film (Figure 2). These results indicate that a reprogramming occurs at the time of the pupal ecdysteroid peak. After reprogramming, the BAGs become competent to make adult proteins when placed in an adult environment. The data from the transplantation experiments and the stathmokinetic experiments indicate that the ecdysteroid peak which acts at 4-5 pupal days to accelerate cell cycles might also cause a change in commitment. All of our morphological and biochemical data indicate that the onset of terminal differentiation does not occur until 8 pupal days or older, well after ecdysteroid levels have fallen. If we consider the time scale of the effect on mitoses and the putative affect on differentiation, it appears that ecdysteroids acting on the BAGs have two short-term effects (on cell cycling and on competence) which later lead to the long-term effects (terminal differentiation).

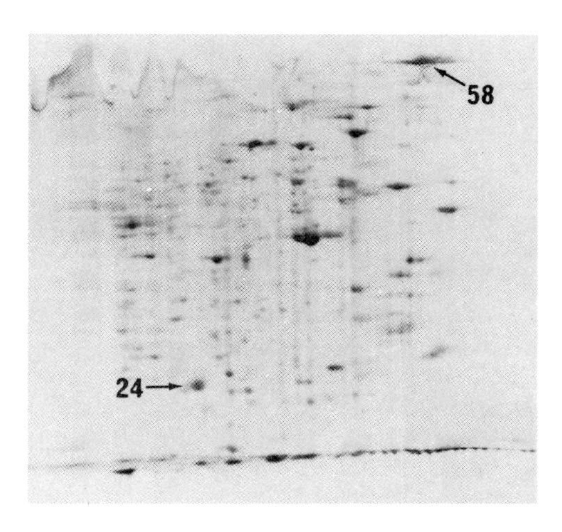

FIGURE 2 Two-dimensional fluorograph showing ^{3}H-leucine incorporation into proteins of BAGs, derived from 5 day pupae and implanted for 9 days into female adults. Incorporation took place in vitro in Landureau's S-20. Nine glands have been pooled for this gel.

ISOLATION AND CHARACTERIZATION OF A SPERMATOPHORIN

Two-dimensional fluorography after leucine incorporation is a cumbersome technique for scoring differentiation. To obtain better probes to study differentiation of the BAGs, our laboratory has prepared monoclonal antibodies (12,13). We have used these antibodies to detect antigens that are common to the spermatophore and the BAG and that also have potential as markers for differentiation. We have recently isolated one of the adult-specific antigens which promises to be a suitable marker for detecting the onset of terminal differentiation. This antigen is very similar (perhaps identical) to a structural protein of the spermatophore, a class which we term "spermatophorins".

Monoclonal antibodies were produced by immunizing Balb-c mice with homogenates of the secretory plug of the BAG and screening the hybridoma clones in an ELISA against water-insoluble fractions of the spermatophore. One of the resulting clones, designated PL21.1, recognized an antigen of 23 kd, This antigen migrated on two dimensional gels like spot #24 that we had previously used to detect terminal differentiation. With western blots of organ homogenates, we found that PL21.1 antigen is organ-specific (found only in the BAGs and not in other parts of the male tract or in the fat body) and is differentiation specific (absent from 0 and 5 day pupae, detected in 8 day pupae, and increasing rapidly thereafter). Using immunocytochemistry, PL21.1 antigen was localized in secretory granules of type 4 cells of the BAGs and in a narrow zone within the multilayered wall of the spermatophore (Fig. 3).

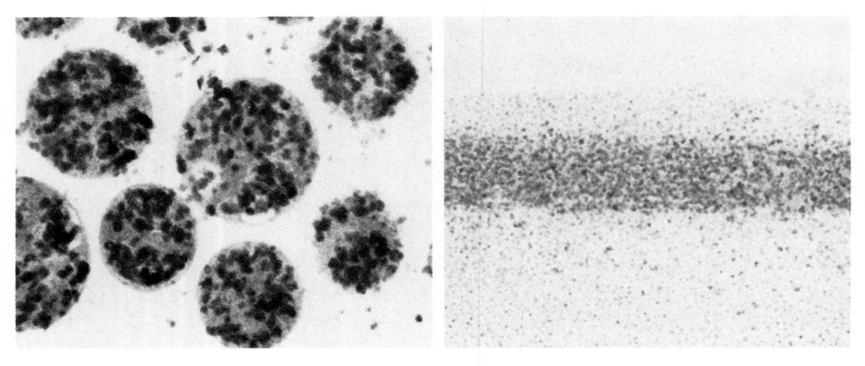

FIGURE 3. Electron immunocytochemical localization of PL21.1 after coupling with primary antibody, biotinylated secondary antibody, and avidin-biotin-peroxidase followed by diaminobenzidine (Vectastain). In type 4 cells of the BAG, the secretion is confined to the secretory granules (X34,000). In the spermatophore (right), the staining is restricted to a discrete zone of the outer wall (X18,000).

With the aid of PL 21.1 antibody, we isolated the corresponding antigen. BAGs were homogenized and the soluble fraction was applied to a Sephadex G-150 column. Fractions containing antibody were identified by dot-blotting, pooled and applied to an antibody affinity column made by coupling PL21.1 to Affigel-10. After washing thoroughly with binding buffer and high salt, a peak was eluted with 0.2 N acetic acid. The antigen was absent from the low salt and high salt washes and was present in the acid-eluted peak. When this peak was run on an SDS gel, there was a major band at 23 kd and also, a minor one at lower molecular weight. Elution of the 23 kd band from a preparative SDS gel gave a single band that was recognized by the PL21.1 antibody (Figure 4).

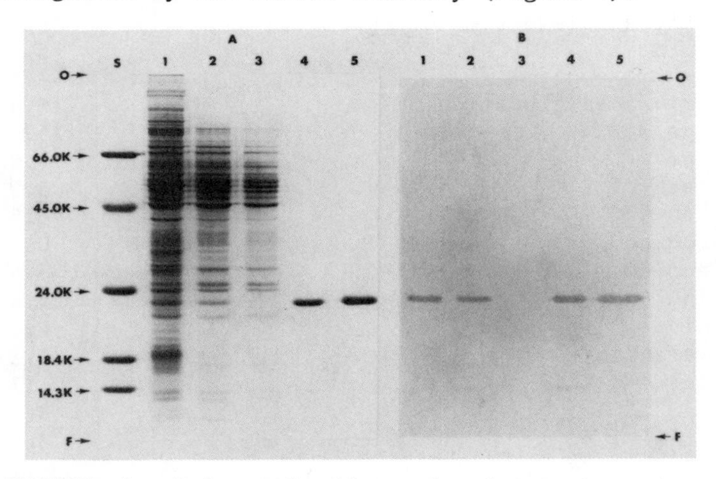

FIGURE 4 Polyacrylamide gel electrophoresis with Coomassie staining of the gel (left) and electroblotting and visualization with PL21.1 antibody, secondary antibody conjugated with peroxidase, and o-diansidine (right). S, standard. Lane 1, crude homogenate; Lane 2, positive fractions from Sephadex G-150; Lane 3, non-binding fractions from PL21.1 affinity column; Lane 4, acid eluant from PL21.1 affinity column; Lane 5, the 23 kd band from another gel (acid eluant of affinity column) which has been eluted and rerun.

Electrophoresis of homogenates on non-denaturing pore-limiting (4% - 20%) gels indicated a molecular weight for this antigen of 370 kd, suggesting that there are 16 monomers in a native protein molecule.

The amino acid composition of the isolated protein was determined after acid hydrolysis (Table 1). The antigen corresponding to PL21.1 is distinguished by its high glutamic acid/glutamine content (15.4%) and even higher proline content (25%). Methionine was absent. Preliminary sequence analysis (23 of 30 N-terminal amino acids) shows the presence of six prolines but does not suggest their involvement in an obvious repeat such as is found in collagen.

TABLE 1
AMINO ACID COMPOSITION OF PL21.1 ANTIGEN

AMINO ACID	Mol per cent*
Aspartic acid/asparagine	6.9
Threonine	3.9
Serine	3.5
Glutamic acid/glutamine	15.4
Proline	25.2
Glycine	5.7
Alanine	5.9
Valine	4.5
Methionine	0
Isoleucine	5.4
Leucine	4.0
Tyrosine	4.0
Phenyalanine	3.2
Histidine	1.7
Lysine	3.2
Arginine	3.5
Cysteic acid**	4.1

* Data from duplicate samples hydrolyzed for 24 hr in 6 N HCL *in vacuo*, 110° C.

** Duplicate samples oxidized with performic acid before hydrolysis.

PL21.1 antigen from the BAG is an unusual protein by virtue of its extraordinarily high proline content (25%). High proline content suggests a relatively rigid structure which has a low α-helical content. This characteristic may be shared by many other spermatophorins of *Tenebrio*, since it was earlier shown that hydrolysates of the wall of the spermatophore of *Tenebrio* are high in proline (14).

In its high proline content, PL21.1 antigen may bear some resemblance to the proteins of the vitelline membrane of Drosophila [18.3% proline (15)] and some cuticle proteins of cecropia [17.9% proline (16)]. Like many cuticle proteins and unlike chorion ones, methionine is absent from Sp23. Unlike cuticle proteins and like some chorion proteins, there is a relatively high content of cysteine in PL21.1 (16, 17).

Spermatophores, the most elaborate products of male accessory glands, are utilized for sperm transfer in diverse insect species, including many agricultural pests and disease vectors. Spermatophorins assemble in extracellular space to form the layers of the wall and the core of the spermatophore of Tenebrio. At least in broad outline, the assembly pattern we have seen for Tenebrio seems to be representative of other species.

According to our evidence from gel electrophoresis and immunochemistry, PL21.1 antigen from the BAGs is indistinguishable from a similar antigen in the wall of the spermatophore. Since the PL21.1 antigen from the spermatophore is a spermatophorin, this 23 kd protein should be designated Sp23. The antigen from the gland appears to be a precursor to Sp23, in the sense that vitellogenins are precursors to vitellins. Further research is required before we can judge whether Sp23 is representative of spermatophorins of Tenebrio or the analagous proteins of other species.

* * * * * *

The accessory glands of Tenebrio have proved to be convenient materials with which to demonstrate the short-term transcription-dependent effects of ecdysteroids on cell cycles. At the same time that high levels of ecdysteroids accelerate cell cycles, the BAGs become competent to mature further and thus to make adult-specific proteins. The reprogramming is expressed days later in the long-term effects - increased expression of genes for spermatophorins. With antibodies as probes, we plan to isolate the cloned cDNAs corresponding to spermatophorin messenger RNAs and to begin to study the genes for these proteins. With the aid of molecular biology, we hope to understand better the differences between the short-term and long-term effects of ecdysteroids and the linkages among them.

ACKNOWLEDGEMENTS

I thank many individuals for contributing data which are summarized in this short communication, including T. Yaginuma and H. Kai for studies on the cell cycle, H. Shinbo for the isolation and partial characterization of PL21.1 antigen, C. Yuncker for the transplantations and the electrophoresis, K. A. Grimnes for production of monoclonal antibodies, and C. S. Bricker for the immunocytochemistry.

REFERENCES

1. Dumser JB (1980). Ann Rev Entomol 25:341.
2. Happ GM (1984) In King RC, Akai H (eds): "Insect Ultrastructure" Vol 2, New York: PLenum, p.365.
3. Chen PS (1984) Ann Rev Entomol 29:233.
4. Happ GM, Happ CM (1982) J. Morphol 172:97.
5. Happ GM, Yuncker C, Dailey PJ (1982). J Exp Zool 220:81.
6. Happ GM, MacLeod BJ, Szopa TM, Bricker CS, Lowell TC, Sankel JH, Yuncker C (1985). Develop Biol 107:314.
7. Szopa TM, Lenoir Rousseaux JJ, Yuncker C, Happ GM (1985). Develop Biol 107:325.
8. Besson-Lavoignet MT, Delachambre J (1979). Develop Biol 83:255.
9. Black PN, Landers MH, Happ GM (1982). Develop Biol 94:106.
10. Puck TT, Steffan J (1963) Biophys J 3:379.
11. Stevens B, Aloareg CM, Bohman R, O'Connor JD (1980). Cell 22:675.
12. Grimnes KA, Happ GM (1986). Insect Biochem, in the press.
13. Grimnes KA, Bricker CS, Happ GM (1986). submitted to J Exp Zool.
14. Frenk E, Happ GM (1976). J Insect Physiol 22:891.
15. Petri WH, Wyman AR, Kafatos FC (1976). Develop Biol 49:185.
16. Willis JH, Regier JC, Debrunner BA (1981). In Bhaskaran G, Friedman S, Rodriguez (eds) "Current Topics in Insect Endocrinology and Nutrition" New York: Plenum p. 27.
17. Regier JC, Kafatos FC, Hamodrakas SJ (1983). Proc Nat Acad Sci 80:1043.

Molecular Entomology, pages 443–452

THE REGULATION AND ORGANIZATION OF OOTHECIN GENES IN THE COCKROACH *PERIPLANETA AMERICANA*

Richard N. Pau

Insect Chemistry and Physiology Group,
Agricultural and Food Research Council,
University of Sussex, Brighton BN1 9RQ,
England

ABSTRACT We are studying the oothecin genes of the cockroach *Periplaneta americana* to investigate the mechanism by which juvenile hormone regulates the expression of genes during reproduction. Oothecins are small structural proteins which are synthesised in the left colleterial gland of the adult female cockroach. They form the major protein components of the egg case (ootheca). Juvenile hormone, but not ecdysone, is essential for oothecin and oothecin mRNA synthesis. The coding sequences of the most abundant oothecins (C oothecins) have been determined from cDNA sequences. The derived amino acid sequences of the C oothecins are highly homologous with silkmoth chorion protein sequences. They have a tripartite structure, consisting of a short central region flanked by two very similar 'arms', whose principal feature is a repeated pentapeptide Gly-Tyr-Gly-Gly-Leu. There are about 11 abundant glycine-rich oothecins. Sequence analysis of cDNAs, translation of hybrid-selected mRNA, and cross-hybridization of cDNAs to transcripts indicate that the glycine-rich oothecin genes constitute a multigene family. Segments of chromosomal DNA containing C oothecin genes have been isolated. Two segments cloned in the vector λ EMBL4 have been partially characterized. Each contains a single oothecin C gene. One of the segments also contains a second oothecin gene which is not closely linked to the C oothecin gene. Like chorion genes, the C oothecin genes are interupted by an intron in the region which codes for the carboxy-terminal end of the signal peptide.

INTRODUCTION

Juvenile hormone plays a central role in the control of reproduction of most insects. The majority of studies on the gonotropic role of juvenile hormone have been concerned with the vitellogenesis and the synthesis of precursor yolk proteins (vitellogenins) in the fat body (1). The regulation of vitellogenin genes provides a convenient system for the analysis of juvenile hormone action at the molecular level, and this is being studied in *Locusta* (2). During the last stages of their formation in the ovary, insect eggs are invested by a chorion (egg shell) which is secreted by the follicular epithelium. After ovulation the eggs are surrounded by secretions from the oviduct or from accessory glands. These secretions serve as adhesives to attach the eggs to the surface, or as a protective coat. In cockroaches and mantids batches of eggs are surrounded by a proteinaceous egg case which is secreted by a pair of accessory (colleterial) glands which, together with the oviducts, open into a modified part of the rear end of the abdomen where the egg case is formed. In cockroaches both vitellogenesis and the synthesis of proteins in the left colleterial gland, which secretes the bulk of the egg case, is controlled by juvenile hormone (3). We have named the structural proteins of the cockroach egg case oothecins. Their synthesis provides an attractive system for investigating, in molecular terms, how juvenile hormone regulates gene expression. In this paper we summarise our studies on the juvenile hormone-induced synthesis of oothecins and on the oothecin genes in the cockroach *Periplaneta americana*.

RESULTS AND DISCUSSION

The oothecins. The structural proteins of the cockroach egg case are synthesised by highly specialized cells and secreted into the lumen of the gland tubules where they are stored until ovulation, when they are extruded from the gland and moulded around the eggs to form the egg case. The egg case consists of submicroscopic protein spheres, and calcium oxalate crystals, imbedded in a protein matrix (4). SDS gel electrophoresis reveals 12 abundant proteins in both the left colleterial gland (fig 1) and in the newly formed egg cases (5). They range in molecular weight from about 13 to 39k. Two-dimensional gel electrophoresis shows that there is a single major protein of MW 39k. This protein has a high content of valine and proline, and so is unrelated to the remaining structural proteins which are all characterized by high contents of glycine, leucine and tyrosine (4). The glycine-rich oothecins can be divided into a few major size classes with average molecular weights about 13k ,16k, 21k, and 28k. For convenient reference we have named the glycine-rich oothecin classes: A to E, and the 39k oothecin: F (fig 1). The glycine-rich oothecins of classes C and D are the major components of the egg case and form the submicroscopic spheres. The class of oothecins of molecular weight 28k (class D) is quantitatively very minor.

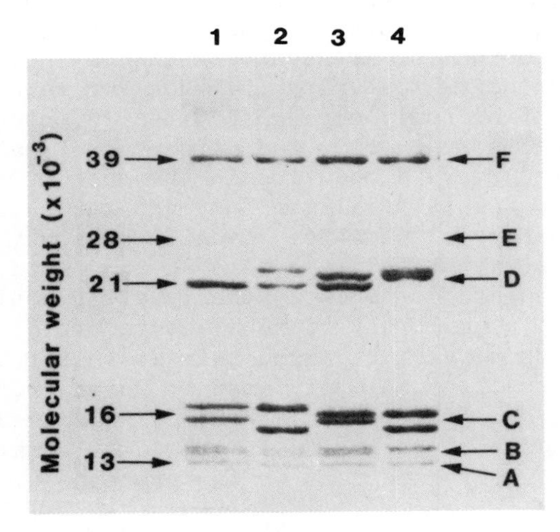

Figure 1. SDS gel electropherogram of the proteins of four individual left colleterial glands (1-4) of Periplaneta americana showing the variants of the oothecins, the various size classes, and the names we have assigned to them. Because of their low abundance class E oothecins are not visualized. The proteins are stained with Brilliant blue R.

Except for oothecin A, each class of glycine-rich oothecins consists of a number of variants which differ from each other in mobility on SDS gel electrophoresis. Individual cockroaches synthesise at least one of each of the oothecin classes, but may differ from each other in the particular variants of classes B to E which they synthesise (fig 1). Each of the oothecin variants present in an individual is secreted synchronously throughout the ovarian cycle (5).

Hormonal regulation of oothecin synthesis. Synthesis of oothecins in the left colleterial gland begins about two days after adult emergence, and the rate of synthesis increases rapidly and coordinately from the fourth to the seventh day. The first egg case is formed on the eighth day after emergence. From adult emergence to the deposition of the first egg case the profile of the rate oothecin synthesis closely parallels the profile of the rate of juvenile hormone synthesis in the corpora allata (6,7). After the first ovulation the rate of juvenile hormone synthesis is markedly cyclical and in phase with the ovarian cycles (7). Measurement of the rates of oothecin synthesis at different stages of a single ovarian cycle does not show a corresponding wide variation in the the rate of oothecin synthesis; the oothecins are synthesised continuously at a high rate after sexual maturity (5). However removal of the corpora allata from newly emerged adults prevents the induction oothecin synthesis, and allectomy of mature adults results in a very marked redution in the rate of oothecin synthesis. The effects of allatectomy are reversed by replacement therapy with either juvenile hormone, or an analogue (6). At about the same time as the rapid rise in juvenile hormone and oothecin synthesis after adult emergence there is also a peak in the titre of ecdysteroids in the haemolymph. But in ovariectomized animals this peak in ecdysteroid titre is completely absent, while protein synthesis in the left colleterial gland is induced normally (8). The synthesis of oothecins in the adult female therefore appears to be solely dependant on the presence of juvenile hormone.

By the use of a cDNA probe which, under standard hybridization conditions, hybridizes only with mRNA encoding C oothecins, we have shown that the hormone-dependent increase in the rate of C oothecin synthesis during sexual maturation is correlated with a rise in the amount of the corresponding mRNA. Allatectomy of newly emerged adults prevents the rise of the specific mRNA, and in adults results in the fall mRNA concentration (6). These results are consistent with the suggestion that juvenile hormone acts like steroid hormones by regulating the rate of specific gene transcription (9), though they do not provide direct proof of this.

Coding sequences of C oothecins. We initiated studies aimed at analysing glycine-rich oothecin genes by determining the complete coding sequence of C oothecins from the sequences of nine overlapping cDNAs (10). The 5' end of the

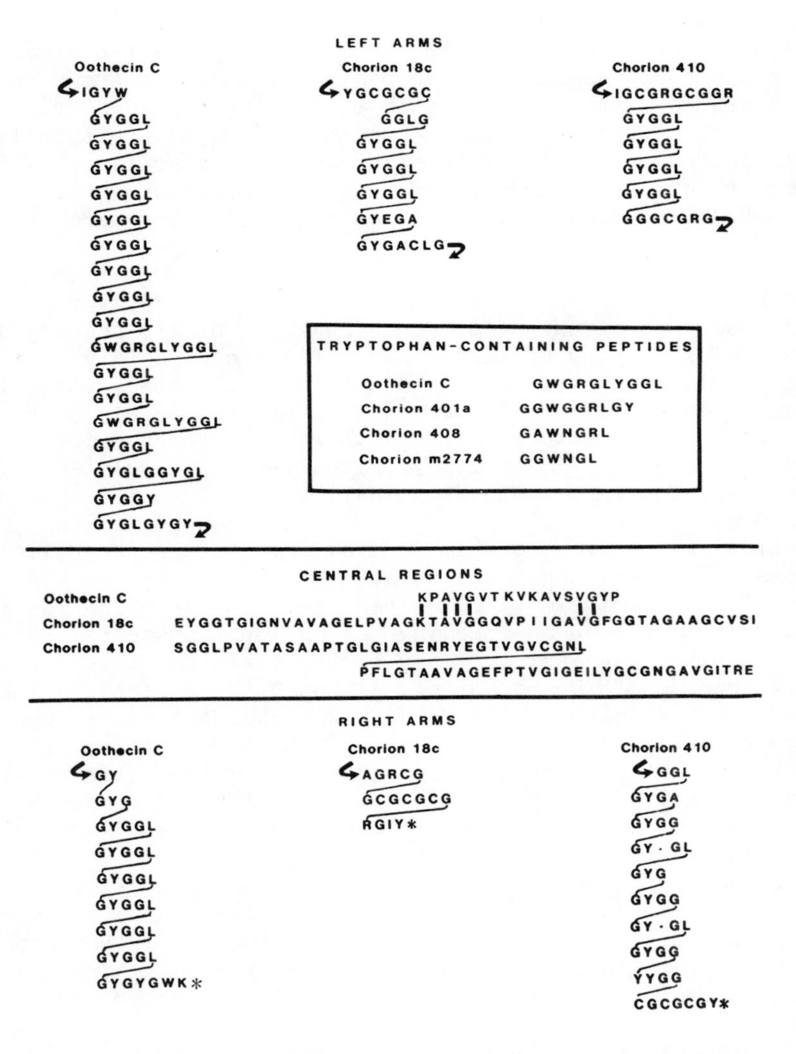

Figure 2. Comparison of the derived amino acid sequence of a C oothecin with the sequences of a silkmoth chorion protein of the A-family (18c from _Antherea polyphemus_), and of the B-family (410 from _Bombyx mori_). One-letter symbols are used for the amino acids, and the sequences read from left to right and top to bottom. The signal sequences have been omitted. The first amino acid of the oothecin is the 23rd from the amino-terminus. Common residues in the best alignment of the central regions of the oothecin and the

(Figure 2, continued)
A-family chorion protein are shown by bars. The repeated tryptophan-containing peptide of the left arm of the oothecin is compared with the tryptophan-containing peptides from the left arms a number of chorion proteins (inset). Dots indicate gaps introduced to maximise alignment and asterisks show the carboxy-termini. The data for the chorion sequences are taken from Rodakis et al (11), Tsitilou et al (12) and Kafatos (13).

coding sequence encodes a signal peptide. The derived amino acid sequence of the mature C oothecin is highly homologous with the sequences of silkmoth chorion proteins (Fig.2). The amino acid sequences of both can be divided into a central region flanked by two 'arms' of similar structure. The salient common feature in the arms is the presence of a repeated pentapeptide: Gly-Tyr-Gly-Gly-Leu, or related peptides. The left (amino-terminal) arm of both the oothecin and many chorion proteins possess tryptophan-containing peptides which are probably derived from the pentapeptide repeat (Fig. 2). Segmental mutations observed in the right arm of different cloned cDNAs result in the deletion (or insertion) of sequences coding for one or two exact pentapeptide repeats. They may be the basis for the polymorphism of the oothecins. Although the central region of the C oothecins (15 amino acids) is much shorter than the highly conserved central regions of the two families of silkmoth chorion proteins (ca. 48 and 58 amino acids for the A and B-families respectively) there is limited homology between the central regions of the C oothecin and A-family chorion proteins (fig.2). Furthermore, despite the differences in sequence in the central regions, plots of structural measurements for both are similar, and indicate a predominantly β-sheet structure. The oothecins differ from chorion proteins in the absence of Gly-Cys-containing peptides at the amino and carboxy-terminal ends of the mature proteins.

Glycine-rich oothecins are encoded by a multigene family. Evidence that the glycine-rich oothecins are coded for by a family of related genes is provided by sequence analysis of oothecin cDNAs, translation of hybrid-selected mRNAs, and the cross-hybridization of cDNAs to abundant transcripts. Sequence analysis of colleterial gland cDNAs provides examples of oothecin sequences which show the general features of the C oothecins, yet differ from them. Thus the derived amino acid sequence of cDNA 610 indicates that it codes for part of the central region and right arm of an oothecin which is different from the C oothecin. The sequence of the fragment of the central region shows limited homology with the central region of B-family chorion proteins (6). Translation of hybrid-selected mRNAs shows that a cDNA coding for the left arm of a C oothecin hybridizes with mRNAs for both C and A oothecins (6). On the other hand, under stringent conditions, a cDNA sequence which codes for the central region and right (carboxyterminal) arm of a C oothecin hybridizes with only C oothecin mRNAs of size 1.2 kb (14), but under less stringent conditions hybridizes with an abundant transcript of approximately 2.6 kb in size. The cross-hybridization of the oothecin cDNAs therefore provides us with probes for at least three classes of oothecin.

Structure and organization of oothecin genes To determine the organization and structure of oothecin genes, clones containing C oothecin genes have been isolated from libraries of *Periplaneta* chromosomal DNA cloned in a cosmid vector, TBE (15), and in the phage λ EMBL4. The results of the preliminary analysis of the clones Pa02 and Pa013 are presented here (fig.3). Each contains a single, but distinct copy of a C oothecin gene. Sequence analysis shows that both of the C oothecin genes are interupted by an intron in the region which codes for the carboxy-terminal end of the signal peptide. Silkmoth chorion genes have a single intron at the same location. The restriction sites on either side of the C oothecin genes in Pa02 and Pa013 differ, suggesting that their flanking sequences may not be homologous. To detect the presence of oothecins which belonged to different classes cDNAs were used as probes for Southern analysis under moderately stringent hybridization conditions. No oothecin gene other than the single oothecin C gene was detected in clone Pa013. In clone Pa02 the single copy of the C oothecin gene lies at least 4 kb away from another oothecin gene. This gene may be tentatively identified as coding for an A oothecin because it hybridizes with a cDNA probe which hybridizes A and C oothecins, and it lacks the EcoRV restriction site characteristic of C oothecins, Thus far our analysis suggests that the oothecin genes are not closely linked. In this respect their organization may differ from that of silkmoth chorion genes. The latter are organized in divergently transcribed pairs of non-homologous genes belonging to different gene families (A and B) which are tandemly repeated (13). The 5' ends of the paired genes are separated by short DNA segments (300 ± 50 base pairs), while the repeated pairs are seperated by segments many kilobases long.

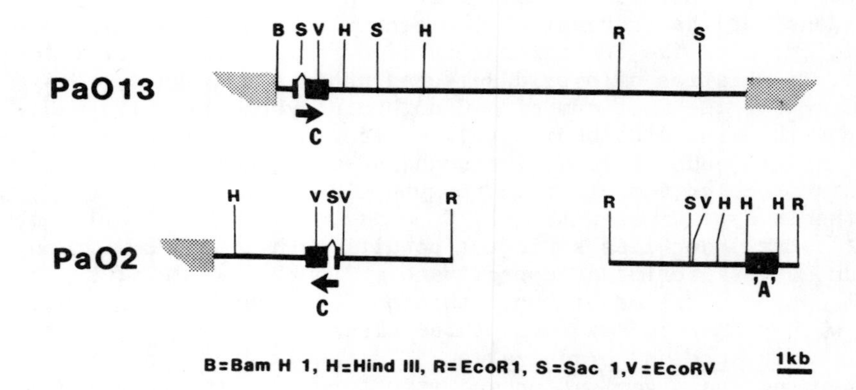

Figure 3. Maps of the cockroach DNA segment in the recombinant PaO13, and of two fragments from the recombinant PaO2. The positions of the oothecin genes (C and the putative A gene are indicated by solid bars. Vector sequences are shown by dotted areas. The positions of the introns and the directions of transcription of the C oothecins are shown.

Concluding remarks. Our preliminary characterization of two fragments of cockroach chromsomal DNA covering more than 30 kb has identified two distinct C oothecin genes and another gene which is probably an A oothecin. They provide promising material for infering presumptive regulatory elements by comparison of nucleotide sequences, as has been done by analysis of the flanking sequences of Drosophila chorion genes (16). They are also suited for determining the functional properties of their flanking DNA by in vitro modification of the DNA followed by reintroduction of the modified genes into either cultured cells, or Drosophila, via P-element mediated transfection. Such transfection between distantly related insect orders has been sucessful in the case of silkmoth chorion genes (17).

But besides the opportunities which they provide for elucidating the mode of action of juvenile hormone, the oothecin genes are also of interest because of their homologies with silkmoth chorion genes, which strongly suggest that the two gene families are related. In contrast to the oothecins, the more numerous chorion genes are expressed in a precisely timed developmental program which procedes in the absence of concurrent hormonal signals. They are also expressed in a different tissue: the follicular epithelium. Our very extensive knowledge of the structure, evolution and developmental expression of the silkmoth chorion multigene family comes from studies of the genes from two silkmoth families which diverged about 60 Myr ago (14). Cockroaches antedate moths about 200 Myr. Further analysis of the structure and organization of oothecin genes therefore promises new insights into the evolution of a superfamily of genes which, in the course of evolution, have changed both tissue and developmental specificity.

ACKNOWLEGEMENTS

I am grateful to K Edwards-Jones for assistance, E. Matsakis for his work on the isolation of the lambda clones, and R.J. Weaver for studies on the regulation of oothecin synthesis.

REFERENCES

1. Englemann F (1983). Vitellogenesis controlled by juvenile hormone. In Downer RGH, Laufer, H (eds): "Endocrinology of Insects," New York: Alan R. Liss, p 259
2. Wyatt GR, Kanost MR, Locke J, Walker VK (1986). Juvenile hormone and the regulation of gene expression in insect fat body. Archives Insect Biochem Physiol: in press 1:suppl 1
3. Scharrer B (1946). The role of the corpora allata in the development of Leucophaea maderae (Orthoptera). Endocrinology 38:35
4. Pau RN, Brunet PCJ, Williams MJ (1971). The isolation and caracterization of proteins from the left colleterial gland of the cockroach Periplaneta americana (L.). Proc. Roy. Soc. Lond. B 177:565
5. Weaver RJ, Pau RN (1986). Absence of sequential or cyclic synthesis of oothecal iso-proteins during the reproductive cycle of female Periplaneta americana Manuscript in preparation.
6. Pau RN, Weaver RJ, Edwards-Jones K (1986). Regulation of cockroach oothecin synthesis by juvenile hormone. Archives Insect Biochem Physiol: in press 1: suppl 1
7. Weaver RJ, Pratt GE (1977) The effect of enforced virginity and subsequent mating on the activity of the

the corpus allatum of *Periplaneta americana* measured *in vitro*, as related to changes in the rate of ovarian maturation. Physiol Entomol 2:59

8. Weaver RJ, Strambi A, Strambi C (1984) The significance of free ecdysteroids in the haemolymph of adult cockroaches. J Insect Physiol. 30:705
9. Wyatt GR, Dhadialla TS, Roberts PE (1984). Vitellogenin synthesis in locust fat body: juvenile hormone-stimulated gene expression. In Hoffmann J, Porchet M (eds): "Biosynthesis, Metabolism and Mode of Action of Invertebrate Hormones." Berlin Heidelberg: Springer-Verlag, p 475
10. Pau, RN (1986). Sequence analysis of cockroach oothecin cDNAs: Homology between oothecins and silkmoth chorion proteins. Manuscript in preparation.
11. Rodakis GC, Moschonas NK, Kafatos FC (1982). Evolution of a multigene family of chorion proteins in silkmoths. Mol Cell Biol 2:554
13. Tsitilou SG, Rodakis GC, Alexopoulou M, Kafatos FC, Ito K, Iatrou K (1983). Structural features of B family chorion sequences in the silkmoth *Bombyx mori*, and their evolutionary implications. EMBO J 2:1845
14. Kafatos FC (1983). Structure, evolution and devolpmental expression of the chorion multigene families in silkmoths and *Drosophila*. In Subtelny S, Kafatos FC (eds): "Gene Structure and Regulation in Development. 41st Symp Soc Dev Biol." New York: Alan R. Liss, p 33.
14. Pau RN (1983). Cloning of cDNA for a juvenile-hormone regulated oothecin mRNA. Biochim Biophys Acta 782:422
15 Grosveld FG, Lund T, Murray EJ, Mellor AL, Dahl HHM, Flavell RA (1982) The construction of cosmid libraries which can be used to transform eucaryotic cells. Nucl Acids Res 10:6715
16. Wong Y-C, Pustell J, Spoerel N, Kafatos FC (1985). Coding and potential regulatory sequences of a cluster of chorion genes in *Drosophila melanogaster*. Chromosoma 92:124
17. Mitsialis SA, Kafatos FC (1985). Regulatory elements controlling chorion gene expression are conserved between flies and moths. Nature 317:453

Molecular Entomology, pages 453–462

CELL AUTONOMOUS AND HORMONAL CONTROL OF SEX-LIMITED GENE EXPRESSION IN DROSOPHILA[1]

Douglas R. Cavener, Michael Murtha, Christopher Schonbaum, and David Hollar[2]

Department of Molecular Biology, Vanderbilt University
Nashville, Tennessee 37235

ABSTRACT We have studied the effect of the temperature sensitive *ecdysoneless-1* (*ecd-1*) mutant in *Drosophila melanogaster* on the expression of four genes: glucose dehydrogenase (GLD), alcohol dehydrogenase (ADH), sorbitol dehydrogenase (SODH), and alphaglycerophosphate dehydrogenase (GPDH). For three of these enzymes, the activities are affected in a sex-limited manner. GLD and SODH expression is depressed in *ecd-1* males at the restrictive temperature. In contrast ADH mRNA and enzyme expression is dramatically increased in *ecd-1* females at the restrictive temperature. Temperature shift-up and shift-down experiments indicate that the *ecd-1* effect on GLD is nearly irreversible whereas the effect on ADH is considerably more reversible. Imaginal disc transplantation experiments show that the *ecd-1* effect on GLD is tissue autonomous.

INTRODUCTION

Sterility is the only apparent adult phenotype associated with two ecdysteroid deficient mutants in *Drosophila* (1,2). Adult viability is not visibly affected in these mutants. This suggests that a major function of ecdysteroids at the adult stage is to modulate genes exhibiting sex-limited patterns of expression. We have chosen to examine the effect of ecdysterone on the adult

[1]This work was supported by NSF Grant PCM-8417724.
[2]Present address: Department of Genetics, North Carolina State University.

expression of four gene-enzyme systems which represent a range of sexual dimorphism. Glucose dehydrogenase (GLD) is expressed in both sexes during development but becomes male limited at the adult stage (3). GLD is secreted by the male ejaculatory duct and is subsequently transferred to females during copulation. Sorbitol dehydrogenase is largely restricted to male reproductive organs but is also found in the thoracic muscles of both sexes (4). A more modest sex difference is observed for alcohol dehydrogenase (ADH) (5). Finally, alphaglycerophosphate dehydrogenase (GPDH) activities are equivalent in males and females (6). Prior to the studies described herein, no evidence had been presented that any of these four enzymes are influenced by ecdysterone. In order to examine the possible effect of ecdysterone on these four enzymes we have used the ecdysoneless-1 (ecd-1) temperature sensitive mutant which conditionally expresses very low levels of ecdysteroids at specific times during development and at the adult stage (1). Both sexes of the ecd-1 mutant are sterile at the restric- tive temperature. In addition, we have tested for general hormonal influences on the expression of GLD via the method of imaginal disc transplantation (7).

RESULTS

In all the experiments with the ecd-1 mutant described below the flies were reared at the permissive temperature (22°C) until the pharate adult stage. Test flies were then shifted to the restrictive temperature (29°C) while control flies were retained at the permissive temperature. The MW1 wild type strain and the st ca strain (the strain from which the ecd-1 mutant was derived) were used as non mutant controls for these experiments. GLD, SODH, ADH, and GPDH activity assays were performed on homogenates of whole flies using standard spectrophotometric analyses.

The most remarkable finding of these experiments was that GLD activities were substantially reduced in adult ecd-1 males incubated at 29°C in comparison with those incubated at 22°C (Fig. 1). By 360 hours after eclosion a 3-4 fold difference in GLD activity was observed between wild type males and ecd-1 males incubated at 29°C. A similar but less dramatic effect of ecd-1 on the expression of male SODH was observed (Fig. 2). Initially, the SODH activity differential between ecd-1 males incubated at 29°C and the three other controls was approximately 70%. This

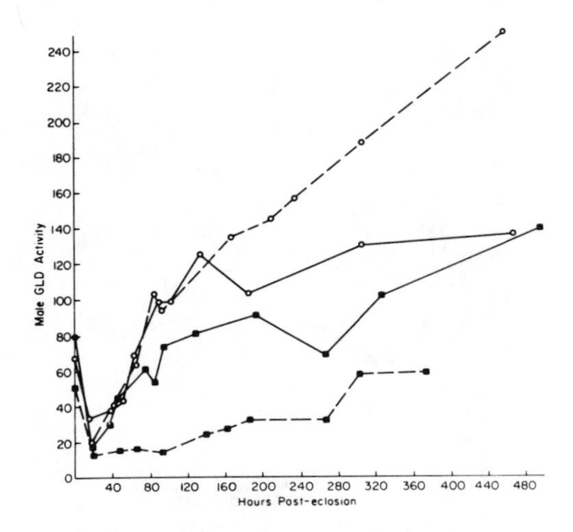

FIGURE 1. Adult male GLD activity profile. Open circles: MW1 wild type strain; closed squares: ecd-1 strain; solid lines: 22°C; dashed lines: 29°C. GLD activity units are expressed as micromoles of DCIP reduced min^{-1} $individual^{-1}$ $x\ 10^{-5}$.

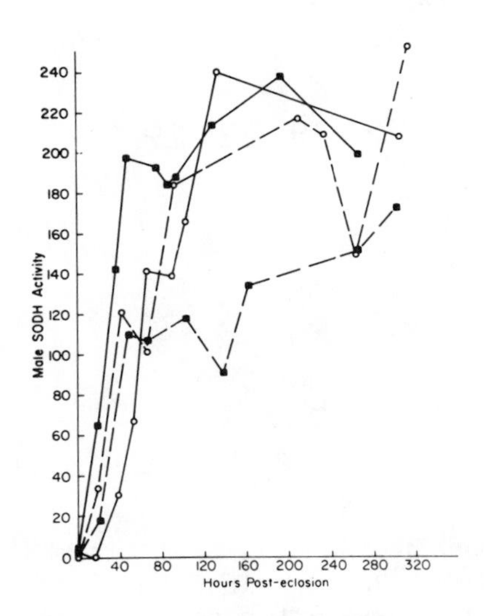

FIGURE 2. Adult male SODH activity profile. Open circles: MW1 wild type strain; closed squares: ecd-1 strain; solid lines: 22°C; dashed lines: 29°C. SODH activity units are expressed as micromoles of NAD^{+} reduced min^{-1} $individual^{-1}$ $x\ 10^{-5}$.

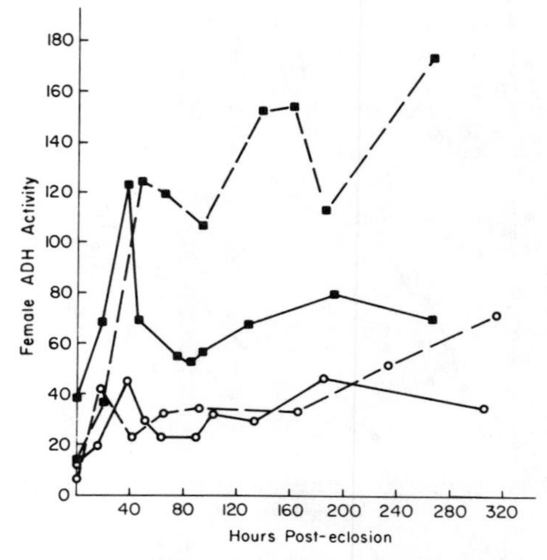

FIGURE 3. Adult female ADH activity profile. Open circles: MW1 wild type strain; closed squares: ecd-1 strain; solid lines: 22°C; dashed lines: 29°C. ADH activity units are expressed as micromoles of NAD^+ reduced min^{-1} $individual^{-1}$ $x\ 10^{-4}$.

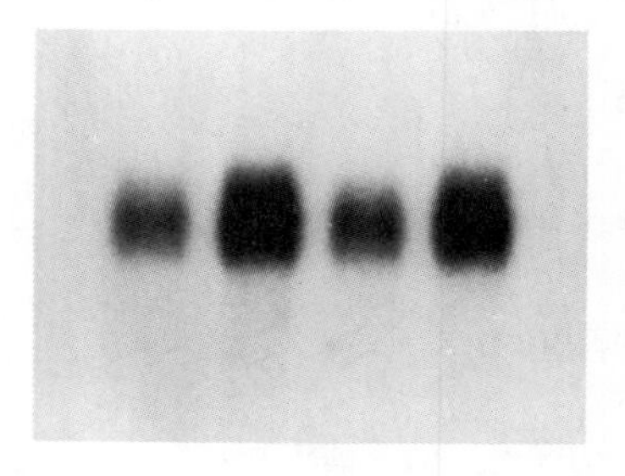

FIGURE 4. Northern blot analysis of Adh mRNA levels in ecd-1 females. Total RNA was isolated (8) from ecd-1 females raised for 9 days after eclosion at the indicated temperature (°C). Three micrograms of total RNA were loaded per lane. The filter was probed with an *in vitro* synthesized complementary RNA specific for Adh.

activity differential declined with time. SODH activity in females is unaffected in the *ecd-1* mutant (data not shown). In contrast to the male limited effects of *ecd-1* on GLD and SODH, a female limited effect was shown for ADH activity and in the opposite direction (Fig. 3). That is, at the restrictive temperature ADH activity was observed to be 2-3 fold higher than the three controls. This elevated ADH activity is observed in the head, thorax, and abdomen implying that the increase is not specific to a reproductive tissue (data not shown). ADH mRNA levels displayed a corresponding pattern (Fig. 4). Densitometry of the ADH Northern indicates that ADH mRNA is approximately 2.6 fold higher in *ecd-1* females incubated at the restrictive temperature.

Temperature shift experiments were performed on *ecd-1* flies in order to see if the observed effects on GLD in males and ADH in females were reversible. *Ecd-1* males incubated for the first four days after eclosion at the restrictive temperature did not regain normal levels of GLD activity after being shifted to the permissive temperature for another four days (Table 1).

TABLE 1
GLD ACTIVITY[a] OF TEMPERATURE SHIFTED *ECD-1* FLIES

Incubation Temperature and Time	Activity (SD)
22°C, 4 days	30.74 (3.67)
29°C, 4 days	9.86 (0.65)
22°C, 4 days; 29°C, 4 days	24.10 (2.21)
29°C, 4 days; 22°C, 4 days	16.96 (1.59)
22°C, 8 days	28.35 (2.51)
29°C, 8 days	12.62 (0.79)

[a]micromoles of DCIP reduced min^{-1} mg^{-1} of soluble protein

In the reciprocal experiment (4 days permissive temperature - 4 days restrictive temperature) it was found that GLD activity was not substantially less than the activity observed in males reared for the entire eight days at the permissive temperature. Both shift-up and shift-down experiments have been repeated using two different incubation periods (data not shown). The results of all of the temperature shift experiments support the interpretation that the effect of ecd-1 on male GLD activity is nearly irreversible. However, this interpretation for the shift-up experiments is somewhat complicated by the unknown stability of the GLD protein. The elevated ADH activity observed in female ecd-1 flies incubated at the restrictive temperature is reversed to a much greater extent (Table 2). Furthermore, the reciprocal experiment demonstrates that the increase of ADH activity in ecd-1 females can occur several days after eclosion.

TABLE 2
FEMALE ADH ACTIVITY[a] OF ECD-1
TEMPERATURE SHIFTED FLIES

Incubation Temperature and Time	Activity (SD)
22°C, 4 days	267.52 (9.04)
29°C, 4 days	573.38 (81.35)
22°C, 4 days; 29°C, 4 days	435.78 (32.58)
29°C, 4 days; 22°C, 4 days	409.66 (27.77)
22°C, 8 days	279.65 (40.17)
29°C, 8 days	686.05 (66.10)

[a]micromoles of NAD^+ reduced min^{-1} mg^{-1} of soluble protein

In order to test the tissue autonomy of the ecd-1 effect upon GLD expression in the male ejaculatory duct we performed imaginal disc transplantation experiments (7) using ecd-1 genital imaginal discs and Gld null mutant hosts. (The ejaculatory duct is one of the three major somatic reproductive organs derived from the male genital imaginal discs.) Genital discs were transplanted during the third instar larval stage. At the pharate adult stage the test animals were shifted to the restrictive temperature and their GLD activity assayed seven days later. The results of these experiments clearly indicate that the effect on GLD activity is autonomous to the ecd-1 derived reproductive tissues regardless of the genetic milieu of its host (Table 3).

TABLE 3
MALE GLD ACTIVITY[a] OF IMAGINAL DISC TRANSPLANTATION EXPERIMENTS

Disc Donor	Host	Temperature (7 days)	Activity (SD)[b]
none	ecd-1	22°C	406.69 (93.52)
none	ecd-1	29°C	140.03 (30.14)
none	Gld null	22°C	16.65 (10.03)
ecd-1	Gld null	22°C	450.10 (170.0)
ecd-1	Gld null	29°C	149.84 (52.55)

[a]micromoles of DCIP reduced min^{-1} mg^{-1} of soluble protein
[b]Each value is the mean of four or more individually assayed males.

DISCUSSION

The major role of ecdysterone is to initiate molting and metamorphosis. Indeed, the _ecd-1_ mutant was selected on the basis of its temperature sensitive inability to pupariate (1). Ecdysterone titers were subsequently found to be extremely low in the _ecd-1_ mutant at the restrictive temperature. Germane to our studies was the finding that _ecd-1_ adults exhibit low ecdysterone titers and low fertility (1,9). Thus, it appears that the _ecd-1_ mutation affects a wide range of processes during development but is restricted to sex related functions at the adult stage. Whether these effects are due to the observed low titers of ecdysterone is still vigorously debated (9,10,11). Moreover the _ecd-1_ gene product and its molecular function have not been identified.

The remarkable feature of our data is that three enzymes (GLD, ADH, and SODH) are affected only in one sex in each case. In addition these three enzymes normally exhibit some quantitative level of sexually dimorphic expression whereas the single unaffected enzyme (GPDH) shows no sexual dimorphism. The _ecd-1_ effect on GLD and SODH occurs only in males whereas the effect on ADH occurs only in females and in the opposite direction. However, the net result of the mutant state of _ecd-1_ is to greatly diminish the normal quantitative sex differences which occur for all three enzymes. An intriguing dissimilarity between the normal expression of GLD, SODH, and ADH in males is that for the former two enzymes the exclusive or higher level of expression occurs in sex-limited tissues whereas the normally elevated ADH expression occurs in non sex limited tissues (almost certainly the fat body). From these considerations we deduced a series of simple hypotheses. Because ADH is expressed in both female and male fat bodies, we propose that ecdysterone plus perhaps other unknown factors repress ADH expression in female fat bodies. The yolk proteins provide examples of fat body genes which are regulated differently depending on both sex and ecdysterone concentration (reviewed in 12). Secondly, we speculate that ecdysterone is directly or indirectly required as a positive regulator of GLD and SODH expression.

Although the experiments that we have conducted at the adult stage do not directly test the hypothesis that the observed _ecd-1_ effects are due to a deficiency in ecdysterone, the temperature shift and imaginal disc transplantation experiments critically bear on this hypothesis. If the observed effects of _ecd-1_ on the three enzymes are due to a simple and direct interaction of the genes encoding them and the ecdysterone-receptor complex, then the effects should be reversible. Thus, the temperature shift data for ADH activity is consistent with this model. However, the temperature shift data for GLD expression are clearly inconsistent with this model. In addition the results of the imaginal disc transplantation experiments for GLD activity indicate that the _ecd-1_ effect is tissue autonomous, again a result inconsistent with the model of GLD expression controlled by circulating ecdysterone. An important caveat to this latter interpretation is that insect reproductive organs have recently been implicated as sites of steroid hormone production (David Denlinger et al.; this volume). Thus for example, the ejaculatory duct may normally receive its sole source of ecdysterone from the paragonia (which is also a genital disc derivative and empties into the ejaculatory duct). If this is true, then transplantation of an _ecd-1_ genital disc into a normal host would not relieve the ecdysterone deficiency in the ejaculatory duct. We have also found that the male _melanogaster_ ejaculatory duct can differentiate normally and express GLD in female hosts of _Drosophila_ _melanogaster_ and _pseudoobscura_ (13). Thus the expression of GLD in the ejaculatory duct does not require any species or sex specific circulating hormone for its expression.

Recently we have embarked upon a series of experiments to test the hypothesis that ADH and GLD are regulated by ecdysterone throughout the life cycle of Drosophila. We have found that ADH and GLD activity and mRNA levels are affected by ecd-1 at the third instar larval stage (unpublished data). These effects are reversible by the application of exogenous ecdysterone. It is interesting to note that this ecdysterone treatment on larvae results in a stimulation of GLD expression and a repression of ADH expression, consistent with our models stated above for the role of ecdysterone on these two enzymes at the adult stage.

ACKNOWLEDGMENTS

We thank Dr. Paul Macdonald, Harvard University, for supplying us with a clone of the Adh gene. We thank Cindy Rae Young for preparing this manuscript.

REFERENCES

1. Garen A, Kauvar L, Lepesant J (1977). Roles of ecdysone in Drosophila development. Proc Natl Acad Sci 74:5099.
2. Wilson TG (1980). Studies on the female-sterile phenotype of 1(1)su(f) ts76a, a temperature-sensitive allele of the suppressor of forked mutation in Drosophila melanogaster. J Embryol Exp Morphol 55:247.
3. Cavener DR, MacIntyre RJ (1983). Biphasic expression and function of glucose dehydrogenase in Drosophila melanogaster. Proc Natl Acad Sci 80:6286.
4. Bishoff WL (1978). Ontogeny of sorbitol dehydrogenase in Drosophila melanogaster. Biochem Genet 16:485.
5. Oakeshott JG (1976). Selection at the alcohol dehydrogenase locus in Drosophila melanogaster imposed by environmental ethanol. Genet Res Camb 26:265.
6. Rechsteiner MC (1970). Drosophila lactate dehydrogenase and α-glycerophosphate dehydrogenase: Distribution and change in activity during development. J Insect Physiol 16:1179.
7. Ursprung H (1967). In vivo culture of Drosophila imaginal discs. In Wilt FH, Wessells MK (eds): "Methods in Developmental Biology," New York: Crowell, p 485.

8. Chirgwin JM, Przybyla AE, MacDonald RJ, Rutter WJ (1979). Isolation of biologically active ribonucleic acid from sources enriched in ribonuclease. Biochemistry 18:5294.
9. Klose W, Gateff E, Emmerich H, Beikirch H (1980). Developmental studies on two ecdysone deficient mutants of Drosophila melanogaster. Wilhelm Roux's Arch 189:57.
10. Redfern CPF, Bownes M (1983). Pleiotropic effects of the ecdysoneless-1 mutation of Drosophila melanogaster. Mol Gen Genet 189:432.
11. Sliter TJ, Bryant PJ (1982). Autonomous effects on imaginal discs caused by the ecd-1 mutation of Drosophila. Amer Zool 22:928.
12. Bownes M (1982). Hormonal and genetic regulation of vitellogenesis in Drosophila. Quart Rev Biol 57:247.
13. Cavener DR (1985). Coevolution of the glucose dehydrogenase gene and the ejaculatory duct in the genus Drosophila. Mol Biol Evol 2:141.

VI. APPLICATIONS OF MOLECULAR ENTOMOLOGY

Molecular Entomology, pages 465–468

WORKSHOP SUMMARY: POTENTIAL FOR MOLECULAR BIOLOGICAL CONTROL OF INVERTEBRATES

Joel B. Kirschbaum

CODON Corp.
430 Valley Dr.
Brisbane, CA 94005

The goal of this 2-hour workshop was to provide a forum for exploring scientific and technological concepts central to developing molecular biological strategies for insect control. The designated participants represented university, governmental and industrial research laboratories, and the areas they presented covered genetic engineering of insects, microbial agents, and plants.

Insects

In selected cases, germline transformation of insects might be profitably used to suppress natural populations by the production and release of suitably altered insects. One obvious example would be developing genetic sexing schemes to produce pure populations of males that could be rendered sterile, by a variety of techniques, prior to release in the environment (S. Miller-USDA/ARS). In principle, germline transformation could be used to introduce a conditional lethal gene under the control of a female-specific promoter, such that females would be eliminated from the transformed population by rearing the insects on a selective medium. The resulting population of pure males could then be sterilized using current X-ray procedures. Although X-irradiation has been used in sterile male-release programs for many years, it has become evident that X-irradiation suffers from a major drawback; namely, X-rays will not only induce sterility, but also can introduce secondary mutations that impair the ability of the sterile males to copulate. Thus a second application of germline transformation could be to introduce one or more additional genes, whose expression would interfere with male gonad development or spermatogenesis. An anti-sense RNA approach might be applicable in this situation.

The success of such approaches to controlling insect populations clearly not only depends on identifying the necessary genes and promoters to drive them, but also on developing methods for introducing the requisite genetic constructs into the insect germline. The identification and characterization of transposable elements in target insects could lead to the development of such specific transformation systems. In the _Anopheles gambia_ complex of mosquitoes, crosses between certain members of the complex are known to lead to sex-ratio distortions and genetic inversions, suggesting the existence of transposable elements. Recent molecular studies within this complex (H. Hagedorn & R. MacIntyre-Cornell) have lead to the cloning of a 14.6 kb fragment that by both Southern analysis and _in situ_ hybridization contains sequences located at several different positions within the mosquito genome. This fragment in turn contains three 1.4 kb repeats that are currently being sequenced to determine what features they have in common with transposable elements. Additional studies will be required to establish whether these regions can be induced to transpose via genetic crosses.

Microbial Agents

The microbial agents in use today were originally obtained as natural isolates purified from moribund insects in the field. Molecular biology affords the possibility of not only improving these natural pathogens to better suit their targets, but also of converting a nonpathogenic microbe to a pathogen, thus creating new weapons for the arsenal against insect pests (D. Miller-Genetics Institute). Designing effective microbial pathogens requires an intimate understanding of the organism itself, its mode of action, and how it interacts with the target insect and environment.

Two groups of microbial insect control agents that are currently receiving considerable attention from the molecular biology community are: 1) baculoviruses, which are active on Lepidoptera, and 2) varieties of the bacterium _Bacillus thuringiensis_ (BT) which, depending on the specific BT variety, synthesize a protein toxin (single gene product) specific for either Lepidoptera, Coleoptera, or Diptera. Although baculoviruses are extremely potent, under normal conditions they are slow to kill the insect. Faster kill is required and might be achieved by engineering the virus to synthesize a toxin molecule early in infection that would quickly incapacitate and kill the insect. A variety of insect-specific toxins exist that might be suitable candidates including, perhaps, the BT toxin itself.

The genes for several BT toxins have been cloned, sequenced, expressed in bacteria, and the minimum genetic sequence required to encode a fully active toxin protein has been defined. The stage is now set for ultimately engineering more potent BT toxins, so that smaller doses can be used in the field in order to offset the currently high usage cost of the product. Also, several groups reported during this meeting having transferred a BT toxin gene into non-pathogenic *Pseudomonas*, and thus converting it to a new microbial agent active against Lepidoptera.

Plants

Engineering plants for insect resistance could one day be a safe and inexpensive alternative to using toxic chemicals for insect control (K. Barton-Agracetus). The development of smaller Ti plasmid vectors containing multiple dominant selectable markers, cloning sites, efficient promoters and polyA signals has greatly streamlined *Agrobacterium*-mediated plant transformation. Additional techniques including microinjection, electroporation, and particle acceleration will most likely make possible the transformation of plants which have thus far resisted transformation by the *Agrobacterium* approach. In addition to transformation *per se*, another formidable challenge will be learning how to regenerate mature plants from transformed cells or protoplasts, in those cases that have to date been refractory.

At the present time, sufficiently good model plant systems exist that can be used to test genetic engineering strategies for insect control. Several groups at this meeting reported the successful introduction of a BT toxin gene into tobacco, resulting in otherwise normal plants resistant to tobacco hornworm. Chitinases are another family of genes that might be introduced into plants to control nematodes and insects, by virtue of the ability of these enzymes to degrade integral protective structures within these organisms. Inserting genes into plants that encode protein inhibitors of insect digestive enzymes is another approach to population control (D. Foard-Bergamo, Italy). One such candidate might be the 70 amino acid Bowman-Birk trypsin inhibitor found in soybeans.

Also, one should not exclude the possibility of engineering plants to produce secondary metabolites lethal or repulsive to insects (D. Jones-U. of Kentucky). In selected cases, engineering a plant by adding 1-2 additional biosynthetic steps to a pre-existing pathway in the plant could result in the synthesis of a protective metabolite. By way of illustration, nicotine is insecticidal but the tobacco hornworm is resistant to the levels of nicotine produced by cultivated tobacco. However, a wild-type relative of the commercial tobacco plant produces a single acetylated derivative of nicotine that is 1000 times more active and is lethal to the hornworm. Screening wild-type relatives of a cultivated plant for resistance to an insect to which the cultivated plant is susceptible could lead to the identification of those additional genes that could be used to develop protection in commercial plants.

Molecular Entomology, pages 469–478

THE TOBACCO-INSECT MODEL SYSTEM FOR GENETICALLY ENGINEERING PLANTS FOR NON-PROTEIN INSECT RESISTANCE FACTORS[1]

Davy Jones, Joseph Huesing, Ernö Zador, and Craig Heim

Department of Entomology, University of Kentucky, Lexington, Ky 40546

ABSTRACT Experiments are described which lead toward the isolation of a gene for an enzyme which acylates nornicotine, a major alkaloid of the cultivated species *N. tabacum*, which is not toxic to *Manduca sexta*. Transfer of the gene for this enzyme from wild *N. stocktonii* into *N. tabacum* should enable the latter to synthesize N-acylnornicotine, which is highly toxic to *M. sexta*.

INTRODUCTION

Fundamental research on the basis for host plant resistance (HPR) to insects has contributed susbstantially to development of concepts in basic disciplines as well as those of agricultural importance. Traditionally plant breeding has not been concerned with the biochemial basis for resistance, relying instead on empirical approaches (12). Possible benefits of understanding the resistance have been articulated with respect to avoiding selection for pest biotypes (1) or minimizing incompatiability of HPR with biological control (3,15) and insecticides (18). However, examples of deliberate commercial implementation of the knowledge of the biochemical basis of plant resistance to insects are rare.

Advances in plant biotechnology have created a new opportunity to make practical use of this knowledge. Instead of determining the basis of resistance after transfer into the plant, we can transfer the gene(s) for resistance which are defined in advance by the situation as the optimal traits for the plant. We have been developing a tobacco-

[1]Supported, in part, by USDA Coop. Agr. 58-43YK-5-0034

insect model system for testing concepts in genetic engineering of plants for insect resistance (11) and in this report we summarize the results obtained up to this time.

RESULTS

Selection of Host Plants for Study

A number of wild species within the genus *Nicotiana* were examined in the field (13) for toxicity to *M. sexta*, *T. ni* and *H. virescens* larvae. Several species consistently induced high mortality to larvae of at least one of these insects (Table 1a-e). Species within the section Repandae (*N. repanda*, *N. stocktonii* and *N. nesophila*) were particularly toxic to larvae of *M. sexta*. In contrast, the species *N. benthamiana* and *N. gossei* (section Suaveolentes)

TABLE 1
% MORTALITY IMPARTED TO LARVAE OF *M. sexta* (AND *H. virescens*) AFTER 96 HRS EXPOSURE TO PLANTS OR FOLLOWING TOPICAL TREATMENT

Treatment	Field[a]	Greenhouse[a]	Topical[b]
a. *N. stocktonii*	100[c] (28)	100 (15)	100 (10)
b. *N. repanda*	90[c] (21)	- (0)	100 (-)
c. *N. gossei*	100[c] (67)	80 (84)	0 (17)
d. *N. benthamiana*	100[c] (-)	- (-)	- (-)
e. *N. tabacum*	10[c] (17)	5 (4)	17 (0)
f. *repanda* x *tabacum*		13 (0)	
g. *gossei* x *tabacum*		3 (61)	
h. nicotine base			4 (0)
i. TLC acylnornicotine[d]			100 (0)
j. etoh			0 (0)

[a]n usually >20; [b]0.5 (0.1) mg leaf exudate, n>10; [c]from ref (13); [d]10 (50) ug TLC semipurified N-acylnornicotine

were toxic to both insects. These plant species were re-examined in the greenhouse, and similar trends were observed (Table 1a-e). We then focused our efforts on identifying and characterizing mechanisms of antibiosis in these species.

Identification of Toxic Plant Compounds

The leaf exudate of each species was removed with acetonitrile, concentrated *in vacuo* and topically applied to sec-

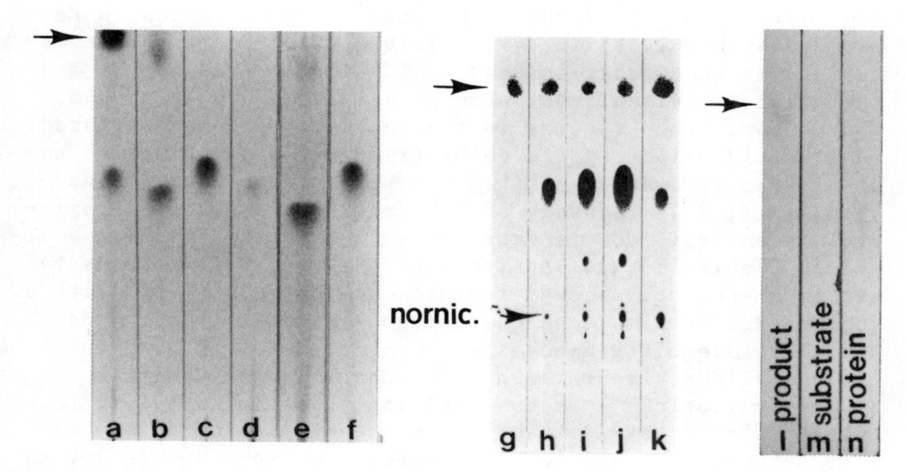

FIGURE 1. a-g leaf surface alkaloids. a- N. repanda (arrow N-acylnornicotine), b-N. repanda x N. tabacum sexual hybrid, c-Nicotine base standard, d- N. gossei x N. tabacum sexual hybrid, e-N. gossei, f- N. tabacum. 1 mg equivalent of exudate for each. g-k Leaf surface alkaloids from leaves fed HOH 24 hrs, then fed 0,1,3 or 5% nicotine, respectively, for 48 hrs. 0.5% equivalent of alkaloids obtained from one leaf for each lane. l-n In vitro production of N-acylnornicotine from nornicotine and myristoyl-CoA by leaf protein.

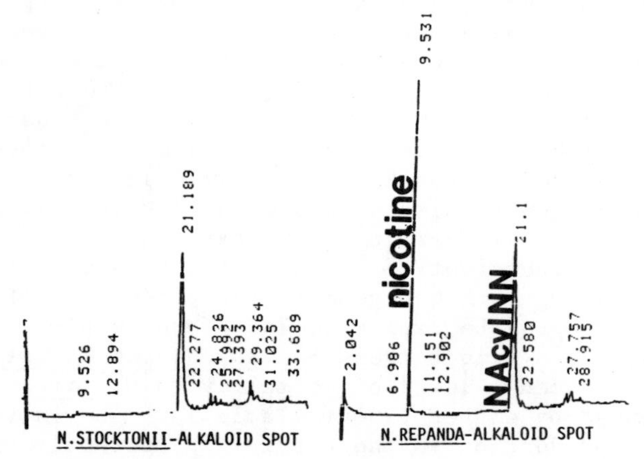

FIGURE 2. Gas chromatography of TLC semipurified N-acylnornicotine from leaf surface of N. stocktonii and repanda. The N-acylnornicotine has a longer elution time than standard alkaloids such as nicotine. (Varian thermionic nitrogen sensitive detector.)

ond instar M. sexta larvae. The exudate from species in Repandae was highly toxic to the larvae (Table la-e). The material was then fractionated on TLC (100:40 CHCl :MeOH) and the alkaloids visualized with para-aminobenzoic acid and CNBr vapor. Each species in the section Repandae possessed an alkaloid which, by its color (red) and Rf, was unlike any alkaloid previously described from fresh Nicotiana tissue (Fig. la,f; 7). Bioassay of TLC fractions showed the toxicity was entirely due to material in the region of the new alkaloid (Table 1;j,k). Analysis of the exudate alkaloids by gas chromatography showed the presence of an alkaloid with a longer elution time than that of conventional nicotinoids, and more resembling N-acylated nornicotine (Fig. 2). The biologically active material was identified as a mixture of N-acylnornicotines, the overwhelmingly predominant form being iso-{N-'(3-hydroxy-12-methyltetradecanoyl)-nornicotine} (14).

Although the species in the section Repandae are not as toxic to H. virescens as to M. sexta, N. gossei and N. benthamiana are. However the leaf exudate of N. gossei was not toxic to H. virescens (Table 1;16). The possible role of alkaloids inside the leaf is being examined.

Expression of Resistance in Conventionally Produced Hybrids

The species in the Repandae are not immediately related to, and have different chromosome numbers than, N. tabacum. Therefore, efforts to obtain hybrids between Repandae spp. and N. tabacum have met with little success. Seeds from one successful effort were germinated and the plants examined for resistance and alkaloid expression. The hybrid plants induced mortality to hornworm larvae greater than that in in the N. tabacum control but less than that caused by the species in the section Repandae (Table la,e,f). The exudate from these species also possessed an alkaloid missing from N. tabacum but similar in color staining properties and Rf to the N-acylnornicotine found in the Repandae (Fig. 1;b). These data suggest that gene(s) for production of N-acylnornicotines can be expressed in a foreign plant background. Tests with a hybrid between N. tabacum and N. gossei showed that it induced a level of mortality to H. virescens between that caused by either parent (Table 1). The expression of alkaloid production in the hybrids is currently under study.

Expression of Resistance in Tissue Culture Produced Hybrids

Advances in tissue culture of Nicotiana have permitted

new approaches to transfer of genes from wild species into cultivated forms. Several hybrids between N. tabacum and the Repandae species or N. benthamiana have been obtained by researchers using embryo rescue. For example, prelimenary studies with a hybrid between N. tabacum and N. benthamiana are testing for expression of resistance to M. sexta and H. vires-cens (Huesing, Jones, Myers, DeVerna, and Collins, in prep).

Approaches to Recombinant DNA and Insect Resistance in Plants

The tissue culture derived hybrids described above provide access to hybrids that were previously difficult or impossible to obtain by conventional means. Still, considerable backcrossing to the cultivated species is required to complete movement of the gene(s) of interest into the proper background. However, gene splicing technology makes it possible, for the first time, to move a gene of interest directly into the background of the cultivated plant. Therefore, we have initiated a project to isolate the gene(s) responsible for synthesis of new, highly active alkaloids from precursors which are conventional nicotine alkaloids found in N. tabacum.

TABLE 2
DISTRIBUTION OF MAJOR NICOTINE ALKALOIDS IN N. stocktonii

	Surface			Interior		
Location	Nicotine	Nor-nicotine	AcylNN	Nicotine	Nor-nicotine	AcylNN
Root	+	-	-	++++	T	-
Stem	-	-	+	+++	+	-
Petiole	T	-	++	+++	T	-
Leaf	T	T	++++	+++	++	-
Trichome (leaf)	-	-	++++	+	++	-

T - trace amount, + = present, - = not detected by TLC

The structure of N-acylnornicotine suggested that a single, terminal acylation step. A tissue distribution study identified nicotine as the primary alkaloid in the root, stem and phloem (Table 2; 19). Nornicotine was essentially absent from these tissues, but was found in abundance in the leaves. N-acylnornicotine was absent from the interior of the leaf and only trace amounts were found inside the leaf trichomes. However, it was very abundant in the trichome exudate (Table 2; 19). These data indicated the biosynthesis of the alkaloid is localized to the trichome.

The biosynthesis of N-acylnornicotine in leaf trichomes was investigated by feeding isolated leaves putative precursors. Feeding leaves unlabelled nicotine increased levels of leaf nornicotine, in a concentration-dependent manner (Fig. 2;g-k). In other experiments an apparent increase in the level of N-acylnornicotine was suggested. Feeding with labelled nicotine resulted in labelling of nornicotine and N-acylnornicotine (Table 1). Feeding with labelled nornicotine resulted in labelled N-acylnornicotine. This latter result is particularly significant because the specific act- of the nornicotine was 20,000 times less than that of the nicotine label, yet the specific activity of the product of nornicotine feeding was greater than that obtained from the nicotine feeding. These data strongly suggest that N-acylnornicotine is synthesized from nicotine via nornicotine and that the product is rapidly secreted into the exudate.

Having determined that the acyltransferase is localized in the trichome, we are currently directing efforts toward purifying the enzyme. It is necessary to assay for enzyme activity during sequential purification steps. To this end, we prepared total ammonium sulfate precipitatable protein from the leaf, and supplied nornicotine and myristoyl-CoA as precursors. In our first attempts, we obtained complete conversion of nornicotine to N-acylnornicotine (Fig. 2;l-n). Thus, the acyltransferase appears to be a highly active and/or a highly abundant enzyme. In order to obtain measurements of reaction velocity more quickly than provided by the TLC procedure, we are developing a partition assay based upon radiolabelled nornicotine substrate.

Our future research on this system will concern sequencing the N-terminus of the enzyme, synthesis of a synthetic DNA probe, and using the probe to locate the gene for the enzyme in a clone bank of *N. stocktonii*. The final goal is to insert this gene into *N. tabacum* and test for production of N-acylnornicotine by the expressed enzyme.

DISCUSSION

The advent of biotechnology and its potential

application to agricultural plants has at least temporarily outrun development of a conceptual framework of criteria by which genes are deliberately selected for insertion into the given plants. The question "What gene?" has also caught insect and plant biochemists off guard, as they have identified few genes at the molecular level which might be candidates for insertion into plants. In the insect-plant interaction, genes from either participant are valid for consideration. Examples are given below, some of which have been expounded upon in the speculative literature to the neglect of the others.

Genes From Insects

Neurotransmitters such as proctolin in the diet affect the insect gut (pers. obsv.), although the results of others suggest that very high levels of such materials in the plant would be required to significantly affect feeding larvae. Administration of neurohormones causes changes in insect physiology or development. Enzymes important in molting and metamorphosis have been considered, such as chitinases or juvenile hormone esterase (JHE). The possible effects of sustained exposure to JH esterase have been observed in larvae of _T. ni_ parasitized by _Chelonus_ spp. In these hosts JHE appears precociously during the penultimate stadium, and is an important component in host precocious metamorphosis (8). _Chelonus_ regulatory allomones, or arthropod venoms directly toxic to the nervous system, may offer alternatives if use of insect genes in plants eventually becomes practical. Use of most insect proteins is probably not practical, at least until ways of affecting the movement of their active fragment across the midgut are developed.

Genes from Plants

Movement of bacterial or foreign plant genes for insect resistance into crop plant offers a more immediate application of advances in molecular biology. The genes most emphasized currently by industrial and academic researchers are those which code for proteins which are, in themselves, toxic to insects (4,11). There are certainly some proteins highly attractive for such an approach. One advantage often cited is that only a single gene need be moved, in contrast to an entire biosynthetic pathway for a nonprotein antibiosis factor. Also, site specific mutagenesis may increase its insecticidal activity or change its insect host range.

However, most naturally occurring plant antibiosis factors are not proteins, and the prevailing concepts tend to define away these factors as not being amenable to plant genetic engineering. Overlooked is that wild relatives of crop plants which contain antibiosis compounds probably make these within the context of biosynthetic pathways common to it and the crop plant. It is our assertion that many antibiosis compounds are but one or a few enzymatic steps away from metabolic products already occurring in the crop plant. Within this conceptual framework, we believe researchers will find the answer to "What gene?" in the results of studies on the routes of biosynthesis of these compounds in the wild relatives. Movement of the one or few genes from the wild species necessary to complete the biosynthetic pathway in the crop plant will enable the latter now to synthesize the insect resistance compound.

The insect-tobacco system described in this paper is particularly suited to test the above concepts. Tobacco has been the primary plant for studies on plant molecular techniques, and expression of foreign genes in tobacco is now almost commonplace, including genes for enzymes. These techniques are now being applied even to distant, unrelated food plants. Furthermore, the biosynthesis of nicotine, the parent compound of the active alkaloids to be discussed below, is well studied, providing a solid foundation upon which to build. _Manduca sexta_ has been a workhorse for insect physiologists and biochemists, as have other tobacco pests such as _Heliothis virescens_ and _Trichoplusia ni_. The toxicology of nicotine alkaloids on these and other insects is well studied (17). If our approach to plant genetic engineering for insect resistance is valid, it should work in this system. The wide species hybrids obtained by conventional or tissue culture means offer additional opportunities unique to this system, including the ability to compare and evaluate the methodology and end results of breeding approaches at sequentially higher levels of sophistication. It has not escaped our attention that in at least our case all three means may result in identical cultivatable plants. This situation would make an interesting test case to challenge those opposed to genetic engineering in agriculture - will they allow the form produced by classical means to be released, but block the identical form produced by biotechnology? There are a number of allelochemics described in the literature whose structures suggest families of compounds suitable for our approach. For example, a solanaceous plant, the potato, produces beta-farnesene, an aphid alarm pheromone (5).

This compound is very similar to farnesol, a metabolite along the steroid synthesis pathway common to all plants. Movement of the gene(s) for conversion of farnesol to beta-farnesene into crop plants may allow them to synthesize the aphid alarm pheromone. Also, some phytoecdysones differ from less active ones by a single hydroxylation (6).

It should be emphasized that a system in which the missing enzyme is that for a terminal enzymatic step is not the only possible context in which our approach should be useful. It may be that the crop plant is missing an enzyme for an intermediate step in the potential biosynthesis pathway. For example, Bowers (2) described a biologically active juvenile hormone mimic from sweet basil whose structure suggests its synthesis involves a condensation between a monoterpene and a cinnamyl moiety. Both of these are metabolites common to many plants. It is also possible to envision that deletion of the gene for an enzyme may prevent the plant from synthesizing an attractant or a secondary metabolite which acts as a feeding stimulant (1).

ACKNOWLEDGEMENTS

We thank G. W. Pittarelli and Dr. V. Sisson for providing seed used in this study. Drs. G. Collins, J. Myers and J. DeVerna provided the embryo rescue material. Dr. L. Bush provided the GC analysis in Fig. 2. Drs. M. Jackson and R. Severson provided valuable interaction and data during the study. Published with the approval of the Director of the Kentucky Agricultural Experiment Station (86-7-75).

REFERENCES

1. Beck SD, Schoonhoven, LM (1980). Insect behavior and plant resistance. IN Maxwell FG, Jennings PR (eds): "Breeding Plants Resistant to Insects," New York: John Wiley & Sons, p 115.
2. Bowers WS, Nishada R (1980). Juvocimenes: potent juvenile hormone mimics from sweet basil. Science 209:1030-1032.
3. Campbell BC, Duffey SS (1979). Tomatine and parasitic wasps: Potential incompatibility of plant antibiosis with biological control. Science 205:700-702.
4. Foard DE, Murdock LL, Dunn PE (1983). Enginering crop plants with resistance to herbivores and pathogens: An using primary gene products. Plant Mol Biol 2:223-233.
5. Gibson RW, Pickett JA (1983) Wild potato repels aphids by release of aphid alarm pheromone. Nature 302:608-609.

6. Horn DHS (1971). The Ecdysones. In Jacobson M, Crosby DG (eds): "Naturally Occurring Insecticides," Marcel Dekker, New York, p 333.
7. Huesing J, Jones D (1986) A new form of antibiosis in Nicotiana (submitted).
8. Jones D, (1985). Parasite regulation of host insect metamorphosis: a new form of metamorphosis in pseudoparasitized larvae of Trichoplusia ni. J Comp Physiol 155:583-590.
9. Jones D, Jones GA, Hagen T, Creech E (1983). Wild species of Nicotiana as a new source of tobacco resistance to the tobacco hornworm Manduca sexta. Entomol Exp Appl 38:157-164.
10. Jones D, Jones G, Jackson DM (1984). Tobacco and insects: A model system for investigating the potential of plant genetic engineering for insect resistance. In Gustafson R (ed): "Gene Manipulation in Plant Improvement," New York: Plenum Press, p 656.
11. Kaufman RJ (1986). The development and risk assessment of a genetically engineered microbial insecticide. J Cellular Biochem, Supp 10C, p 94.
12. Maxwell FG (1980). Future Opportunities and Directions. In Maxwell FG, Jennning PR (eds) "Breeding plants resistant to insects," New York: John Wiley & Sons, p 535.
13. Parr, JC, Thurston, R (1968) Toxicity of Nicotiana and Petunia species to larvae of the tobacco hornworm. J Econ Entomol 61:1525-1531.
14. Severson RF, Huesing JE, Jones D, Arrendale RF, Sisson VA (1986). Identification of a tobacco hornworm antibiosis factor from the cuticulae of Nicotiana section Repandae. J Chem Ecol (submitted).
15. Thurston R, Fox PM (1972). Inhibition by nicotine of emergence of Apanteles congregatus from its host, the tobacco hornworm. Ann Entomol Soc Amer 65:547-550.
16. Thurston R, Parr JC and Smith WT (1966). The phylogeny of Nicotiana and resistance to insects. Proc 4th Int Tob Sci Congr, Athens, Greece, p 424-430.
17. Yammamoto I, Kamimura H, Yamamoto R, Sakai S, Goda M (1962). Studies on nicotinoids as an insecticide. Part I. Relation of structure to toxicity. Agr Biol Chem 26: 709-716.
18. Yu SJ, Berry RE, Terriere LC (1979). Host plant stimulation of detoxifying enzymes in a phytophagous insect. Pest Biochem Physiol 12: 280-284.
19. Zador E, Jones D (1986) Biosynthesis of a novel nicotine alkaloid in the trichomes of N. stocktonii. (submitted).

Index